# SEMIMETALS & NARROW-BANDGAP SEMICONDUCTORS

D.R.Lovett

Applied physics series

Series editor H.J.Goldsmid

1 **Structure and properties of magnetic materials** D.J.Craik

2 **Solid state plasmas** M.F.Hoyaux

3 **Far-infrared techniques** M.F.Kimmitt

4 **Superconductivity and its applications** J.E.C.Williams

5 **Physics of solid state devices** T.H.Beeforth and H.J.Goldsmid

6 **Hall generators and magnetoresistors** H.H.Wieder

7 **Introduction to noise analysis** R.W.Harris and T.J.Ledwidge

8 **Thermal conductivity of solids** J.E.Parrott and Audrey D.Stuckes

9 **Sound waves in solids** H.F.Pollard

Series editors H.J.Goldsmid and D.W.G.Ballentyne

10 **Semimetals and narrow-bandgap semiconductors** D.R.Lovett

# SEMIMETALS & NARROW-BANDGAP SEMICONDUCTORS

D.R.Lovett

Pion Limited, 207 Brondesbury Park, London NW2 5JN



ISBN 0 85086 060 1

Printed in Great Britain

# Preface

The layout of this book has been arranged intentionally in two distinct parts. Part I deals with the preparation, electrical properties and applications of semimetals and narrow-bandgap semiconductors in general terms. Part II is devoted in turn to the various elements, compounds and alloys constituting these groups of materials. Some materials have been investigated much more thoroughly than others and for these data are therefore available in the literature in very much greater quantity and detail. Despite this, an attempt has been made to give a balanced account for each material, although as a consequence readers may consider the sections devoted to materials which have a long history of investigation somewhat short. Although part I is intended as an introduction and explanation of what is discussed in part II, it is hoped that it will stand on its own as an introduction to these classes of materials. There is a strong emphasis on the materials aspects, and this emphasis should prove of particular interest to engineers and to physicists who often require to know more about the growth and chemical physics of semimetals and semiconductors in addition to their applications.

I am grateful to the authors and publishers who granted permission for the use of particular diagrams and graphical data. Most figures that are included in this book are in a modified form to produce a uniform format, but sources are quoted with the individual captions.

Contents

## List of main symbols

|  |  |
|---|---|
| $A$ | area |
| $a_0$, $b_0$, $c_0$ | lattice parameters |
| $B$, $\boldsymbol{B}$ | magnetic induction |
| $c$ | velocity of light |
| $C$ | number of components in a thermodynamic system; heat capacity per unit volume |
| $C(x)$ | concentration |
| $d$ | thickness |
| $D$ | diffusion coefficient |
| $D^*$ | detectivity |
| $E$ | energy |
| $E$, $\boldsymbol{E}$ | electric field |
| $E^*$ | reduced energy ($= E/k_B T$) |
| $E_F$ | Fermi energy |
| $E_F^*$ | reduced Fermi energy ($= E_F/k_B T$) |
| $e$ | electronic charge |
| $F$ | number of degrees of freedom in a thermodynamic system; merit factor |
| $F_m(x)$ | Fermi–Dirac integral |
| $f$ | distribution function |
| $f_0$ | equilibrium distribution function (usually Fermi–Dirac) |
| $g$ | Landé splitting factor |
| $\mathrm{g}(x)$ | density-of-states function |
| $H$ | Hamiltonian |
| $h$ | Planck's constant (also $\hbar = h/2\pi$) |
| $\boldsymbol{h}$ | the unit vector $\boldsymbol{B}/B$. heat flow |
| $I$ | total current; intensity of thermal radiation |
| $\mathcal{I}(x)$ | integral |
| $I_{ij}$ | transport integral |
| $j$ | current density |
| $K$ | thermal conductivity ($K_L$, lattice component; $K_{el}$, electronic component); anisotropy factor |
| $\boldsymbol{K}_n$ | reciprocal lattice vector |
| $k$ | distribution coefficient ($k_0$ effective distribution coefficient) |
| $\boldsymbol{k}$ | wave vector |
| $k_B$ | Boltzmann's constant |
| $L$ | length |
| $L_0$ | Lorenz number |
| $\ell$ | mean free path; direction cosine |
| $m_0$ | free electron mass |
| $m^*$ | effective electron mass |
| $m_c^*$ | conductivity effective electron mass |
| $m_d^*$ | density-of-states effective electron mass |

| | |
|---|---|
| $N_A$ | concentration of acceptors |
| $N_D$ | concentration of donors |
| $N_v$ | number of valleys |
| $n$ | carrier concentration (usually electrons) |
| $P$ | matrix element in Kane theory; number of phases in a thermodynamic system; Ettingshausen coefficient; power |
| $p$ | pressure |
| $\mathrm{p}$ | hole concentration |
| $\boldsymbol{p}$ | momentum vector |
| $Q$ | Nernst coefficient |
| $R$ | electrical resistance; thermal resistivity |
| $R_H$ | Hall coefficient |
| $\mathcal{R}$ | responsivity |
| $r$ | scattering index |
| $\boldsymbol{r}$ | position vector |
| $S$ | Righi–Leduc coefficient |
| $T$ | absolute temperature (K) |
| $t$ | time |
| $V$ | voltage |
| $V(\boldsymbol{r})$ | periodic potential |
| $v$ | velocity |
| $\boldsymbol{W}$ | thermal current density |
| $w$ | width |
| $x$, $x_g$ | $E/k_BT$, $E_g/k_BT$ |
| $z$ | figure of merit |
| $\alpha$ | Seebeck coefficient |
| $\zeta$ | quantum efficiency |
| $\lambda$ | wavelength |
| $\mu$ | mobility ($\mu_n$ electron mobility, $\mu_p$ hole mobility, $\mu_H$ Hall mobility, $\mu_0 = e\tau/m^*$); magnetic moment |
| $\nu$ | frequency |
| $\rho$ | density, electrical resistivity |
| $\Pi$ | Peltier coefficient |
| $\sigma$ | electrical conductivity; surface energy |
| $\sigma_x$, $\sigma_y$, $\sigma_z$ | Pauli matrices |
| $\tau$ | relaxation time |
| $\tau_0$ | equilibrium relaxation time |
| $\chi$ | susceptibility |
| $\omega$ | angular frequency |

# Part 1

# General preparation and properties

# 1

# Introduction

The term *semimetal* originated in mediaeval times when it was used to describe bismuth which, because of its brittle nature, was considered to be a 'half-metal'. Today, the term is used to describe those solids which have a small overlap between the valence and conduction bands. Pure semimetal crystals should contain equal numbers of conduction electrons and holes. Although there should be sufficient electrons to fill all the available states of the uppermost valence band, some of these spill over into the bottom of the conduction band, leaving behind an equal number of holes. Thus the properties of these materials are rather like those of a semiconductor with a large number of carriers; however, at ~0 K a semimetal still exhibits this large carrier concentration, whereas a semiconductor should tend to an insulating state. On the other hand, carrier densities in semimetals are well below those for metals.

Whether a solid is likely to be a metal, semiconductor, or insulator can be estimated from the number of electrons per unit cell for, if there are $N$ unit cells in a particular volume, then there will be $2N$ electrons per band. Hence solids with one free electron per unit primitive cell (e.g. Li, Na, Cu, Ag, Au) will always be metals, and so will solids with an odd number of electrons per cell (e.g. Al, Ga, In). In calculating the number of electrons it is important to take into account the number of atoms per unit cell. Thus As, Sb, and Bi have five electrons per atom, but as they have two atoms per unit cell there are in fact ten electrons almost filling five bands. These three elements are typical semimetals since there is a small overlap of the fifth and sixth bands. This results in the fifth band being not quite filled and in there being some electrons in the sixth band. These electrons in the sixth band and the holes in the fifth band carry the current. Other examples of semimetals are graphite, mercury telluride, and mercury selenide.

A semiconductor will have a filled uppermost band at zero temperature separated by a bandgap from the next empty band, this bandgap being sufficiently small for thermal excitation of electrons from the filled valence band to the empty conduction band to occur at reasonable temperatures. The term *narrow-bandgap* as applied to semiconductors is a less specific one. It usually refers to those semiconductors which have an energy gap between the conduction and valence bands of the order of ten times or less the room-temperature value of $k_B T$ (0·026 eV), where $k_B$ is Boltzmann's constant and $T$ the absolute temperature. This value for the limiting bandgap will be used as the guideline but will not be regarded as a 'hard-and-fast' rule. For instance, there will be some mention of lead sulphide, which has a bandgap of 0·286 eV, because its properties are similar to those of lead selenide and lead telluride which have smaller bandgaps. But, more generally, semiconductors with bandgaps greater than approximately 0·26 eV at room temperature will be omitted.

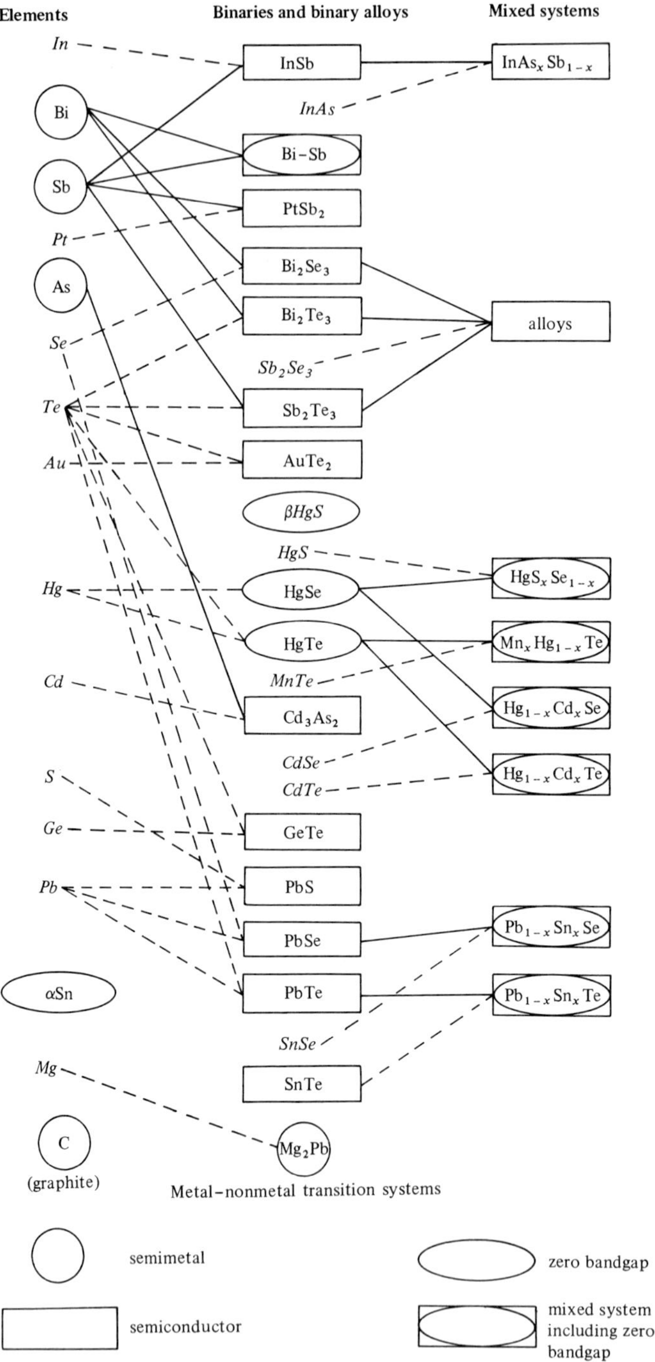


**Figure 1.1.** Semimetals and narrow-bandgap semiconductors considered in this text. (Elements and compounds in italics are not considered.)

Between the semimetals and narrow-bandgap semiconductors is an interesting series of materials exhibiting a *zero bandgap*. Grey tin has such a zero bandgap, but more usually the property is exhibited by mixed crystal systems where alloying is chosen to produce such a condition for a particular temperature. The two most widely investigated systems are the mixed-crystal systems PbTe and SnTe, and also HgTe and CdTe.

In figure 1.1 the elements, compounds, and alloys presented in ordinary type-set constitute a definitive list of the semimetals, narrow-bandgap semiconductors, and zero-bandgap semiconductors described in this book; the materials shown in italics are constituent components falling outside these categories. Obviously, in a book of this length, some eligible but rather obscure materials have been excluded.

The book refers to materials possessing a crystalline structure and usually they will be required in single-crystal form. In the next chapter there is a discussion of techniques of single-crystal growth and of purification. Preparative techniques, particularly for alloys, depend on the form of the relevant phase diagram, and the chapter begins with sections on phase diagrams and distribution coefficients.

Electrical and other properties of the materials depend very much on the band structure, and so chapter 3 has been devoted to crystal symmetry and band properties. Some readers may find this chapter particularly difficult as it assumes a more extensive background knowledge than other chapters. In particular, it involves some knowledge of group theory. However, it is possible to omit this chapter and to assume the expressions for the forms of the band structure, such as are given in table 3.4 for the Kane model.

Because carrier concentrations in semimetals and also in degenerate narrow-bandgap semiconductors are neither as large as those in metals nor as small as those in nondegenerate semiconductors, the transport properties generally require statistics specific to the particular carrier concentration and Fermi energy. Classical statistics tend to be inapplicable because carrier concentrations, and hence the density of filling of the quantum states in energy or momentum space, are too high; yet certain limiting features of Fermi–Dirac statistics for high carrier concentrations are also not applicable. Consequently, part I includes a discussion of intermediate statistics and their use in the Boltzmann transport equation.

Finally, an introduction to the theory and materials requirements of those devices which involve the use of semimetals and narrow-bandgap semiconductors is included in this section.

The emphasis in part II, which constitutes the major part of the book, is on the preparation, electrical properties, and applications of the selected materials, and optical properties are only mentioned in passing. Even with these restrictions, it has been necessary to be very selective in a book of this size. Nevertheless, it is hoped that the coverage provides an interesting and homogeneous description of the more general properties of semimetals and also semiconductors having a bandgap less than 0·26 eV.

# 2

# Preparative techniques

## 2.1 Introduction

In chapter 1 it was stated that the range of semimetals and narrow-bandgap semiconductors to be considered would be limited to crystalline materials. Although polycrystalline material can often be used for making physical measurements and for device applications, single-crystal material is often preferred or is essential. In this chapter, the main preparative techniques for the semimetals and semiconductors will be discussed, leaving any very specific methods for discussion in part II under the heading of the element or compound in question. The problems associated with producing mixed crystals will be discussed in some detail.

Often there are a number of methods of growing crystals of a particular material. Sometimes one method may be significantly better than another; for example one method may give material with the highest purity or best stoichiometry, an alternative method may provide the larger single crystals. Choice of method of growth is made according to the physical properties of the material and to the size, etc., of crystal required. Usually, semimetals and narrow-bandgap semiconductors do not have especially high melting points (graphite is an exception), in which case problems of contact with the crucible do not exist. Hence growth from the melt (for example by the Bridgman–Stockbarger method or the Czochralski method) is particularly common. If the material has a high vapour pressure below the melting point, then vapour deposition is likely to be a suitable method. On the other hand, a high vapour pressure makes certain methods of growth less suitable. Zone melting is certainly more involved in cases where the material has a high vapour pressure since the vapour pressures above the material can be controlled to prevent sublimation of the molten zone (see section 2.6.5). If a material has a solid–solid phase change below the melting point it is often necessary to grow single crystals below the phase change. Again vapour deposition is the likely answer, provided the vapour pressure is sufficiently high. If the methods are suitable for the materials in question, the Bridgman technique or crystal pulling (Czochralski method) are good choices as these methods produce large crystals—particularly crystal pulling. The Bridgman method can be a good method of obtaining seed crystals since pulling often requires a single crystal to initiate growth.

There are a number of introductory texts on the growth of single crystals. For instance, Lawson and Nielsen (1958), Gilman (1963), and Laudise (1970) have discussed the complete range of methods available, whereas Brice (1973) has discussed growth from solution. Once grown, the crystals often have to be etched or chemically polished. A review of the chemical polishing of semiconductors has recently been given by Tuck (1975).

## 2.2 Phase diagrams

In the preparation of binary and ternary compounds and alloys it is imperative that information on the possible phases which can exist for different composition and preparation conditions be available. This information is contained in the relevant phase diagram, and a particular state of equilibrium can be characterised by the number and identity of the phases. The phase rule relates the number of phases, $P$, the number of components, $C$, and the number of degrees of freedom $F$; thus

$$P+F = C+2 . \qquad (2.1)$$

The division into *phases* includes the division into states of matter—solid, liquid, or gas—but also includes division of, for instance, the solid phase into different groupings of matter having identifiable properties. Strictly, a phase should be homogeneous, but in alloys there may be compositional variation within the alloy. The number of *components* is the smallest number of independently variable constituents necessary for expressing the composition of each phase within the complete system. In an alloy system it is therefore the number of constituent metals (or semimetals). In addition there are a number of other conditions which have to be stated in order to define the state of the system. These conditions can be altered independently, and in the statement of the phase rule above they are assumed to be temperature, pressure, and composition, and are called *degrees of freedom*. If other variables such as magnetic, electric, and gravitational fields are included, then the value of 2 in equation (2.1) must be increased by the number of new variables. If, as in metallic systems, the vapour pressures of the liquid and solid are negligible, then pressure can be considered as constant and equation (2.1) becomes

$$P+F = C+1 . \qquad (2.2)$$

### 2.2.1 One-component systems

If one phase is present, two degrees of freedom are possible; i.e. pressure and temperature can be varied independently without any change of phase. When two phases are present and in equilibrium, as along AX, BX, and CX, only one degree of freedom is possible (see figure 2.1). Thus, on BX, if a particular temperature is chosen, this also establishes the vapour pressure of the liquid. For three phases to be present at the single position X (the triple point), the pressure and temperature must have a fixed value, and there is no degree of freedom. In addition the diagram may show metastable equilibria (as along XY) where there is equilibrium between supercooled liquid and vapour, solidification having not occurred because of a lack of nucleation centres. The solid phase may be further divided if the component exists in more than one crystallographic form.

**Table 2.1.** Values of $P$ and $F$ for a one-component system ($C = 1$).

| $P$ | 1 | 2 | 3 |
|---|---|---|---|
| $F$ | 2 | 1 | 0 |

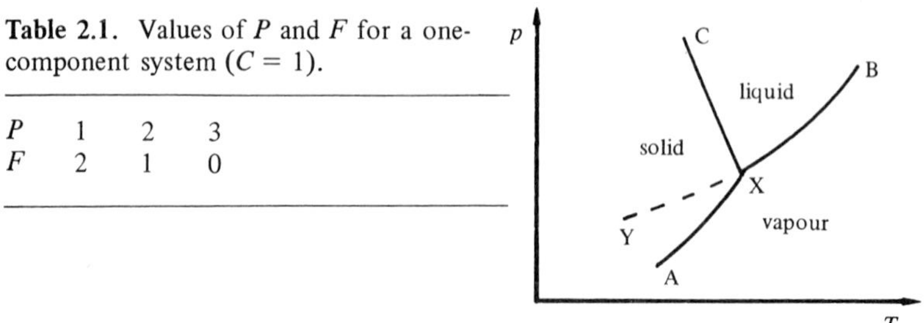


**Figure 2.1.** $p$–$T$ diagram for a one-component system.

**2.2.2 Two-component systems; binary phase diagrams**

With temperature, pressure, and composition all variable, it would require a three-dimensional diagram to represent the phases. It is usual to fix one of the variables, say pressure, and show the phases graphically as a function of the other two variables, temperature and composition. Figure 2.2 shows various binary phase diagrams that are possible and the features of these will now be discussed. For fixed pressure no diagram can show more than three phases coexisting at a point. The binary systems to be discussed in this book include most of the features of these diagrams.

**Table 2.2.** Values of $P$ and $F$ for a two-component system ($C = 2$).

(a) Pressure variable

| $P$ | 1 | 2 | 3 | 4 |
|---|---|---|---|---|
| $F$ | 3 | 2 | 1 | 0 |

(b) Pressure fixed

| $P$ | 1 | 2 | 3 |
|---|---|---|---|
| $F$ | 2 | 1 | 0 |

*Figure 2.2a.* This diagram shows a system where A and B are completely miscible with each other in the solid and liquid states. Where A and B are together in the solid phase it is called a solid solution. As there is no real difference between solvent and solute, the term mixed crystal is also used. The liquidus curve shows the temperature at which solidification of the liquid occurs (it is a function of composition), whereas the solidus curve shows the temperature at which melting of the solid occurs. These two curves can be connected by horizontal tie lines which then connect liquid and solid of compositions which are in equilibrium. One consequence of this type of diagram is that freezing tends to reject the lower melting component; this will be discussed later (section 2.3.3). Between the liquidus and solidus there exist therefore two phases—solid and liquid. This type of relationship tends to exist for components of similar atomic size. A similar type of diagram can be drawn between liquid and vapour when the corresponding boundary curves are liquidus and vaporus, and for the liquid/vapour case in particular it is also possible to draw a similar pressure/composition diagram for constant $T$.

I. Complete miscibility in the liquid phase.

(i) Complete miscibility in the solid phase.

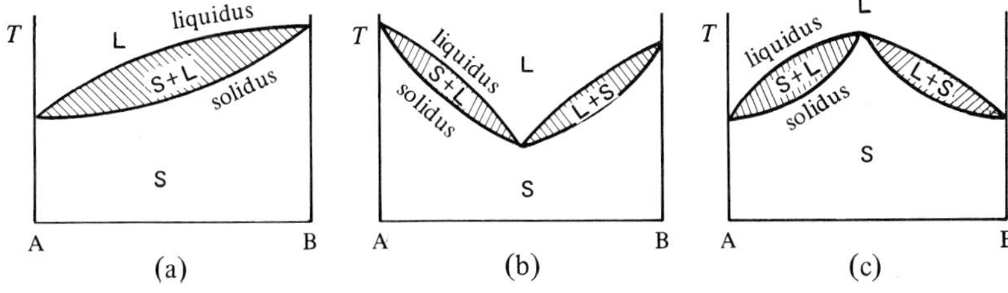


(ii) Limited miscibility in the solid phase.

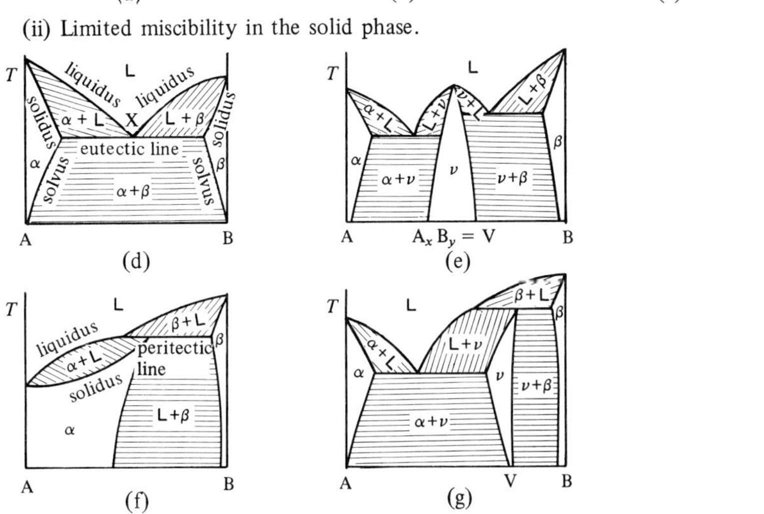


(iii) Complete immiscibility in the solid phase.

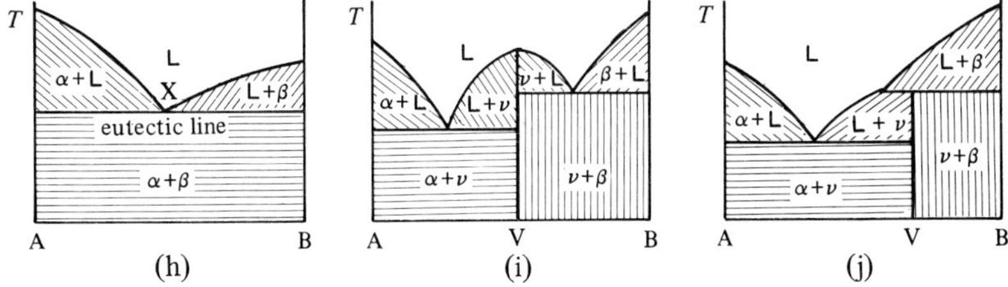


II. Partial miscibility in the liquid phase, complete immiscibility in the solid phase.

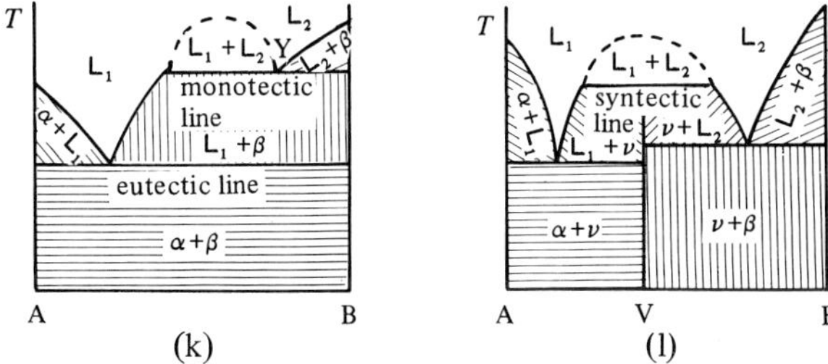


**Figure 2.2.** Examples of phase diagrams.

*Figures 2.2b and 2.2c.* Even if A and B are completely miscible, local maxima and minima may occur in the liquidus and solidus at certain temperatures. At the extreme points the liquidus and solidus coincide so that the solid melts isothermally at these compositions.

*Figures 2.2d and 2.2h.* These are examples of eutectic phase diagrams. Figure 2.2d shows an example of the case where A and B are miscible over restricted composition ranges and form phases $\alpha$ (composed mainly of A) and $\beta$ (composed mainly of B). Between the composition ranges where phases $\alpha$ and $\beta$ exist there is the two-phase region $(\alpha+\beta)$, and this region is often called the miscibility gap. Again, there are liquidus and solidus curves which can be connected by horizontal tie lines. Figure 2.2h shows no solubility of A and B, so that two phases exist over the complete composition range. Between the solidus and the eutectic line (see below) there are solid and liquid phases coexisting; this effect occurs also in figure 2.2d. In both cases, if a melt is cooled, solidification occurs once the liquidus temperature is reached, the crystallised phase being composed entirely of $\alpha$ or $\beta$ depending on which side of the phase diagram crystallisation is occurring, unless solidification takes place at the minimum X, in which case joint crystallisation of $\alpha$ and $\beta$ occurs. When the latter happens, large numbers of interwoven crystals of the two phases are produced, and the resultant material is very different from that produced when solidification occurs further up the solidus. Material of this composition is said to have eutectic composition (see section 2.3.4), and this is the only composition at which melting occurs at a sharply defined temperature. Any alloy cooled through the eutectic line will show an arrest in the cooling curve.

*Figures 2.2e and 2.2i.* The forms of these phase diagrams are similar to those of figures 2.2d and 2.2h respectively, except that A and B form a compound constituting phase $\nu$ and showing *congruent* melting; when the compound melts it forms a liquid of the same composition.

*Figure 2.2f.* This is an example of a so-called peritectic phase diagram which may be compared with the eutectic phase diagrams shown in figure 2.2d etc. The features of the peritectic diagram resemble those of the eutectic. It has extensive two-phase regions and, in particular, a composition for which a slight temperature change results in a single phase changing into two phases (point Y). Whereas the eutectic point is the lowest point of a particular liquid phase, the peritectic point is the highest melting point of a particular solid phase. If cooling of an alloy takes place through the peritectic, there will be an arrest in the cooling curve in the same way as in the eutectic.

*Figures 2.2g and 2.2j.* Here both compound formation and a peritectic exist. The compound $A_x B_y$ (or V) does not melt to form a single liquid phase of the same composition, hence this is called *incongruent* melting.

*Figures 2.2k and 2.2l.* It is fairly common for the solid in a binary system to melt into two immiscible liquids which differ in chemical composition and physical properties. However, the region of immiscibility tends to be restricted in composition and temperature. The usual case occurs when a melt reacts with a crystal to form a second melt which is immiscible with the first, for example

$$\beta + \mathrm{L}_1 \underset{\text{cooling}}{\overset{\text{heating}}{\rightleftarrows}} \mathrm{L}_2 \,. \tag{2.3}$$

This type of reaction, called monotectic, is illustrated in figure 2.2k where Y is the monotectic point and is an invariant point. The less usual case occurs when a solid melts directly into two liquids, for example

$$\nu \underset{\text{cooling}}{\overset{\text{heating}}{\rightleftarrows}} \mathrm{L}_1 + \mathrm{L}_2 \,, \tag{2.4}$$

as illustrated in figure 2.2l; this is called a syntectic system.

The diagrams showing eutectic features are widely applicable. For instance, Pb–Te shows this feature as also does Pb–Se which, in the high Se composition range, possibly illustrates immiscibility in the liquid phase. The systems Bi–Se and Mg–Pb illustrate peritectic features.

### 2.2.3 Lever-arm principle

If a solution of any given composition is frozen out, it is possible from the phase diagram to determine the relative quantities of the phases present in the alloy. In figure 2.2a tie lines were drawn from the liquid to the solid phases. Now consider a liquid corresponding to composition $C_a$. It will start to solidify at temperature $T_a$ and the composition of the solid will correspond to point $C_b$ (figure 2.3). If cooling is continued down to temperature $T_b$, the composition of the solid is fixed by point $C_s$ and the composition of the liquid is fixed by point $C_l$; hence, as there is an overall average composition $C_a$, the relative proportions of solid and liquid must be fixed.

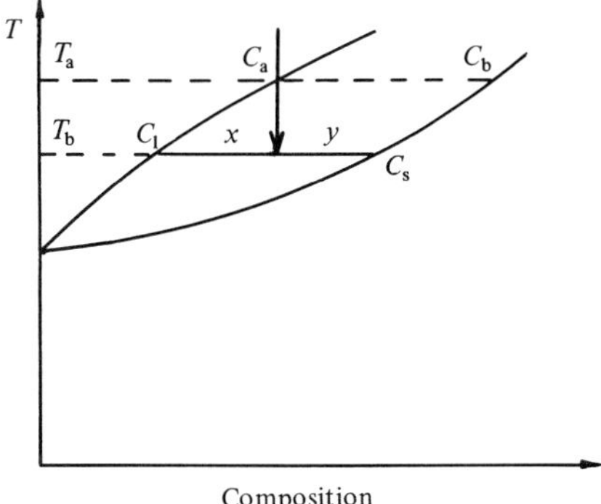


**Figure 2.3.** Illustration of the lever-arm principle.

Let $L$ be the amount of liquid and $S$ be the amount of solid then,

$$LC_1 + SC_s = (L+S)C_a ,$$

giving (see figure 2.3),

$$\frac{S}{L} = \frac{C_a - C_1}{C_s - C_a} = \frac{x}{y} ,$$

or

$$Sy = xL , \qquad (2.5)$$

where $x$ and $y$ are the lever arms after which the principle is named. Hence the fraction of solid is

$$\frac{S}{L+S} = \frac{C_a - C_1}{C_s - C_1} , \qquad (2.6a)$$

and the fraction of liquid is

$$\frac{L}{L+S} = \frac{C_s - C_a}{C_s - C_1} . \qquad (2.6b)$$

What is more important is that the approach can be extended to solid–solid transformations and be used over the range of phase diagrams illustrated. Construction of the phase diagrams assumes thermodynamic equilibrium so that slow changes are assumed. Phase changes may be suppressed if quenching is carried out.

### 2.2.4 Three-component systems; ternary phase diagrams

If the pressure is considered to be fixed, the phase rule for a ternary material gives the values of $P$ and $F$ listed in table 2.3.

Hence a three-dimensional model is required to represent the phases. It is common practice to use a triangular prism, with temperature plotted vertically and the sides of the equilateral base used to plot the binary systems AB, BC, and AC (figure 2.4). Thus the corners A, B, and C represent the three components; points on the sides are binary alloys, while ternary alloys are represented by points within the prism.

**Table 2.3.** Values of $P$ and $F$ for a three-component system ($C = 3$).

| $P$ | 1 | 2 | 3 | 4 |
|---|---|---|---|---|
| $F$ | 3 | 2 | 1 | 0 |

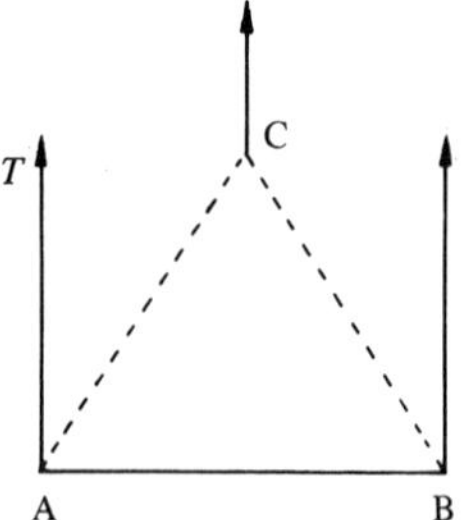


**Figure 2.4.** Axes for a three-component system.

It is usual to plot composition along the sides AB, BC, and CA as weight percentages. Hence, if an isothermal section is taken, lines corresponding to equal weight percentages for one component can be drawn parallel to each base line. The composition for any point within the triangle can be obtained by reading off the base lines (see figure 2.5). An alternative but less frequently used method is to take the altitude of the triangle for the respective component as 100% and measure off composition as a percentage of this height. In this case, composition is given by the length of the perpendicular from the point X to each side. Another alternative is to use a right-angled triangle.

Analogous to the lever rule, there is for these ternary phase diagrams the triangle rule. For binary alloys the tie line is horizontal; that is, it is an isothermal line. Similarly, for ternary alloys, there is a tie triangle which corresponds to the isothermal plane and can be represented on the

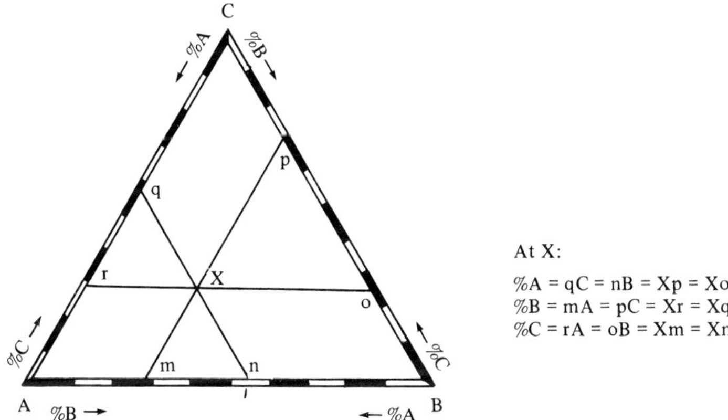


**Figure 2.5.** Isothermal section through a ternary phase diagram.

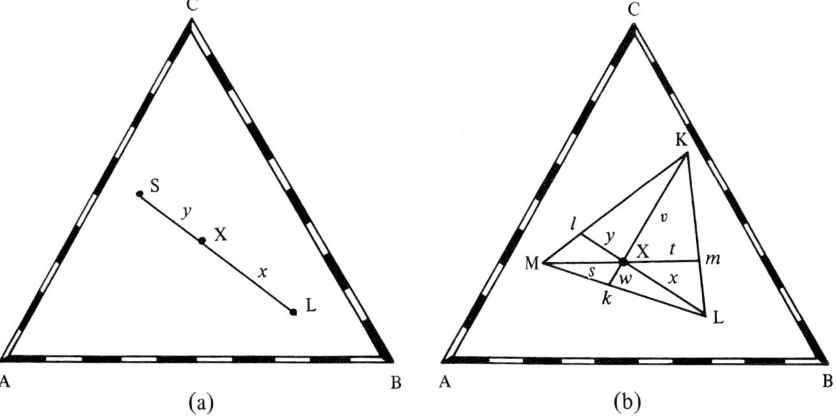


**Figure 2.6.** The triangular rule for ternary systems.

single triangular diagram. First, consider an alloy of composition X melting into solid and liquid, corresponding to compositions S and L (figure 2.6a). Point X always lies on the SL line. As with tie lines for binary alloys, what is usually known is the composition at S and L, and what is required is a knowledge of the quantitative ratios of S and L. By analogy with the binary case,

$$Sy = xL \qquad \text{[equation (2.5)]} .$$

Now consider the more general case of three phases, K, L, and M (figure 2.6b). The compositions of these three phases are known, in addition to the composition of X. By analogy with equations (2.6a) and (2.6b) for binary alloys, we have for the fractions of K, L, and M:

$$K = \frac{w}{w+v} , \qquad L = \frac{y}{x+y} , \qquad M = \frac{t}{s+t} . \tag{2.7}$$

In this case, triangle KLM may be considered as supported on a fulcrum at X with masses at the corners proportional to the amounts of the alloys.

Besides using isothermal sections, it is common to use projected diagrams of, for instance, the liquidus or solidus surfaces, which are projected on to the base plane. Vertical plane sections can also be taken. Examples of ternary diagrams arise in such cases as Pb–Sn–Te, Pb–Sn–Se (section 7.3) and Hg–Mn–Te (section 7.5). Ternary diagrams are discussed in more detail in many texts (for instance, West, 1965; Tamás and Pâl, 1970; Reisman, 1970).

## 2.3 Distribution coefficients

As can be seen from the phase diagrams discussed in the previous section, the presence of a solute or, alternatively, an impurity in a solvent will affect the melting point of the solvent; it may raise the melting point or, more often, lower it. This will then affect the proportion of solute or impurity being solidified, and hence will either affect the composition in the case of a compound or mixed crystal, or affect the purity of the solid. A distribution coefficient $k$ is defined such that it is the ratio of the solute concentration in the freezing solid to that in the liquid (coefficients can also be defined between other phases). If the solute raises the freezing point, its concentration will be greater in the solid than in the liquid and $k$ is greater than unity; if the solute lowers the freezing point, there will be a lower concentration in the solid than in the liquid, and $k$ is less than unity. $k$ can range through values from $10^{-3}$ to values greater than 10. The fact that $k$ is usually less than unity means that growth techniques often automatically lead to purification; however, in the case of mixed crystals it increases the difficulty of producing material of an exactly specified composition. It also leads to problems of constitutional supercooling which will be discussed later.

During the actual freezing there may not be complete equilibrium within the liquid phase, and between the liquid and solid phases. Consequently $k$ is mainly used to refer to the effective value of the distribution coefficient—a value which depends on the conditions of the freezing; whereas $k_0$ is used for the equilibrium value—the value which is obtained from the phase diagram. Figure 2.7 shows the constitution diagram for the case where $k_0$ is less than unity. XY represents the liquidus line; at temperatures above this line the solute is entirely liquid whatever the concentration. XZ represents the solidus line; at temperatures below this line the solute is entirely solid. Now consider a point P on the liquidus. If freezing occurs here, corresponding to a temperature $T_1$ and solute concentration in the liquid of $C_L$, the frozen solute concentration corresponds to Q. Then, by definition of the equilibrium distribution coefficient, the solute concentration at Q is $k_0 C_L$. Similarly, if solid material of solute concentration $C_S$ is allowed to melt at $T_2$, the concentration of solute in the liquid will be $C_S/k_0'$ (one cannot assume that the distribution coefficient remains constant with variable solute concentration).

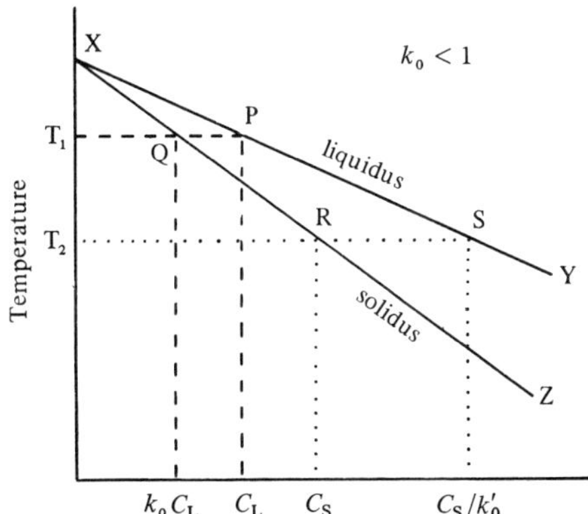


**Figure 2.7.** Constitution diagram, $k_0 < 1$.

### 2.3.1 Normal freezing

First consider the application of the equilibrium distribution coefficient to normal freezing as typified by the Bridgman–Stockbarger growth technique, which is discussed later in section 2.6.1. An ingot is gradually frozen from one end. Let $g$ be the fraction of the ingot which has been frozen out at any particular time (figure 2.8). It is assumed that at the solid–liquid interface the concentrations of impurities in the two phases are related by $k_0$. That is, the impurity concentration is shown by curve a in figure 2.8, with the solute uniformly distributed in the liquid by diffusion; no diffusion of impurity in the solid will be assumed. Thus the rate of freezing must be small compared with the rate of diffusion in the liquid, but large compared with the rate of diffusion in the solid. If $C(x)$ is the impurity concentration in the solid at position $x$ (since no diffusion in the solid is assumed, this concentration remains fixed at this position $x$) and $C_0$ is the impurity concentration in the liquid at commencement of crystal growth, then

$$C(x) = k_0 C_0 (1-g)^{k_0 - 1} \tag{2.8}$$

(see, for instance, Lawson and Nielsen, 1958, page 71). Typical curves are shown in figure 2.9. In fact, the equation cannot hold for the entire range of $g$; $k_0$ will vary as the impurity concentration changes, and should a eutectic or peritectic be produced then the equation breaks down completely.

If freezing is so fast that rejected solute builds up in the liquid in front of the advancing solid–liquid interface, then the distribution of the solute takes the form shown by curve b in figure 2.8.

Complete mixing does not usually take place unless growth is carried out at a very low speed or there is forced or natural convection that causes mixing. The form of curve b can be calculated if the value of $k_0$ is assumed constant and convection in the liquid is negligible.

Because the solute concentration will vary only along the ingot and not in the radial direction, conservation of solute atoms may be described by the one-dimensional continuity equation

$$D\frac{\partial^2 C(x)}{\partial x^2} + v_i \frac{\partial C(x)}{\partial x} = 0 , \tag{2.9}$$

where $D$ is the diffusion coefficient for the solute in the liquid and $v_i$ is the velocity of propagation of the interface. The first term represents the diffusion coefficient (as in Fick's Law, $\partial C/\partial t = D\partial^2 C/\partial x^2$, which is equal to the rate of change of concentration when this arises from diffusion alone). This is balanced by growth at the solid–liquid interface, there being an effective flow of solute in the direction $-x$.

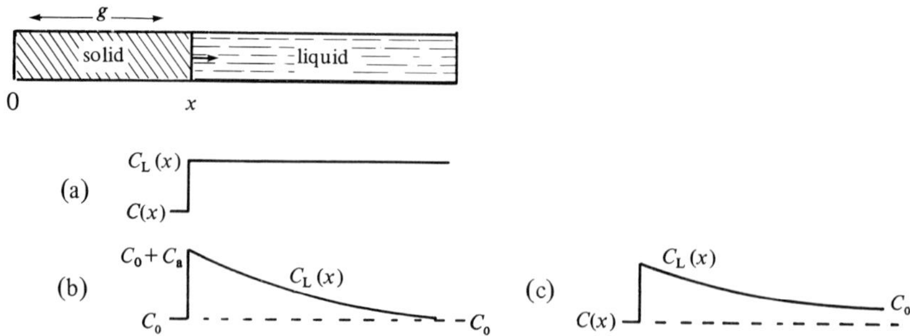


**Figure 2.8.** Impurity concentration for normal freezing, $k_0 < 1$.

The solution of equation (2.9) takes the form

$$C_L = C_a \exp\left(-\frac{v_i x}{D}\right) + C_0 \tag{2.10}$$

(Tiller *et al.*, 1953), where $C_a + C_0$ is the total solute concentration in the liquid at the interface and $x$ is the distance in front of the interface. Hence the distribution of solute in the liquid in front of the interface decays exponentially with a decay constant $v_i/D$. $D$ is approximately 1–10 $cm^2$ per day for liquid metals, whereas $v_i$ can be varied. The concentration of solute in the solid rises from an initial value of $k_0 C_0$ to its equilibrium value of $C_0$ which is maintained through the remaining solidification process. (The amount of solute leaving the liquid during the advance of the interface equals the amount entering the solid and, hence, the concentration of solute in the solid must equal $C_0$.) On applying steady state conditions, equation (2.10) becomes

$$C_L = C_0\left[1 + \frac{1-k_0}{k_0}\exp\left(-\frac{v_i x}{D}\right)\right] . \tag{2.11}$$

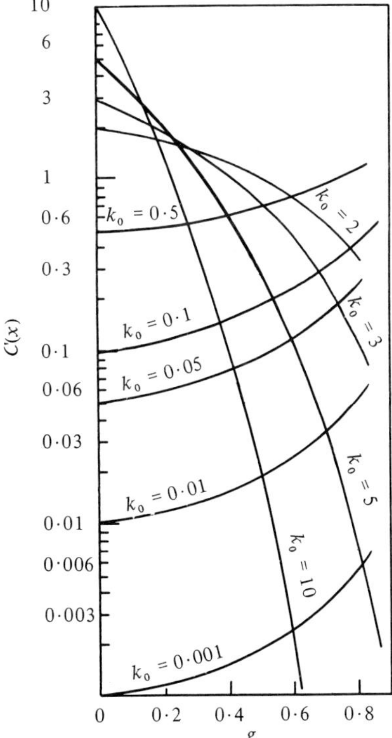


**Figure 2.9.** Variation of impurity concentration along the length of a normally-frozen ingot.

If the speed of growth is changed during a run, bands of higher or lower concentration are produced before the concentration returns to its equilibrium value.

In practice, the distribution of solute in the liquid and the concentration of solute in the solid will take values between cases a and b, as shown by curve c in figure 2.8. An effective distribution coefficient is sometimes used to describe the situation and this distribution coefficient will be a function of $v_i$.

Equation (2.11) is particularly useful when considering constitutional supercooling.

### 2.3.2 Zone melting

This technique consists of passing a molten zone down the length of an ingot (see section 2.6.5). As in the case of normal freezing, a variation of solute concentration occurs down the sample once it has been traversed by a molten zone. Even if one assumes the idealised case of complete diffusion of solute in the liquid phase and no diffusion in the solid phase, the calculation of the resultant solute concentration is more complicated than for normal freezing and depends on such additional factors as the ratio of molten zone length $l$ to ingot length $L$ and the number of traverses.

The nett amount of impurity within the molten zone will depend on the total impurity content of the portion of the ingot liquified, A + B, during passage of the zone, less the amount of impurity left behind in the solidified portion A (see figure 2.10). With $n$ denoting the number of the pass in progress, the concentration of impurity in portion C will be $C_{n-1}(x)$ and in A will be $C_n(x)$. Hence

$$C_n(x) = \frac{k_0}{l}\left[\int_0^{x+l} C_{n-1}(x)\,\mathrm{d}x - \int_0^x C_n(x)\,\mathrm{d}x\right], \tag{2.12}$$

(where the last part of the ingot which is solidified by normal freezing is neglected). Although a limiting distribution can be easily found by putting $C_n(x) = C_{n-1}(x)$, full solution of the equation requires numerical methods (see Pfann, 1953, for typical curves). The limiting concentration

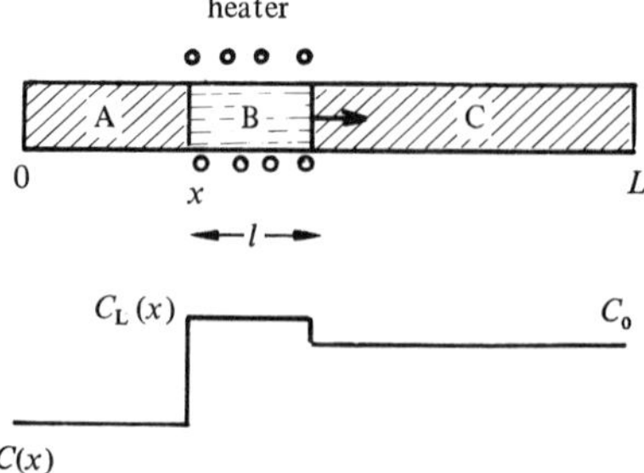


**Figure 2.10.** Impurity concentration for zone melting, $k_0 < 1$.

for the purified part of the ingot is lower for a narrow zone than for a wide zone. However, more traverses are required with the narrow zone to obtain this improved distribution. For a few traverses only, a wider zone can give better results. Ideally, it would be better to pass a wide zone through a few times followed by a number of passages of a narrower zone.

### 2.3.3 Constitutional supercooling

If there is not complete mixing of the solute or impurity within the liquid phase such that a concentration profile shown in figure 2.11a results, then the equilibrium solidus temperature will also vary, as shown in figure 2.11b. When growth is occurring within a furnace, the temperature profile will not necessarily follow that of the solidus temperature. Suppose it varies linearly with distance as shown in figure 2.11c. If b and c are superimposed, as shown in figure 2.11d, then any difference in temperature corresponds to positive or negative supercooling of the liquid. In the case shown the hatched area represents positive supercooling called constitutional supercooling (Rutter and Chalmers, 1953). This encourages cellular dendritic growth and should be avoided by using a steep temperature gradient or a slow growth rate.

However, if the effective distribution coefficient $k_0$ is used (as assumed in the previous section) then the conditions for the onset of constitutional supercooling can be derived mathematically. Provided the amount of solute is small, the equilibrium temperature $T_e$ can be expressed in the form

$$T_e = T_m - mC_L \ . \tag{2.13}$$

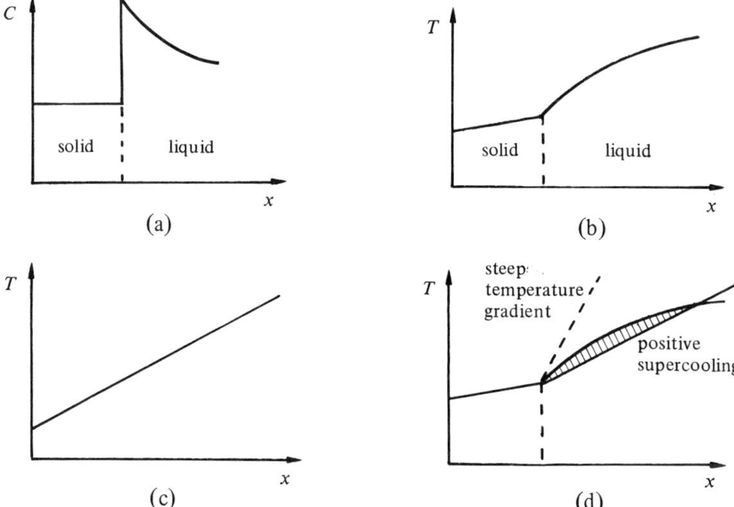


**Figure 2.11.** Constitutional supercooling.

Here, increasing the amount of impurity depresses the solidus temperature from the value $T_m$ for the pure solvent. For small amounts of solute the depression can be taken as proportional to the solute concentration (figure 2.12). At the interface the concentration of the impurity is $C_0/k_0$ so that its temperature is given by

$$T_i = T_m - \frac{mC_0}{k_0} . \tag{2.14}$$

If $dT/dx$ is taken to be constant in the liquid for a small distance $x$ from the interface, then

$$T = T_m - \frac{mC_0}{k_0} + x\frac{dT}{dx} . \tag{2.15}$$

For no constitutional supercooling,

$$\frac{dT}{dx} \geqslant \frac{dT_e}{dx} , \qquad \text{at } x = 0 .$$

Using equations (2.13) and (2.11) gives

$$\frac{dT_e}{dx} = -m\frac{dC_L}{dx} = mC_0\left(\frac{1-k_0}{k_0}\right)\frac{v_i}{D} . \tag{2.16}$$

Hence the condition for no supercooling becomes

$$\frac{dT}{dx} \geqslant mC_0\left(\frac{1-k_0}{k_0}\right)\frac{v_i}{D} . \tag{2.17}$$

Hence if $D$, $k_0$, and $m$ are known, suitable values of $dT/dx$ and $v_i$ can be chosen.

A usual consequence of the constitutional supercooling is the formation of a cellular substructure in the solid. Up to now, a plane interface between solid and liquid has been assumed, but with constitutional supercooling a perturbation of the surface enters the supercooled region and continues to grow. An analysis of the situation now involves lateral

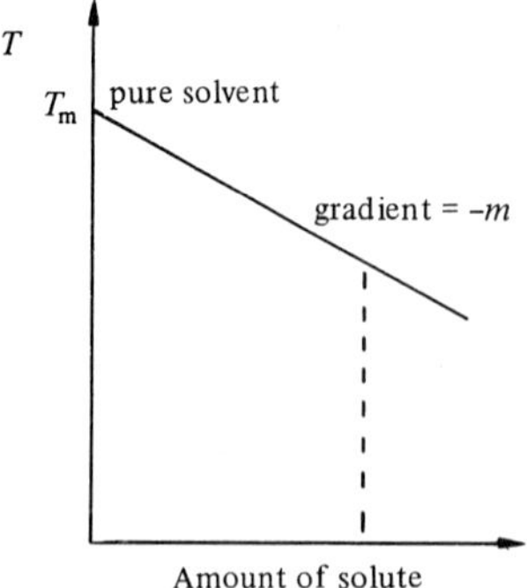


**Figure 2.12.** Solidus temperatures for addition of solute.

as well as longitudinal diffusion of the solute or impurity. The resultant structure is called cellular because of the 'honeycomb' appearance of a transverse section of the solid.

The fact that the cellular structure is stable is thought to be due to the fact that this morphology reduces the amount of constitutional supercooling. Solute-rich liquid solidifies in the grooves between the cells well back from the main interface so that the build up of solute at the interface is restricted. This effect is important not only for growth from solutions containing impurities but particularly for growth of binary and ternary alloys. The effect can be studied more easily in binary alloys because growth rates are slower than in pure metal systems; growth rates in binary alloys are limited by diffusion rather than by heat-transfer rates.

The conditions under which the planar surface is replaced by cellular growth have been studied in some detail, particularly by Mullins and Serkerka (1964) and Serkerka (1967); reviews are given by Serkerka (1968) and Woodruff (1973). The approach used investigates the time dependence of a perturbation to the interface. This perturbation is represented as a Fourier series of sinusoidal functions subject to the effect of diffusion and surface tension. Steady-state thermal and solute diffusion has been assumed, and Serkerka has concluded that this is a good approximation. Constitutional supercooling is a necessary but not a sufficient condition for instability and, hence, cellular growth. It is found that the gradient of the supercooling must exceed a specified amount, but the stability condition depends on a number of other factors, including rate of propagation of the interface $v_i$, thermal conductivities of the solid and liquid phases, and the latent heat of fusion. Experiments by Verhoeven and Gibson (1971) on the melting of Sn–Sb and Sn–Bi alloys showed planar interface stability for high temperature gradients, with a thin layer of constitutionally superheated solid present. However, the melting interface was found to be unstable at low temperature gradients, contrary to theory. This instability was considered to be due to heterogeneous nucleation within the constitutionally superheated region. It is difficult to predict the shape of the cellular surface. From experiment, cell size is proportional to $v_i^{-1/2}$ and largely independent of temperature gradient; the former dependence agrees with theory but the latter is difficult to explain theoretically.

When there is sufficient constitutional supercooling and enough latent heat evolved to produce a negative temperature gradient, dendritic growth occurs. It is a common phenomenon in solidification of alloys and can also occur in growth from the vapour. The name originates from the Greek for tree-like, and arises from the branched effect shown by many dendrites. However, the term is now used rather more generally for growth under these particular temperature conditions.

As with cellular growth analysis, the procedure is to apply a perturbation to the surface and consider the time evolution of the shape.

One different condition is that in dendritic growth it is not necessary for the absolute magnitude of the perturbation to increase, merely that the nonsphericity of the nucleus should increase. Thermal diffusion tends to increase the nonsphericity by allowing the protuberances to grow, whereas capillarity effects arising from surface tension tend to prevent the protuberances from growing. Materials which are able to facet tend to be more stable than those which are not. When formed, dendrites tend to show symmetries associated with their crystal structure; cubic materials show orthogonality of their growth directions, implying four-fold symmetry. Critical radii can be calculated from stability criteria, but the presence of small amounts of impurity leads to instability at smaller radii. No detailed explanation regarding the branching of dendrites is available.

### 2.3.4 Eutectic growth

Occasionally, eutectics of semimetal/semiconductor materials are of direct interest; for instance, Elliott and Hiscocks (1969) have studied the $Cd_3As_2$–NiAs eutectic. Even where the eutectic itself is not directly important, many of the relevant phase diagrams involve a eutectic, so that the morphology of growth of a eutectic is of concern.

As can be seen from figure 2.2 (for instance diagram d), the eutectic consists of two phases $\alpha$ and $\beta$. If the eutectic liquid is cooled unidirectionally, then the two phases may solidify separately, growing as rods or lamellae in the direction of solidification. Hunt and Jackson (1966) suggested that the structure of eutectics be classified according to whether or not the separate phases form facets. Eutectics between nonfaceting phases are the more understood. In this case the most common structure is the lamellar one. It is probable that one phase nucleates first, and then as growth occurs this produces enrichment of the other phase in the liquid whereupon this phase also nucleates, presumably heterogeneously at the liquid–solid interface. However, the lamellae do not originate randomly as they are always oriented in the same crystallographic direction, except for small misorientations. Their spacing is a function of growth rate and decreases as the growth rate increases. Where growth rates are small, the total free energy will be kept down by keeping the $\alpha$–$\beta$ interface small; however, as growth rates increase, diffusion of $\alpha$ and $\beta$ between the spacings becomes significant and a reduction in the lamellar spacing is required. Jackson and Hunt (1966) have given a quantitative analysis. Rod growth, rather than lamellar growth, can occur under certain conditions. Obviously rods, rather than lamellae, will be formed if they can grow under a smaller undercooling. Hunt and Chilton (1962) have found that if $\sigma_L$ is the average interfacial free energy between the $\alpha$- and $\beta$-phases for the lamellae and $\sigma_R$ is the similar quantity for the rods, then the condition for growth of rods occurs when the phase incorporated in the rods has volume less than $(\sigma_L/\sigma_R)^2/\pi$.

For isotropy of the interfacial surface energy $\sigma_L = \sigma_R$, so that for eutectics where there are large ratios between the volumes occupied by the two phases rods will occur; otherwise lamellae are obtained.

For growth of mixed composition $\alpha+\beta$ away from the eutectic point, two possibilities exist:

(i) the primary phase nucleates and grows dendritically into the melt; then, as the liquid between the dendrites becomes enriched with the second phase, eutectic growth can occur;

(ii) eutectic growth occurs below the eutectic horizontal, the volume fractions of the $\alpha$- and $\beta$-phases adjusting to give the correct composition.

Because it is easier for excess solute to diffuse laterally between adjacent lamellae in case (ii) than to diffuse into the bulk liquid in case (i), case (ii) should become dominant at fast growth rates. Figure 2.13 shows typical growth rates for the eutectic and for dendrites of the two phases superimposed on corresponding parts of the phase diagram. Whether or not only a eutectic is produced depends on the composition, the temperature, and the growth rate (Hunt and Jackson, 1967). The analysis used did not take into account temperature-gradient effects, although some work has been done on this.

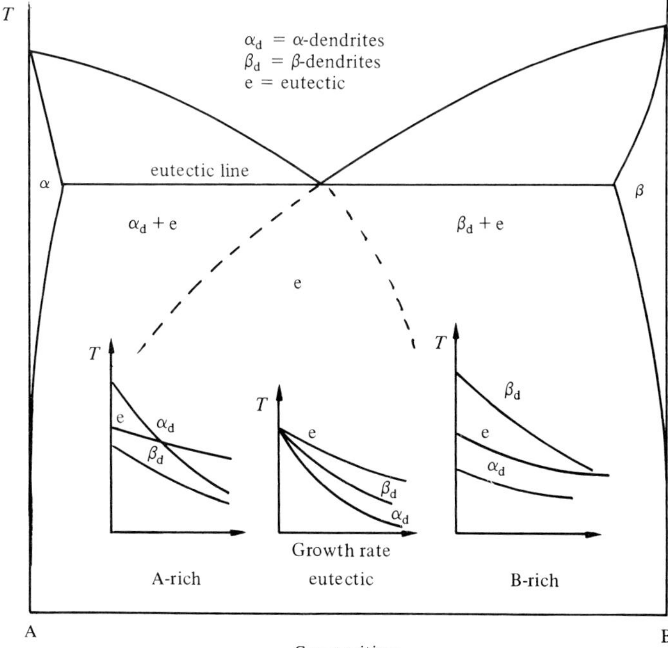


**Figure 2.13.** Eutectic and dendritic growth. (After Hunt and Jackson, 1967.)

If both phases do not tend to facet, more complex structures are possible and the lamellar structure may be broken or show varying degrees of regularity. General reviews of eutectic growth are given, for instance, by Hunt (1968) and (on alloys involving silicon and germanium) Hellawell (1970), and a chapter on eutectic growth is included by Woodruff (1973).

## 2.4 Starting materials and purification

Most of the materials discussed in this text are available commercially in single-crystal form with high purity. Where the compounds are not available, they can usually be obtained by direct reaction between stoichiometric amounts of the constituent elements. Most of the elements are available in 5N or 6N purity (i.e. 99·999% or 99·9999% purity), although great care must be taken to prevent oxidation before actual use. Unless one of the elements reacts with silica (almost without exception the elements relevant here do not, but the presence of oxide may affect the situation) or unless the reaction temperature is above 1200°C, silica vessels are suitable. It is essential that the vessels are thoroughly cleaned. Where a material reacts with silica, a carbon coating is often satisfactory (this can be produced by cracking a hydrocarbon such as toluene within the tube) or carbon can be used as a crucible material. Alumina is available as tubes and can be used up to 1850°C. Platinum crucibles can be heated to 1500°C in air, but there is some risk of recrystallisation.

Either the primary materials can be purified by vacuum distillation, zone melting, etc., or the final material can be purified prior to crystal growth. All the methods of purification are based on a distribution equilibrium between two different phases. At equilibrium, the relation between the concentration of the foreign atoms in the various phases is given by $C_1^n/C_2^m = k$ in which $n$ and $m$ are small integers or simple ratios $(\frac{1}{2}, \frac{1}{3}, ...)$. If $n = m$ (which is usually approximately the case), $k = C_1/C_2$. For the solid/liquid case (the important case) the ratio is $C$(solid)/$C$(liquid), and, as impurities usually segregate in the liquid phase, $k$ is then less than 1. Thus greater purification then occurs where $k$ is smaller. In fact most methods of crystal growth also act as methods of purification at the same time. As has been discussed already in section 2.3.2, zone melting is particularly important as a method of purification.

## 2.5 Growth from the vapour

### 2.5.1 Bulk growth

This method is suitable for the growth of crystals of materials having a large vapour pressure and is an especially useful method for cases where the material has a phase change below the melting point. Both static methods in a closed tube and dynamic methods in either an inert or a reducing gas flow are used. Sometimes the process includes chemical reaction with a carrier gas. The classic example of the method was that used for CdS by Piper and Polich (1961), although it is suitable for a wide range of other compounds.

Figure 2.14 shows a cross section of the furnace and crucible used; also shown is the temperature profile applicable to growing CdS; this profile must be suitably adjusted for other compounds. The profile is achieved by using suitably tapped platinum–10% rhodium wire wound on an alumina muffle. Nucleation is achieved at the conical end of the growing crucible, and usually only one single crystal grows even though the charge is in sintered-powder form. The cone is gradually moved from the hot zone to the cooler region by moving the mullite tube. Hence the length of crystal grown need not be limited by the temperature profile of the furnace. The method is analogous to the Bridgman method used for growth from the liquid and discussed in section 2.6.1. The mullite tube is closed at one end and is gas-tight so that argon can be passed very slowly through the tube. However, once the charge is heated it seals the tube from one end of the growing crucible and so the argon flow was not considered as part of the transport process (but see below). The quartz rod attached to the tip of the crucible is present merely to conduct away heat while the crystal is growing.

Ballentyne *et al.* (1970) considered the growth kinetics when preparing single crystals in sealed tubes, as by the Piper and Polich method (although they used a vertical arrangement). Even though the inert gas is not participating directly in the transport process, its presence does affect the growth conditions when mass transport is considered. They showed that

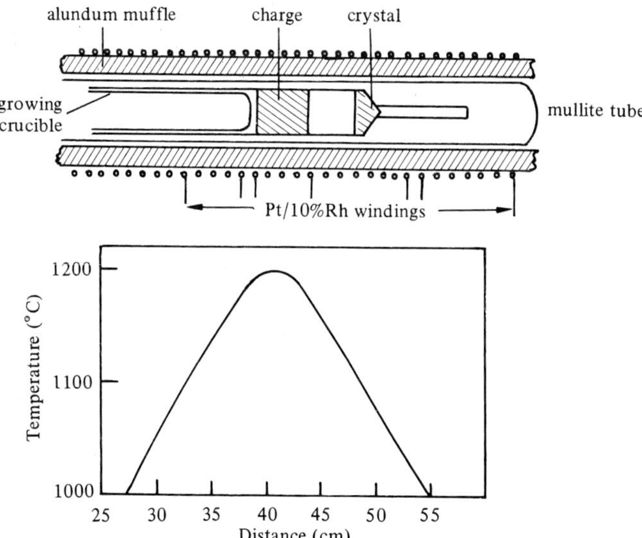


**Figure 2.14.** Furnace arrangement and temperature gradient used by Piper and Polich (1961) for the vapour growth of CdS single crystals.

good quality crystals of CdS were not likely to be obtained from growth in a stoichiometric vapour. For growth from a nonstoichiometric vapour or under an inert gas atmosphere, the growth rate should depend on either the degree of nonstoichiometry or on the partial pressure of the inert gas, as well as on the charge temperature, transport length, and pulling speed. The temperature gradient at the growing surface determines the highest attainable growth rate once charge temperature and partial pressures have been set. However, with these conditions chosen, including the partial pressure of the inert gas, the pulling speed must be chosen within a restricted range of values for good quality crystal boules to result.

Often a two-zone furnace is used to establish vapour transport from the hot zone to a nucleating point in the cold zone. If control of the partial pressure of one of the constituents is required then a further zone is necessary, with one of the constituents held in a reservoir. Such a system was developed by Prior (1961) for the growth of PbSe and is illustrated in figure 2.15. The charge of melt-grown PbSe crystals was placed at A

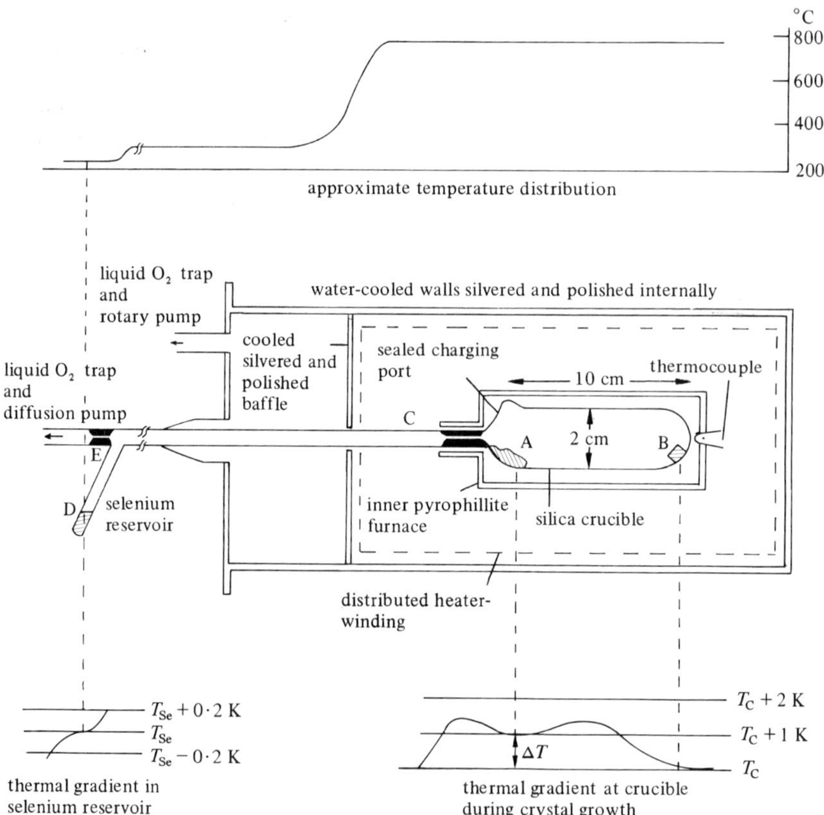


**Figure 2.15.** Furnace arrangement and temperature gradient used by Prior (1961) for the vapour growth of PbSe single crystals.

in the silica crucible and sublimed through the small temperature drop of 1 K to point B which was 10 cm away, the overall temperature being 775°C. The selenium reservoir D was connected to the crucible via a capillary tube C whose diameter was small enough to keep the loss rate of PbSe low and yet to allow control of the selenium partial pressure. The capillary tube E was large enough to allow impurity gases to be pumped out and yet sufficiently small to prevent a reduction in the selenium pressure. Over a period of two to three days, crystals of $\sim 0 \cdot 2$ cm$^3$ were grown and material of a low carrier concentration produced.

Growth in open tubes has been carried out by many workers and carrier gases have included Ar, $N_2$, $H_2S$, $H_2$, or $H_2 + H_2S$, with flow rates from a few cm$^3$ min$^{-1}$ to 300 cm$^3$ min$^{-1}$, or more. Corsini-Mena *et al.* (1971) have grown cadmium chalcogenide single crystals using high flow rates of argon and nitrogen and with the modification that they packed the charge between plugs of quartz wool. The crystals were grown on an alumina rod placed along the axis of the crystallisation chamber.

Reaction transport methods are less applicable to the narrow-bandgap semiconductors, although II–V compounds including InSb, have been grown by closed-tube reversible reactions with the use of, in particular, halogens as the transport reagent. One particular advantage of the method is that it allows control of the concentration profile. Concentrations of dopant in the gas phase can be altered abruptly either by opening a valve or by altering the temperature at which the reservoir of dopant is held; in this way, for instance, an n–p junction can be produced.

### 2.5.2 Thin film growth and sputtering

Most of the results to be discussed will have been obtained from bulk specimens, so little will be said about the preparation of thin films. The usual method of preparing thin films by epitaxial growth is to use a vacuum evaporator (figure 2.16a) in which evaporation of the material is carried out by use of a resistance heater nearby. The substrate can also be heated if desired and a baffle can be positioned between charge and substrate. Very high vacuums of $10^{-9}$ to $10^{-12}$ Torr are possible with very clean conditions (less than monolayer absorption on the surfaces). In practice, a vacuum of $10^{-6}$ Torr is often suitable.

Sputtering is more usually used for the preparation of polycrystalline and amorphous films but growth of thin films is possible. Volatilisation is electrical rather than thermal so that it can occur at lower temperatures. Cathodic sputtering is illustrated in figure 2.16b. A DC potential of several kV is maintained between cathode and anode, the cathode being made of the material to be sputtered. The method is restricted to monocomponent systems, but in reactive sputtering a reactive gas is added and this combines with the species from the cathode to produce a deposit of the compound formed from reaction between the gas and the cathodic material. Voltages of a few kV are again used, together with pressures of

$10^{-1}$ to $10^{-3}$ Torr. An inert gas can be present in addition to the reactive gas. A further technique, that of ion implantation, involves the bombardment of a substrate (the cathode) by ions accelerated in a DC field. The method allows the production of doping levels not normally attainable by thermal diffusion techniques.

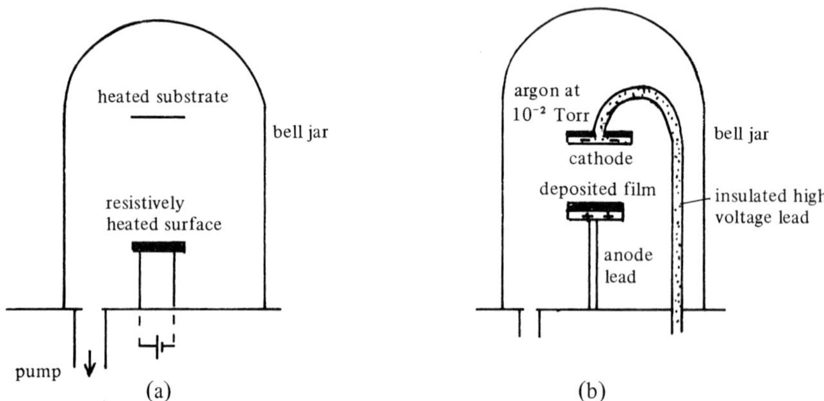


**Figure 2.16.** Arrangements for preparing thin films by (a) evaporation and (b) sputtering.

## 2.6 Growth from the liquid

Growth from the liquid is usually the preferred method of crystal growth. It is used commercially for the growth of a wide range of crystals and the Czochralski method of crystal pulling, in particular, can be used for growing very large crystals. Methods of growing from the liquid are not suitable if at the vapour–liquid transformation the required polymorphic form is not obtained, if the material has a very high vapour pressure at the melting point or if it sublimes, if melting is incongruent (see page 10), or if the melting point is so high that containment of the melt becomes impossible (but see the cold crucible adaptation).

### 2.6.1 Bridgman–Stockbarger method

This method is also called the directional freeze method and, as this description implies, it involves the gradual cooling of the melt from one end. Bridgman (1925) and Stockbarger (1936) used slightly different forms of the method and the general technique has come to be named after them. The cooling can be achieved either by moving the molten charge into the cooler part of the furnace or by lowering the overall temperature of the furnace in such a way that the melt is progressively cooled from one end. In the former variation, either advantage can be taken of the natural drop in the temperature towards the outside of a single-zone furnace (Bridgman variation, figure 2.17a) or the furnace can be of two zones, often separated by a baffle, with a steep temperature gradient between the zones (Stockbarger variation, figure 2.17b).

The charge is contained in a cylindrical capsule, usually of silica, and is held in the hotter part of the furnace which is maintained at a sufficiently high temperature for the charge to be molten. If the melting point of the material is low (below 600°C) then Pyrex can be used for the crucible. At higher temperatures, alumina, graphite, and the noble metals can be used. The capsule with its molten contents is lowered slowly into the cooler zone of the furnace so that solidification progresses from the lower tip. This tip is made either pointed or of a capillary or with a constriction, so that single crystals are selectively produced. Lowering of the capsule is usually achieved by fabricating a hook on the upper end of the capsule and suspending it via suitable high-melting-point wire to a chain or cord which is played out from a synchronised motor. The capsule is often counterweighted with a suitable weight suspended outside the furnace. Constitutional supercooling can be prevented by having a large temperature gradient at the liquid–solid interface; hence a preference for the two-zone method, with a heat baffle between the zones to enhance the temperature gradient. It is advantageous for the capsule to possess a low thermal conductivity. The likelihood of constitutional supercooling is also reduced by keeping the rates of movement slow, and rates actually used for the method range from approximately 1 mm $day^{-1}$ to 1 cm $h^{-1}$.

In the second case the arrangement is often horizontal, with the charge held in a pointed boat (figure 2.18). The temperature of the furnace is raised to melt the charge and then lowered so that the isotherms sweep across the sample from the pointed end of the boat inwards. The method has the advantage that, without movement of the charge, no vibrations are set up. However, it does not provide such large temperature gradients at the liquid–solid interface and so constitutional supercooling can be a problem. Movement of the boat past a stationary furnace can be used in

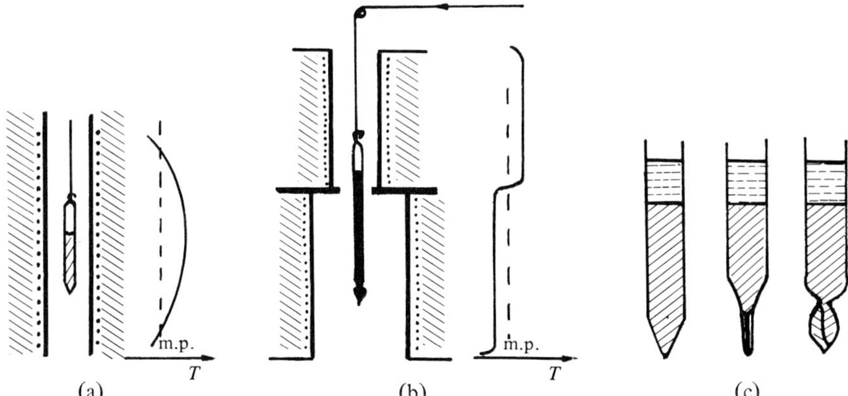


**Figure 2.17.** Vertical Bridgman–Stockbarger method; (a) single zone furnace; (b) two-zone furnace; (c) selection of a single-crystal region.

this horizontal method, but it is often easier and better to keep the boat fixed and move the furnace, thus preventing vibration of the charge.

The Bridgman–Stockbarger method has the advantage that it does not require a seed to initiate growth. Indeed, it is difficult to arrange the temperature and crucible position precisely enough for a seed material to be used since it is necessary to melt the tip of the seed without melting the whole. Thus the method can be used for obtaining initial crystalline material for seeding the much larger crystals which can be grown by the Czochralski method.

The space inside the capsule above the charge can be evacuated or an inert atmosphere can be used. If the vapour pressure of the charge is very high and the melting point in excess of ~800°C, a layer of boric oxide may be included above the charge and the inert atmosphere applied at pressure.

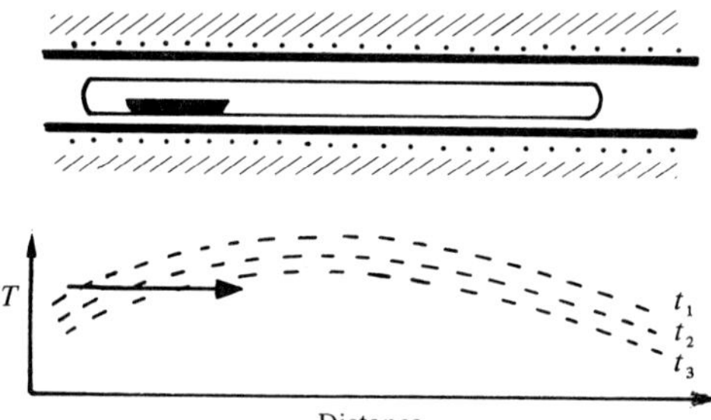


**Figure 2.18.** Horizontal Bridgman–Stockbarger method.

### 2.6.2 Czochralski method

This method of crystal pulling enables the production of large dislocation-free crystals. The technique was originally used by Czochralski (1917) for pulling crystals of metals with low melting points, but it is now used for growing a wide range of materials, including semiconductors and crystals used for optical work.

The melt is contained within a crucible and is usually heated by induction by means of an r.f. coil. Commonly the crucible is made of graphite which acts as the susceptor; i.e. the currents induced in the graphite crucible provide the heating. Otherwise the crucible must be placed on a graphite susceptor. The melt and crucible are electrically and thermally insulated from the rest of the apparatus. So that the temperature can be closely regulated, a thermocouple is inserted in the melt. A seed crystal is suspended from a chuck fitted to a rotating shaft (figure 2.19). This shaft can be raised or lowered, and it is possible for the shaft to be raised very slowly and for the rate of rise to be closely controlled so that the crystal can be grown at any desired rate. Although it is possible to grow some crystals at a rate of up to 10 cm $h^{-1}$, better quality crystals are usually obtained at lower rates. To initiate growth it is preferable to use a seed crystal which is correctly oriented in the best growth direction and the tip of which is just dipped into the melt. If a seed crystal is not available, then it may be possible to commence growth on the tip of a wire or in the end of a capillary. If growth is polycrystalline, it may be possible by changing the pulling rate to neck down the crystal thus limiting it to a single-crystal region, and then allow it to increase in diameter again so as to produce single-crystal growth (figure 2.20). Rotation of the crystal during pulling helps to ensure cylindrical symmetry of the isotherms around the crystal in the melt. Otherwise it is difficult to produce a straight uniform crystal. Even with rotation, the grown crystals are often well-facetted. The diameter at which the crystal grows is dependent both on the temperature of the melt, which affects the viscosity of the melt, and on the rate of pulling. Variation of these parameters enables crystals of the required dimensions to be grown.

Many modifications and improvements to these basic requirements are possible. In order to reduce the concentration of dislocations and other defects in the crystals, it is often desirable to have 'after-heating', which entails placing a second heating element above the main r.f. coil. This heats the crystal as it is withdrawn and so keeps the vertical change of temperature to a minimum. It also tends to reduce the radial temperature gradient. This after-heater often can be merely a ceramic shield which retains the heat.

Usually the pulling arrangement is surrounded by a chamber which allows growth to be carried out in a vacuum or in an inert gas. Part of the chamber can be made of Pyrex so that it remains possible to view the crystal directly as it is grown. In apparatus for high-temperature melts it may be necessary to have water-cooling coils around the chamber.

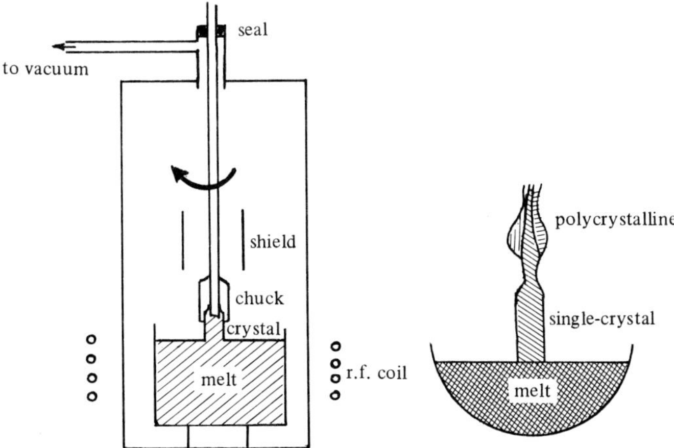


**Figure 2.19.** Crystal pulling (Czochralski) method.

**Figure 2.20.** Necking down to produce single-crystal growth.

It is only possible to grow crystals of constant diameter if there is a balance between the temperature of the melt, the rate of pulling, and the heat loss from the crystal rod. When a material is opaque to infrared and is of low thermal conductivity, it is difficult to achieve the balance. One alternative is to cool the crystal rod immediately above the melt by means of a stream of cool air.

### 2.6.3 Refinements to the Czochralski method

#### 2.6.3.1 *Liquid encapsulation*

A particularly important modification is liquid encapsulation (Mullin *et al.*, 1968) in which the use of boric oxide, $B_2O_3$, enables the method to be extended to materials having a high vapour pressure at the melting point. Boric oxide is a transparent inert glass which, when used in the temperature range of about 800–1350°C, is viscous enough to provide an encapsulating film around the crystal as it is pulled from the melt. Provided a high pressure is applied on the outside, the high vapour pressure is suppressed. After growth the boric oxide is removed by washing. Addition of small quantities of alkaline ions has brought the usable range for boric oxide down from 800°C to below 500°C (it is thought that the addition of these ions breaks the molecular chains and reduces the melting point). Boric oxide is slightly acidic and has the advantage that it tends to remove 'crud'.

#### 2.6.3.2 *Sealed pullers*

When high pressures are necessary, the difficulty is to construct a satisfactory seal around the shaft of the puller so that rotation and pulling are possible through the seal. Various arrangements are possible. At low temperatures rubber seals can be used. At higher temperatures the use of metal bellows may be possible. One arrangement that has been used is the magnetic suspension of the puller, with both the puller and the charge retained within an enclosed capsule.

#### 2.6.3.3 *'Cold' crucible*

Materials that are electrically conducting can be contained in a 'cold' crucible. The actual crucible is made of copper so that it also is electrically conducting; it consists of an array of vertical segments each of which is separately water-cooled and is electrically isolated from its neighbours. Eddy currents are induced both in the segments of the crucible and in the charge thus causing magnetic levitation of the charge. Whereas the segments of the crucible are kept cool by the water cooling, the charge which is no longer in contact with the crucible is heated to the required temperature, which can be considerably in excess of the melting point of copper. The method is particularly useful for high-melting-point semiconductors.

2.6.3.4 *Automatic control*

Automatic control of cross-sectional area has been achieved by Bardsley *et al.* (1972; 1974) by weighing the growing crystal with an industrial weighing cell. This method requires no constraints at the point where the pull rod enters the growth chamber, and so a gas bearing is used here. The rod is rotated by a low-friction pin-and-fork arrangement. By comparing the signal from the weighing cell with that from a rectilinear potentiometer driven from the leadscrew nut, any difference in amplitude can be used to adjust the heating power in the required direction. The desired growth diameter can be set by adjusting the potentiometer output voltage per unit length of rod pull, and initial growth can be automatically controlled by using a nonlinear element in series with output from the potentiometer. Surface tension and buoyancy affect the magnitude of the force measured by the weighing cell and so first and second time derivatives are also used to give more sophisticated servocontrol. This method has the particular advantage that there are no optical components on which condensation could otherwise occur.

A number of workers have controlled crystal diameter by using sophisticated systems incorporating television and electronic device gates (Gartner *et al.*, 1972) or a digital computer (O'Kane *et al.*, 1972), and the diameter has also been controlled with the use of reflections of a laser beam vertically incident onto the free-melt surface (Gross and Kersten, 1972) and by means of Peltier cooling (Vojdani *et al.*, 1974).

**2.6.4 Kyropoulos method**

This is a variation on the Czochralski technique (Kyropoulos, 1926; 1930) being similar in that a seed crystal is dipped into a melt, but different in that the seed is not then withdrawn or rotated. Instead, the temperature of the melt is allowed to drop so that crystalline growth occurs on the seed (figure 2.21). The decrease in temperature of the melt can be produced by the heat leak through the seed crystal, by moving the seed together with the crucible through a temperature gradient, or by reducing the heat input to the melt. Whereas the Czochralski technique is best suited to producing crystals of a large length-to-diameter ratio, the Kyropoulos method tends to produce crystals of a large diameter-to-length ratio.

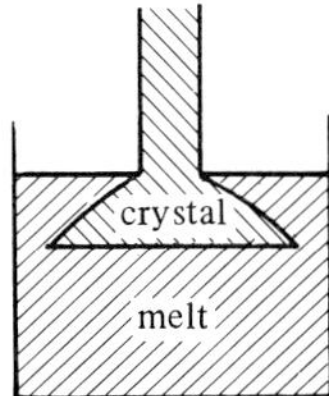


**Figure 2.21.** Kyropoulos method.

**2.6.5 Zone melting method of growth** (see also section 2.3.2)
Somewhat similar to the Bridgman–Stockbarger method, this method involves placing a seed crystal at one end of a long crucible which otherwise contains polycrystalline material. A small zone is melted and moved up the crucible. Usually this zoning is started away from the seed crystal, the zone is moved sufficiently close to the seed crystal for the seed to be wetted, and then the zone movement is reversed. Often it is necessary to control the vapour pressure of one of the constituents by having a reservoir of the constituent kept at a fixed temperature (figure 2.22). One refinement is to carry out the zone melting in a magnetic field. This is found to reduce temperature fluctuations along the sample. Temperature fluctuations should be avoided, particularly in semiconductors where these fluctuations can result in sharply oscillating magnetoresistance along the sample. Zone melting is a very powerful method of purification. A horizontal arrangement is commonly used and the hot zone is produced by means of an additional wire-wound heater. However, vertical arrangements exist and other forms of heating are then common. An r.f. heater can be used, or focused arc lamps, or even a powerful laser beam. In these cases it is usual to have a rod of material inserted top and bottom in chucks which are simultaneously rotated so that the heating is uniform. The whole arrangement is then allowed to move past the heater.

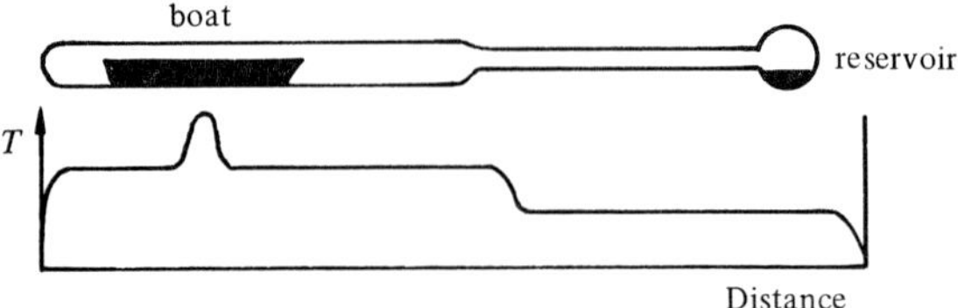


**Figure 2.22.** Zone melting.

## 2.7 Growth by other methods

Outlined here are methods which have been used to a limited extent for the production of crystals of semimetals and narrow-bandgap semiconductors.

**2.7.1 Solution growth**
This is growth from supersaturated solutions and growth rates are largely controlled by the overall temperature and the degree of supersaturation, which can be changed either by decreasing the overall temperature or by evaporation under contolled conditions. Seed material is generally necessary and growth can be very dependent on the quality of the seed material; however, it is possible to improve growth by reseeding and regrowing a number of times. The growth of grey tin from a solution of tin in mercury is a classic example of this method and is described in section 6.1.1.

### 2.7.2 Hydrothermal growth

Various solvents can be used in solution growth in order to obtain a system with suitable solubility, and temperature can also be adjusted to establish the degree of solubility. Another possibility is to carry out growth at high pressure in order to increase solubility, and this method has, for instance, been used to grow arsenic crystals from a solution of hydrogen iodide. The method does not as a rule produce large crystals. Because of the high pressures involved, an autoclave (figure 2.23; Laudise and Nielsen, 1961) is required. This is made of steel or a special high-temperature alloy. Conditions are such that it is cooler at the top in order that the solution will become supersaturated, thereby enabling growth. Replenishment is accomplished by convection currents produced by the temperature differential.

Methods such as the Verneuil method (involving melting of the charge as it falls through an oxyhydrogen flame), most chemical and electrochemical reactions, and solid–solid growth are not of particular interest here.

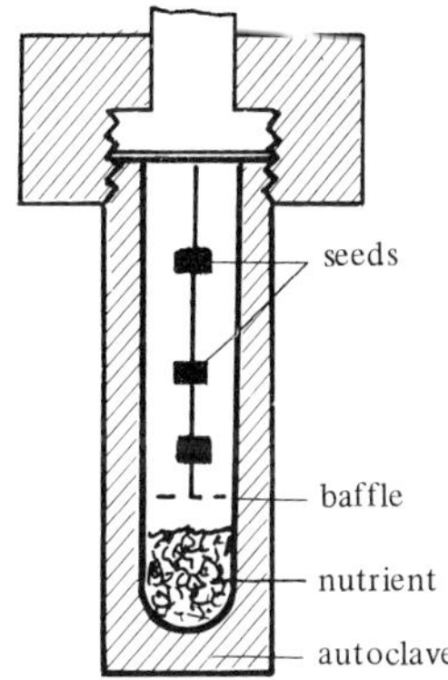


**Figure 2.23.** Hydrothermal growth.

## 2.8 Annealing and phase equilibria

It is usual to reduce the carrier concentration in as-grown material by annealing, i.e. by heating for a suitable length of time at a constant temperature that is lower than the growth temperature. In compound material excess metal and excess nonmetal defects introduce donor and acceptor levels, and annealing alters the concentration of these defects. Samples can be equilibrated under different overall temperature conditions, or it is possible to equilibrate under the excess pressure of one of the components, or samples can be equilibrated with an alloy powder of known composition. The concentrations of the various defects are interrelated and can be calculated from a statistical–thermodynamic approach. The subject is dealt with at length by Kröger (1964) and a good introduction is given by Swalin (1972), although the method was originally proposed by Wagner and Schottky (1931).

### 2.8.1 Types of defect

Primary defects occurring in elements are vacancies (vacant lattice sites), interstitials (atoms occupying interstitial sites), and foreign atoms.

In stoichiometric compounds, although the same types of defect occur, there are five identifiable types of disorder because of the relationship between defect concentrations to maintain stoichiometry. For a compound MX, these are:

(i) Equal numbers of vacancies on the X sublattice, $V_X$, and interstitial atoms of species X, $X_i$ ($[V_X] = [X_i]$); this type of defect is called Frenkel disorder.

(ii) The same as (i) but species M ($[V_M] = [M_i]$).

(iii) Equal numbers of vacancies on the M and X sublattices ($[V_M] = [V_X]$); this is known as Schottky disorder.

(iv) Equal numbers of M and X interstitials ($[M_i] = [V_i]$).

(v) Substitutional disorder with M occupying X sites and X occupying M sites ($[M_X] = [X_M]$); this is called antistructure disorder.

Various defect complexes are also possible, particularly if foreign atoms are present producing, for instance, vacancy pairs. But of particular relevance is the electronic disorder which can occur. The latter can exist without the presence of atomic disorder; with atomic disorder there can be direct interaction. However, interaction between defects becomes less important the more metallic the material, since the mobile electrons in a metal can group around a defect and electrically shield it from other defects.

Provided the defects are present in very low concentrations, the properties associated with them are additive with respect to their concentration. This is really saying that they obey Henry's law, which states that the partial pressure due to a volatile solute A in a nonvolatile solvent is proportional to the concentration of solute atoms, i.e. $p \propto x_A$. Once a compound deviates considerably from stoichiometry, the concentration of defects increases and Henry's law becomes invalid. Even more important, it is no longer necessary for pairs of defects to occur. However, charge neutrality must be preserved at all times. Another point is that when classical statistics apply, both Henry's law and the law of mass action are applicable, but when Fermi–Dirac statistics apply then the law of mass action, coupled with Henry's law, cannot be applied (Swalin, 1972). Hence the treatment which follows should strictly be applied to nondegenerate semiconductors only, but in any case the treatment becomes increasingly useful for larger bandgap material.

### 2.8.2 Electronic aspects of defect theory

A defect present in an otherwise perfect semiconductor can contribute electrons to the conduction band or absorb them from the valence band, and often the defects have an energy level associated with them which lies within the bandgap. The presence of a number of types of defect can give rise to a number of energy levels. The concentrations of electrons in the conduction band and holes in the valence band are related by the law of mass action:

$$K = [n][p] , \tag{2.18}$$

where $K$ is a constant (see table 2.4 for a list of defect symbols). Suppose a Frenkel disorder exists in a compound crystal MX and that ionisation can occur. The overall reaction for the creation of Frenkel disorder is

$$M_M \rightleftharpoons M_i + V_M ; \tag{2.19}$$

the concentrations of $V_M$ and $M_i$ are related by a reaction constant, i.e.

$$K_1 = [M_i][V_M] . \tag{2.20}$$

(As the concentration of interstitials on the interstitial sublattice is small, $[V_i]$ can be treated as a constant.) M will be the more electropositive and can ionise as follows:

$$M_i \rightleftharpoons M_i^+ + n , \tag{2.21}$$

$$M_i^+ \rightleftharpoons M_i^{2+} + n , \tag{2.22}$$

thus acting as a donor. There are reaction energies and reaction constants associated with equations (2.21) and (2.22) and the extent of the reactions depends on the energy of the valence electrons in atom M (when it is in the interstitial position) relative to the energies of the conduction and valence bands. Thus it is possible to place the energy levels for the ionised defects on an energy band diagram.

Similarly, it should be possible for the vacancies to ionise:

$$V_M \rightleftharpoons V_M^- + p$$

etc.

**Table 2.4.** Defect symbols.

| | | | |
|---|---|---|---|
| $M_M$ | M atom on M site | $V_M$ | Vacancy on M site |
| $X_M$ | X atom on M site | $V_X$ | vacancy on X site |
| $M_i$ | M atom on interstitial site | $V_i$ | insterstitial vacancy |
| $X_i$ | X atom on interstitial site | n | electron in conduction band |
| $N_M$ | impurity atom on M site | p | hole in valence band |

### 2.8.3 Kröger-Vink diagrams

The extent to which these defects occur will depend on the conditions of partial pressure under which the crystal is equilibrated. Under a low partial pressure of M, an excess of X will occur in the lattice; as the partial pressure is increased stoichiometry is achieved, and finally with the partial pressure of M increased still further an excess of M is incorporated in the crystal lattice. If it is considered that only singly ionised defects are produced, then for the case under discussion there are six possible defects:

$V_M$, $V_M^-$, $M_i$, $M_i^+$, n, and p, and the concentration of these will vary with the applied pressure. This variation can be shown best by a graph of the logarithm of concentration versus the logarithm of partial pressure of M (Kröger and Vink, 1956), as the concentrations all involve equations of the type of equation (2.20) and will give linear relationships when plotted in this way. However, as indicated, electrical neutrality must be conserved and this involves the relationship

$$[n]+[V_M^-] = [M_i^+]+[p] \,. \tag{2.23}$$

Introduction of this condition removes the simple relationships, because whereas the relations between the logarithms of concentration and partial pressure are linear, this neutrality equation involves a sum rather than a product. This was overcome by Brouwer (1954) by the method of equating dominant pairs of defects.

First consider the use of a low partial pressure of M. As $[M_i]$ will be small and hence also $[M_i^+]$, [n] will be small and [p] large. Equation (2.23) now gives

$$[V_M^-] = [p] \,.$$

These conditions are shown in region I of figure 2.24 (called a Kröger–Vink diagram). As the pressure $p_M$ is increased, region II is obtained

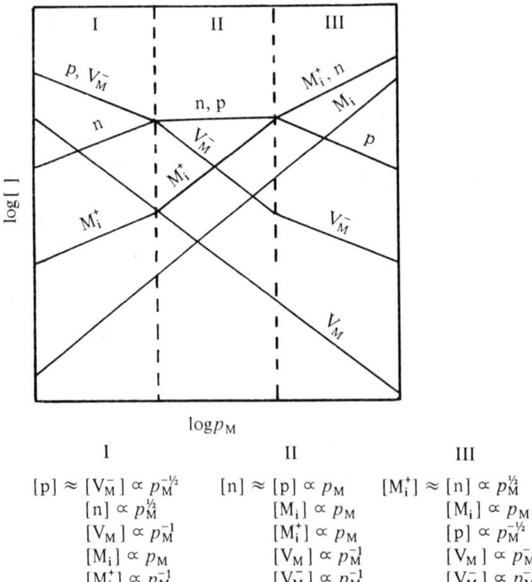


| I | II | III |
|---|---|---|
| $[p] \approx [V_M^-] \propto p_M^{-1/2}$ | $[n] \approx [p] \propto p_M$ | $[M_i^+] \approx [n] \propto p_M^{1/2}$ |
| $[n] \propto p_M^{1/2}$ | $[M_i] \propto p_M$ | $[M_i] \propto p_M$ |
| $[V_M] \propto p_M^{-1}$ | $[M_i^+] \propto p_M$ | $[p] \propto p_M^{-1/2}$ |
| $[M_i] \propto p_M$ | $[V_M] \propto p_M^{-1}$ | $[V_M] \propto p_M^{-1}$ |
| $[M_i^+] \propto p_M^{-1}$ | $[V_M^-] \propto p_M^{-1}$ | $[V_M^-] \propto p_M^{-1/2}$ |

**Figure 2.24.** Relation between the logarithms of concentration [ ] and partial pressure $p_M$ for Frenkel disorder on the M sublattice. (After Swalin, 1972.)

where the dominant condition is

$$[n] = [p]$$

and finally in region III the dominant condition is

$$[M_i^+] = [n] .$$

The proportionalities which are obtained on solving the equations are listed under the diagram.

The exact positioning of all the lines on the graph requires a knowledge of the reaction constants for the particular temperature $T$ in question. For, if actual values are required, a knowledge of the reaction constants is necessary for obtaining both abscissae and ordinates. If the sample is now quenched, the atomic defects are frozen in, whereas the electronic defects remain mobile, equilibrate among the atomic defects, and become much fewer. The dominant defects become after quenching $[V_M]$ $(= [V_M] + [V_M^-]$ before quenching) and $[M_i]$ $(= [M_i] + [M_i^+]$ before quenching).

Similar arguments apply to other defect cases, such as Schottky and antistructure disorder. Additional charge neutrality regions may occur on the Kröger–Vink diagram. From these diagrams it can be seen how

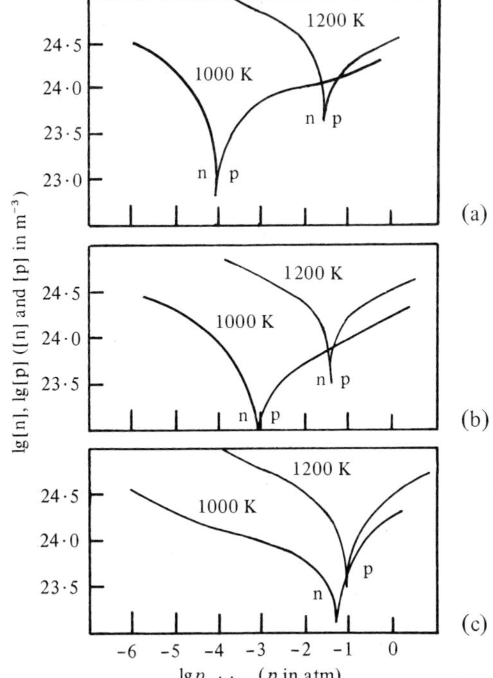


**Figure 2.25.** Electron and hole concentrations in lead sulphide as a function of sulphur pressure. The lead sulphide is (a) doped with silver ($10^{24}$ atoms $m^{-3}$), (b) pure, and (c) doped with bismuth ($10^{24}$ atoms $m^{-3}$). (After Bloem, 1956.)

equilibration of a crystal in a partial gas pressure of one component can control the defect structure. Alternatively, or in addition, the defect structure can be controlled by addition of impurities possessing a charge different from those of the native ions. This system can be analysed in a similar way.

Figure 2.25 shows the effect of a variation of partial pressure of sulphur on the defect structure (concentration of holes and electrons) of lead sulphide (Bloem, 1956). Lead sulphide has a bandgap of 0·286 eV and so lies somewhat outside the area of narrow-bandgap material. The atomic defects here are of Schottky type. Cation vacancies act as acceptors ($V_{Pb} \rightleftharpoons V_{Pb}^{-} + p$) and anion vacancies act as donors ($V_S \rightleftharpoons V_S^{+} + n$), the donor and acceptor levels being close to the conduction and valence bands respectively. When added, silver acts as an acceptor and bismuth acts as a donor.

Although many compounds exist close to stoichiometry, other compounds may exist over a wide range and may not even exist at the stoichiometric composition. This wide range of composition occurs when there is a broad minimum in the plot of Gibbs free energy as a function of composition and is associated with a large amount of intrinsic disorder. However, it is possible that microdomains of different composition exist within a phase and that change of composition corresponds to a change in size of the domains. This probably applies in the case of the vanadium oxides which have similar energies of formation. Plots of free energy against composition take the form shown in figure 2.26 (Eyring and O'Keefe, 1970). Because tangents to the minimum points of the free energy curves for each phase form almost a continuous envelope it is difficult to identify the separate phases (see section 8.5.2).

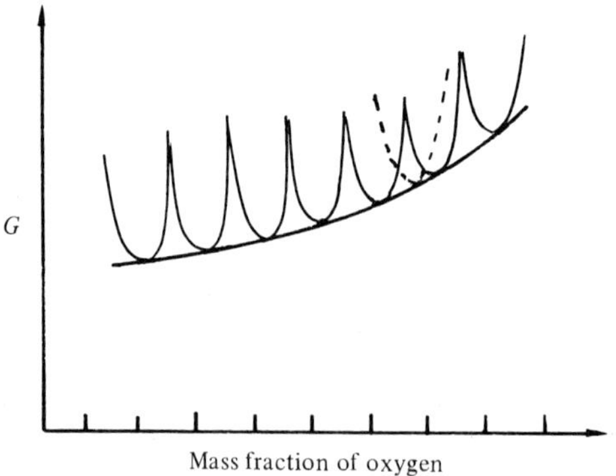


**Figure 2.26.** Form of the free energy as a function of composition for vanadium oxides. (After Eyring and O'Keefe, 1970.)

**References**


Ballentyne, D. W. G., Wetwatana, S., White, E. A. D., 1970, *J. Cryst. Growth,* **7**, 79.

Bardsley, W., Cockayne, B., Green, G. W., Hurle, D. T. J., Joyce, J. C., Roslington, J. M., Tufton, P. J., Webber, H. C., Healey, M., 1974, *J. Cryst. Growth,* **24**, **25**, 369.

Bardsley, W., Green, G. W., Holliday, C. H., Hurle, D. T. J., 1972, *J. Cryst. Growth,* **16**, 277.

Bloem, J., 1956, *Philips Res. Rep.,* **11**, 273.

Brice, J. C., 1973, *The Growth of Crystals from Solution* (North-Holland, Amsterdam).

Bridgman, P. W., 1925, *Proc. Am. Acad. Arts Sci.,* **60**, 305.

Brouwer, G., 1954, *Philips Res. Rep.,* **9**, 366.

Corsini-Mena, A., Elli, M., Paorici, C., Pelosini, L., 1971, *J. Cryst. Growth,* **8**, 297.

Czochralski, J., 1917, *Z. Phys. Chem. (Leipzig),* **92**, 219.

Elliott, C. T., Hiscocks, S. E. R., 1969, *Br. J. Appl. Phys. (J. Phys. D), Ser. 2,* **2**, 1083.

Eyring, L., O'Keefe, M., 1970, *The Chemistry of Extended Defects in Non-metallic Solids* (North-Holland, Amsterdam).

Gartner, K. J., Rittinghaus, K. F., Seeger, A., 1972, *J. Cryst. Growth,* **13**, **14**, 619.

Gilman, J. J. (Ed.), 1963, *The Art and Science of Growing Single Crystals* (Prentice-Hall, Englewood Cliffs, NJ).

Gross, U., Kersten, R., 1972, *J. Cryst. Growth,* **15**, 85.

Hellawell, A., 1970, *Prog. Mater. Sci.,* **15**, 3.

Hunt, J. D., 1968, *J. Cryst. Growth,* **3**, **4**, 82.

Hunt, J. D., Chilton, J. P., 1962, *J. Inst. Met.,* **91**, 338.

Hunt, J. D., Jackson, K. A., 1966, *Trans. Metall. Soc. AIME,* **236**, 843.

Hunt, J. D., Jackson, K. A., 1967, *Trans. Metall. Soc. AIME,* **239**, 864.

Jackson, K. A., Hunt, J. D., 1966, *Trans. Metall. Soc. AIME,* **236**, 1129.

Kröger, F. A., 1964, *The Chemistry of Imperfect Crystals* (North-Holland, Amsterdam).

Kröger, F. A., Vink, H. J., 1956, *Solid State Phys.,* **3**, 307.

Kyropoulos, S., 1926, *Z. Anorg. Chem.,* **154**, 308.

Kyropoulos, S., 1930, *Z. Phys.,* **63**, 849.

Laudise, R. A., 1970, *The Growth of Single Crystals* (Prentice-Hall, Englewood Cliffs, NJ).

Laudise, R. A., Nielsen, J. W., 1961, *Solid State Phys.,* **12**, 149.

Lawson, W. D., Nielsen, S., 1958, *Preparation of Single Crystals* (Butterworths, London).

Mullin, J. B., Heritage, R. J., Holliday, C. H., Stranghan, W. B., 1968, *J. Cryst. Growth,* **3**, **4**, 281.

Mullins, W. W., Serkerka, R. F., 1964, *J. Appl. Phys.,* **35**, 444.

O'Kane, D. F., Kwap, T. W., Gulitz, L., Bednowitz, A. L., 1972, *J. Cryst. Growth,* **13**, **14**, 624.

Pfann, W. G., 1953, *Solid State Phys.,* **4**, 423.

Piper, W. W., Polich, S. J., 1961, *J. Appl. Phys.,* **32**, 1278.

Prior, A. C., 1961, *J. Electrochem. Soc.,* **108**, 82.

Reisman, A., 1970, *Phase Equilibria* (Academic Press, New York).

Rutter, J. W., Chalmers, B., 1953, *Can. J. Phys.,* **31**, 15.

Serkerka, R. F., 1967, *J. Phys. Chem. Solids,* **28**, 983.

Serkerka, R. F., 1968, *J. Cryst. Growth,* **3**, **4**, 71.

Stockbarger, D. C., 1936, *Rev. Sci. Instrum.,* **7**, 133.

Swalin, R. A., 1972, *Thermodynamics of Solids,* 2nd edition (John Wiley, New York).

Tamás, F., Pál, I., 1970, *Phase Equilibria Spatial Diagrams* (Iliffe, London).

Tiller, W. A., Jackson, K. A., Rutter, J. W., Chalmers, B., 1953, *Acta Metall.,* **1**, 428.

Tuck, B., 1975, *J. Mater. Sci.,* **10**, 321.

Verhoeven, J. D., Gibson, E. D., 1971, *J. Cryst. Growth,* **11**, 29, 39.
Vojdani, S., Dabiri, A. E., Ashoori, H., 1974, *J. Cryst. Growth,* **24**, **25**, 374.
Wagner, C., Schottky, W., 1931, *Z. Phys. Chem.,* **B11**, 163.
West, D. R. F., 1965, *Ternary Equilibrium Diagrams* (Macmillan, London).
Woodruff, D. P., 1973, *The Solid-Liquid Interface* (Cambridge University Press, Cambridge).

# 3

# Crystal symmetry and band properties

## 3.1 Symmetry properties and groups

The symmetry properties of crystal lattices have important consequences for the band structure. Group theory can be used to classify the symmetry properties and the notation used will be outlined. Examples of introductory texts on this aspect of group theory are Cracknell (1968) and Wooster (1973), and the subject will not be treated in depth here. Many of the semimetals and semiconductors have the diamond or zinc-blende structure or alternatively a face-centred cubic (f.c.c.) structure (figure 3.1). The first Brillouin zone for each of these structures is a truncated octahedron[(1)] and the notation for the main symmetry points is indicated in figure 3.2. The symbols Γ, X, K, and W stand for points as shown, and Δ, Λ, Σ, Q, and Z stand for axes.

Consider a limiting case in which the periodic potential within the lattice tends to zero (free electron limit) but symmetry properties of the wave function are retained. Expressions for the energy bands are obtained of the form

$$E = \frac{\hbar^2}{2m}(\boldsymbol{k}+\boldsymbol{K}_n)^2 \tag{3.1}$$

where $\boldsymbol{K}_n$ is a reciprocal lattice vector, and these expressions resemble the form of the energy bands in real materials and show the degeneracy as indicated by the symmetry of the points in the Brillouin zone. When a periodic potential is added, much of this degeneracy is removed.

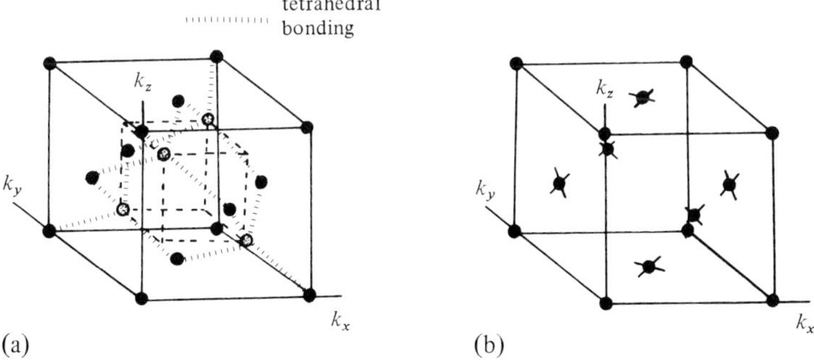


**Figure 3.1.** (a) Zinc-blende and (b) face-centred cubic lattices.

(1) This follows convention. If the parts of the plane shown shaded in figure 3.2 are bulged outwards and the unshaded parts of the same plane correspondingly pushed inwards, the resulting surface would still be satisfactory.

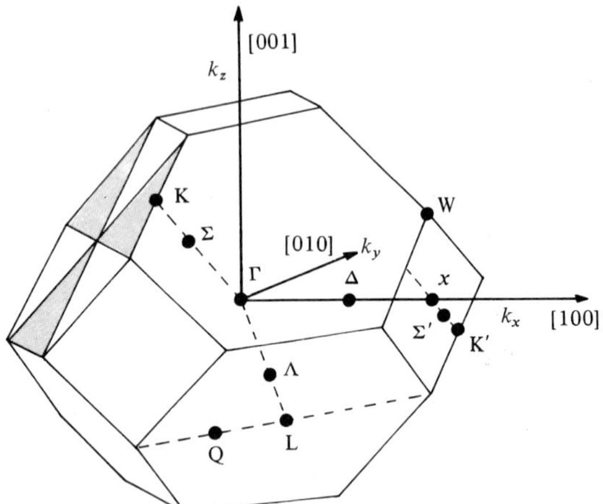


**Figure 3.2.** The first Brillouin zone applicable to the face-centred cubic, diamond, and zinc-blende lattices.

### 3.1.1 Character tables

In order to make actual deductions about the energy bands from symmetry, it is usual to construct a so-called character table. A matrix representation is used for the group of symmetry operations involved and the character of each matrix is the sum of the elements down the leading diagonal. The matrix representations provide linear transformations which can be applied to vectors in a generalised space. There are an infinite number of alternative matrix representations of a group; the one used is an irreducible representation, say $A$, having by definition the property that $M^{-1}AM$ (where $M$ is a unitary matrix) has the same character as $A$. It is possible to deduce the character of an irreducible representation without in fact knowing the matrix itself. For a group there is a finite number of irreducible representations. The number is equal to the number of classes within the group. (Elements which are conjugate are said to be equivalent, and sets of equivalent elements are called classes.) Also, the sum of the squares of the dimensions of the representations equals the number of elements of the group.

Continuing to take as an example zinc blende consider the Γ point at the centre of the Brillouin zone. There are twenty-four symmetry operations in all and these can be divided into classes. The operators involved are the identity operator, $E$, in a class of its own; rotation by $\pm\frac{1}{2}\pi$ about one of the fourfold axes (conventionally taken as the coordinate axes), class $C_4$ (six operations in all); rotation by $\pi$ about the same axis, class $C_4^2$ (three operations); and rotation by $\frac{1}{3}\pi$ about a threefold axis, class $C_3$ (eight operations). In addition, the inversion operator $J$ is a symmetry operator in conjunction with any of the six operators of class $C_4$ and also in conjunction with a rotation of $\pi(C_2)$ about certain axes (six in all).

It was shown previously that all wavevectors of the form $k+K_n$ where $K_n$ is a reciprocal lattice vector, were equivalent. This still applies at a general point but at the points of symmetry there is the more general case

$$\beta k = k+K_n , \tag{3.2}$$

where $\beta$ is an operator within the space group of the lattice. It is now necessary to construct the character table for each of the points of interest in the Brillouin zone. Again as an example consider the point at the centre where group $\Gamma$ applies. States characterised by wavevectors and belonging to the same irreducible representation must have the same energy since the wave functions can be transformed into each other by the operations of the group. Hence, energy bands will have the complete symmetry of the reciprocal lattice. Where there is degeneracy within the point group, there the energy bands will merge. Table 3.1 is the character table for the representations of the single group $\Gamma$.

Deduction of the characters and their association with particular representations will not be given here. The character for $E$ will be the dimension of the representation and gives the degeneracy of the energy level. The bases give the symmetries of the wave functions. Thus the wave functions of the type $\Gamma_1$ have the full symmetry of the point group. $\Gamma_{15}$ is also a simple but triply degenerate type which occurs at low energy, whereas $\Gamma_{12}$ is doubly degenerate with symmetries of the form $x^2-y^2$, and $z^2-\frac{1}{2}(x^2+y^2)$ and with two nodal surfaces. In fact, the four tetrahedral orbitals in zinc blende can be associated with the reducible representations of $\Gamma$. The first two electrons (note however that spin has not been considered yet) are of $\Gamma_1$ type and are the s contribution to the orbitals, whereas the $\Gamma_{12}$ states can be thought of as associated with the d-states.

From the representations and approximate arguments as to which energy levels will be lower it is possible to construct an approximate band structure for zinc blende.

**Table 3.1.** Character table for the representation of the single group $\Gamma$.

| Representation | Basis | $E$ | $3C_4^2$ | $8C_3$ | $6JC_4$ | $6JC_2$ |
|---|---|---|---|---|---|---|
| $\Gamma_1$ | 1 | 1 | 1 | 1 | 1 | 1 |
| $\Gamma_2$ | $x^4(y^2-z^2)+y^4(z^2-x^2)$ $+z^4(x^2-y^2)$ | 1 | 1 | 1 | −1 | −1 |
| $\Gamma_{12}$ | $x^2-y^2, z^2-\frac{1}{2}(x^2-y^2)$ | 2 | 2 | −1 | 0 | 0 |
| $\Gamma_{15}$ | $x, y, z$ | 3 | −1 | 0 | −1 | 1 |
| $\Gamma_{25}$ | $z(x^2-y^2)$ | 3 | −1 | 0 | 1 | −1 |

### 3.1.2 Atomic orbitals

By comparing the bases shown in table 3.1 with the atomic orbitals drawn in figure 3.3 analogy between the two can be seen. The surfaces of the orbitals represent approximate boundary surfaces for the electrons drawn such that, say, 10% of the charge lies outside the drawn surface. Signs are the signs of the electron wave functions. The s orbital is spherically symmetric but the others are asymmetric and show nodal planes (i.e. intermediate regions where the wave functions are zero). Any symmetry operation on an s orbital leaves the orbital unchanged; operation of certain rotations or reflection operations on the p and d orbitals will in some cases restore the orbital to its original position unchanged, and in some cases restore the orbital but with a change of sign for the wave function.

We can extend this to molecular orbitals; a simple way of considering these is by linear combinations of the atomic orbitals (LCAO method). This method will be discussed further (see page 50). Orbitals on adjacent atoms will produce either bonding or antibonding orbitals overall. When overlapping orbitals are of the same sign, there is a high probability of electrons occupying the region and hence bonding.

In diamond-like structures, such as silicon and germanium, the four electrons in an isolated atom occupy two s states and two p states, but in the crystal the lowest states are one s and three p with the orbitals hybridised so as to give four characteristic loops with tetrahedral symmetry and forming the covalent bonds. The angular parts of the wave function

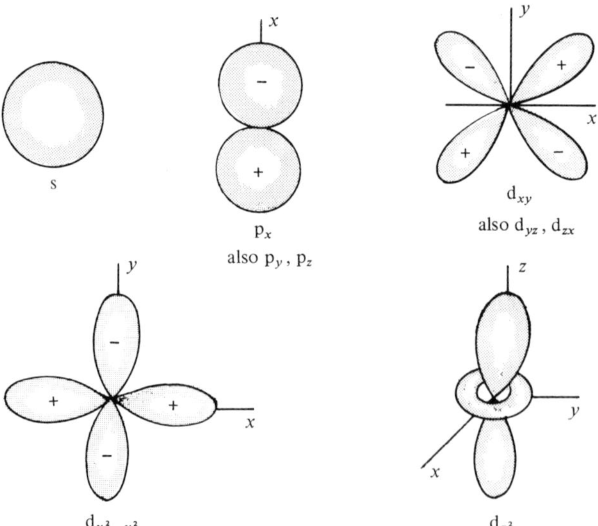


**Figure 3.3.** Atomic orbitals.

(required later) are;

$$s = 1 \text{ and } \quad p_x = \frac{x}{r} = 3^{1/2}\sin\theta\cos\varphi ,$$

$$p_y = \frac{y}{r} = 3^{1/2}\sin\theta\cos\varphi ,$$

$$p_z = \frac{z}{r} = 3^{1/2}\cos\theta ,$$

where $\theta$ and $\varphi$ are the angles used for spherical polar coordinates. The hybridised orbitals have $sp^3$ forms:

$$s + p_x + p_y + p_z ,$$

$$s + p_x - p_y - p_z ,$$

$$s - p_x + p_y - p_z ,$$

$$s - p_x - p_y + p_z ;$$

they then describe the orbital lobes in the four directions and all the valence electrons are identical as far as bonding is concerned. In zinc-blende structures each atom still has an average of four electrons available for forming the bonds but the neighbours have unequal numbers of valence electrons.

However, there appears to be a contradiction in that d-orbital contributions appear to be obtained from the $\Gamma$ representation. In fact, by deriving the wave function from overlapping orbitals for the condition of bonding when the orbitals are of the same sign, orbital shapes of similar form to the d-orbitals are obtained. Hence, the d functions obtained from the $\Gamma$ point group can be used to represent the p-like valence band edge in diamond and zinc-blende type crystals, the functions having the same orthogonality and general shape.

### 3.1.3 Spin-orbit coupling and double groups

So far electron spin has not been included in the arguments although its importance has been implied. Obviously electron spin will introduce extra degeneracies because of time reversal symmetry. At a general point in the Brillouin zone the representation will become doubly degenerate with the presence of spin, but any change will be small. At high symmetry points, however, the addition of spin can cause splitting of the energy levels and to take this into account it is necessary to consider the product of the representation $\Gamma$ and the function $S$ which represents spin ($S$ and $\Gamma$ commute) giving rise to a so-called double group, because inclusion of spin with the point group doubles the number of elements within the group.

Table 3.2 gives the complete table for the double group $\Gamma$. However, the direct product may be reducible. For instance $S^{-1} \times \Gamma_{15}$ gives $\Gamma_7 + \Gamma_8$

in the double group; hence a single energy level splits into two energy levels when spin is taken into account, each energy level having its respective degeneracies. Compatibility between the single and double $\Gamma$ groups is shown in table 3.3. [Note that in diamond, where there is inversion symmetry, there will be $\Gamma_6^-$ and $\Gamma_6^+$, both corresponding to $\Gamma_6$ shown here, etc; often the sign in the superscript is also used for zinc blende but $\Gamma_6^-$ is then the same as $\Gamma_6^+$.]

Similar tables can be constructed for other points in the Brillouin zone of zinc blende (see Parmenter, 1955; Dresselhaus, 1955) and also for any other crystal class.

**Table 3.2.** Character table for the representation of the double group $\Gamma$.

| | $E$ | $\bar{E}$ | $6C_4^2$ or $\bar{C}_4^2$ | $8C_3$ | $8\bar{C}_3$ | $6JC_4$ | $6J\bar{C}_4$ | $12JC_2$ |
|---|---|---|---|---|---|---|---|---|
| $\Gamma_1$ | 1 | 1 | 1 | 1 | 1 | 1 | 1 | 1 |
| $\Gamma_2$ | 1 | 1 | 1 | 1 | 1 | −1 | −1 | −1 |
| $\Gamma_3$ | 2 | 2 | 2 | −1 | −1 | 0 | 0 | 0 |
| $\Gamma_4$ | 3 | 3 | −1 | 0 | 0 | −1 | −1 | 1 |
| $\Gamma_5$ | 3 | 3 | −1 | 0 | 0 | 1 | 1 | 1 |
| $\Gamma_6$ | 2 | −2 | 0 | 1 | −1 | $\sqrt{2}$ | $-\sqrt{2}$ | 0 |
| $\Gamma_7$ | 2 | −2 | 0 | 1 | −1 | $-\sqrt{2}$ | $\sqrt{2}$ | 0 |
| $\Gamma_8$ | 4 | −4 | 0 | −1 | 1 | 0 | 0 | 0 |

**Table 3.3.** Compatibility between single and double $\Gamma$ groups.

| | | | | | |
|---|---|---|---|---|---|
| Single group | $\Gamma_1$ | $\Gamma_2$ | $\Gamma_{12}$ | $\Gamma_{15}$ | $\Gamma_{25}$ |
| Double group | $\Gamma_6$ | $\Gamma_7$ | $\Gamma_8$ | $\Gamma_7+\Gamma_8$ | $\Gamma_6+\Gamma_8$ |

### 3.1.4 Scheme of the band structure for type zinc blende

The energy bands must have the full symmetry of the Brillouin zone and must form continuous curves in $k$-space with continuous derivatives. The nature of the bands in the vicinity of the symmetry points is considered first. There are various possible forms of curves at these points depending on whether the slope is zero or finite. If perturbation theory is used, a band structure as illustrated in figure 3.4 is obtained. In this sketch the spin–orbit splittings are considerably exaggerated. The constant energy surfaces for the bands represented by $\Gamma_6$ and $\Gamma_7$ are spheres, whereas the constant energy surfaces for $\Gamma_8$ are warped. The important point is the top edge of the valence band; it can be seen that, although it is along a principal axis, the valence band maximum can be off the $k = 0$. This is not so in diamond where the maximum occurs at $k = 0$.

The forms of the bands and how they can be determined will be considered in more detail later but figure 3.4 for the band structure of zinc blende will be used as a comparison when discussing the band structures of actual materials in part II.

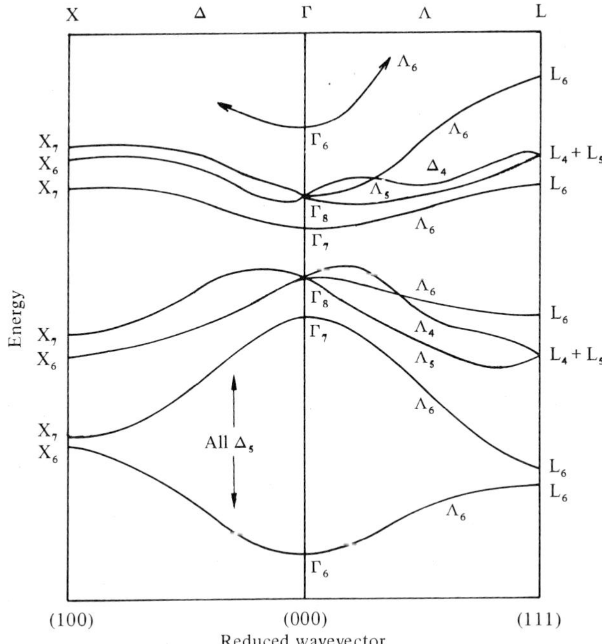


**Figure 3.4.** Schematic diagram of the energy levels for a zinc-blende type structure in which the spin-orbit splittings have been highly exaggerated. (After Dresselhaus, 1955.)

## 3.2 Calculation of band structure

### 3.2.1 Summary of methods

Various approaches have been used in the calculation of the electron states in solids and, hence, of the actual form of the band structure. Most of the methods have been applied first to metals, where they proved most suitable, but have been increasingly applied to semiconductors. The Korringa, Kohn and Rostoker (KKR) and the augmented plane waves (APW) methods have been applied particularly successfully to metals; the orthogonalised plane wave (OPW) method has proved highly suitable for covalent semiconductors. The approach for other solids is very similar. As has been indicated already, the main features of the band structure are determined by the crystal structure, and the nature of the actual atoms becomes a perturbation to the calculation. There is no room here to consider the methods in detail but a brief review of the methods is included so that when band calculations for real materials are referred to in part II the reader will know which method has been used. For further information, the reader is recommended to consult such texts as Quinn (1973) and Callaway (1964) which discuss band-structure methods in detail, or Ziman (1972) which is a general text on solid state physics.

Elementary solid-state theory distinguishes from the beginning between the core electrons which are almost completely localised and the valence or conduction electrons whose states extend throughout the whole crystal and are assumed to be describable by Bloch functions. The tight-binding approximation which starts from wave functions for the free atoms is particularly good for the core electrons. The wave functions are added linearly (hence the so-called LCAO method—linear combination of atomic orbitals method) and is particularly useful in quantum chemistry for obtaining molecular wave functions. However, the method is not very successful for obtaining Bloch functions in solids, although the problem has been overcome to a certain extent, particularly for the alkali metals, by the use of cellular methods. The nearly-free electron model is a good starting point for the conduction electrons. Starting from an equation of the form of equation (3.1), representing the parabolic relationship between energy and wavevector for the free electron, the method satisfactorily shows the presence of energy gaps as a consequence of the periodicity of the lattice which leads to discontinuities at the Brillouin zone boundaries. The problem is one of introducing the potential arising from the ion cores. Until pseudo-potentials were introduced, the strong fields around the ion cores necessitated the introduction of short-wavelength Fourier components and hence of a large number of terms.

3.2.1.1 *Orthogonal plane wave (OPW) method*
Herring (1940) used the simple method of expanding the valence and conduction states as plane waves; here, in addition, the wave functions for the outer states are expanded as sets of plane waves which are orthogonalised to the wave functions of the energies. Hence the method is called the orthogonalised plane wave (OPW) method and overcomes the difficulty of describing the rapid variation of the wave-function near an atomic nucleus. The normalised OPW function can be combined in a similar way to plane waves at points of high symmetry in the Brillouin zone to form the basis functions for the irreducible representation of the crystal space group.

3.2.1.2 *Augmented plane wave (APW) method*
This uses the so-called muffin-tin potentials (figure 3.5) consisting of spherically symmetric potential regions centred at the ion sites with a constant potential in the interstitial regions. The Schrödinger equation for the motion of an electron in the spherically symmetric potential can be solved in spherical polar coordinates. The augmented plane wave after which the method is named is equal to the plane wave outside the regions of spherical potential symmetry, and to a general linear combination of spherical harmonics multiplied by radial functions within the spherically symmetric potential. Coefficients are chosen to meet the requirements for acceptable wave functions, and at the limit of the spherically symmetric potential, the wave functions for the two types of regions must be matched in value and in their logarithmic derivatives. The method was originated by Slater (1937) and computational techniques have been

applied to it in recent years. The number of APW functions required depends on the crystal structure and the type of band involved. It is usual that s and p bands give more rapid convergence than d bands.

**Figure 3.5.** Muffin-tin potentials.

3.2.1.3 *Korringa, Kohn and Rostocker (KKR) method*

This also assumes a muffin-tin potential with a constant potential outside the spherically symmetric potential regions. The wave function is considered as scattered by the potential itself and so Korringa (1947) divided the wave function $\chi_k(\boldsymbol{r})$ into incoming and outgoing components. An integral equation approach was introduced by Kohn and Rostocker (1954). Integration must be carried out avoiding poles. The integration is written as the sum of terms over all the muffin-tin spheres and then transformed to an integral over the muffin-tin sphere around the origin. The KKR method (also sometimes called the Green function method) is in fact very similar to the APW method. In the KKR method it is necessary to sum over (reciprocal) lattice vectors so that a secular determinant in contributions from different spherical harmonics is obtained. In the APW method expansion is made as far as is considered necessary in spherical harmonics and then a secular determinant is obtained which requires solution for contributions from the different (reciprocal) lattice vectors.

3.2.1.4 *Pseudo-potential methods*

Calculation of the muffin-tin potential distribution for a particular crystal can be a tedious process and need not be necessary. Instead, the atomic potential is replaced by a weak potential having in the case of the KKR method the same scattering amplitude for the conduction electrons. This gives rise to the so-called empirical pseudo-potential methods.

3.2.1.5 *The* $\boldsymbol{k}\cdot\boldsymbol{p}$ *method*

This is a method using quantum perturbation theory in conjunction with requirements from crystal symmetry to investigate the wave functions and form of the energy bands at particular points in $\boldsymbol{k}$ space and particularly at $k = 0$. The band structure at a particular point can be determined by obtaining experimentally a limited number of parameters such as bandgaps and electron and hole effective masses. The limitation on the method for obtaining the band structure throughout $k$-space is the detail and accuracy with which the $\boldsymbol{k} \cdot \boldsymbol{p}$ matrix can be determined, and, because the matrix depends on a great many parameters, it is not easy to determine it experimentally in full. The method has been extensively developed by Kane (1957, 1966) and its application leads to the so-called Kane band form which is of considerable importance for narrow-gap materials.

**3.2.2 Application of perturbation theory to a nondegenerate energy band system**

As a consequence of the translational symmetry of the crystal, the Schrödinger equation for an electron in the lattice is

$$H\chi(\boldsymbol{r}) = \left[\frac{p^2}{2m} + V(\boldsymbol{r})\right]\chi(\boldsymbol{r}) = E\chi(\boldsymbol{r}) \tag{3.3}$$

where $\chi(\boldsymbol{r})$ is the wave function for the electron in the lattice (a one-electron model is used as a good first approximation), $H$ the Hamiltonian, $\boldsymbol{p}$ and $m$ the momentum and mass respectively of the electron, $E$ the energy eigenvalue and $V(\boldsymbol{r})$ the periodic potential in which the electron moves in the lattice. $\boldsymbol{p}$ is an operator given by $\boldsymbol{p} \rightarrow \mathrm{i}\hbar\nabla$, so solutions of equation (3.3) take the form (Bloch's theorem)

$$\chi_k(\boldsymbol{r}) = \exp(\mathrm{i}\boldsymbol{k} \cdot \boldsymbol{r})u_k(\boldsymbol{r}) \tag{3.4}$$

where $\chi_k(\boldsymbol{r})$ are called Bloch functions and are indexed by $k$ because they have the periodicity of the lattice; $u_k(\boldsymbol{r})$ also has the periodicity of the lattice. Equation (3.3) ignores interaction between the magnetic moments associated with the electrons and the orbital angular momentum (i.e. assumes no spin–orbit coupling). This will be considered later.

Substituting equation (3.4) into equation (3.3) and using the operator $\boldsymbol{p}$ gives

$$\left[\frac{p^2}{2m} + \frac{\hbar}{m}\boldsymbol{k} \cdot \boldsymbol{p} + \frac{\hbar^2 k^2}{2m} + V(\boldsymbol{r})\right]u_k(\boldsymbol{r}) = E_k u_k(\boldsymbol{r}) . \tag{3.5}$$

The second term in this equation is the $\boldsymbol{k} \cdot \boldsymbol{p}$ term which gives the $\boldsymbol{k} \cdot \boldsymbol{p}$ method its name. Both this term and the third term, the $k^2$ term which is usually very small, can be dealt with by perturbation theory. (Without spin–orbit coupling it is not essential, however, to use perturbation theory.) The solution of the unperturbed equation

$$\left[\frac{p^2}{2m} + V(\boldsymbol{r})\right]u_k^0(\boldsymbol{r}) = H^0u^0(\boldsymbol{r}) = E_k^0 u_k^0(\boldsymbol{r}) \tag{3.6}$$

must of course be known. $H^0$ is the Hamiltonian and $u^0$ the wave function for the unperturbed case $k = 0$.

The $k = k$ case is now considered as a perturbation of the $k = 0$. $H$ can be put in the form $H^0 + H'$, where $H^0$ is the dominant part obtained for $k = 0$ and $H'$ is a much smaller part arising from the perturbation on this original system. In fact it is easier to put

$$H = H^0 + \lambda H' \tag{3.7}$$

where $\lambda$ is a variable parameter such that when $\lambda = 0$ there is no perturbation and when $\lambda = 1$ the fully perturbed situation exists. Considering the $x$ direction only, it is necessary to solve [from equation (3.3)] an equation of the form

$$Hu_k(x) = E_k u_k(x)\,. \tag{3.8}$$

Assuming initially that for each value of $u_k(x)$ there is a distinct value of $E_k$ (i.e. assuming that $E_k$ is nondegenerate), we can expand the eigenfunctions and eigenvalues as power series of the variable parameter $\lambda$:

$$u_k = u_{0k} + \lambda u_{1k} + \lambda^2 u_{2k} + \ldots\,, \tag{3.9a}$$

$$E_k = E_{0k} + \lambda E_{1k} + \lambda^2 E_{2k} + \ldots\,. \tag{3.9b}$$

Here, successive terms represent smaller and smaller corrections to the unperturbed terms. Equations (3.7) and (3.9) can now be substituted into equation (3.8) to give a polynomial in $\lambda$. As $\lambda$ is a variable, the coefficients of each power of $\lambda$ in the polynomial must be equal. This then gives

$$H^0 u_{0k} = E_{0k} u_{0k} \qquad \text{(unperturbed case)}, \tag{3.10a}$$

$$H^0 u_{1k} + H' u_{0k} = E_{0k} u_{1k} + E_{1k} u_{0k}\,, \tag{3.10b}$$

$$H^0 u_{2k} + H' u_{1k} = E_{0k} u_{2k} + E_{1k} u_{1k} + E_{2k} u_{0k} \quad \text{etc.} \tag{3.10c}$$

As $\lambda \to 0$,

$$u_k \to u_{0k} \equiv u_k^0\,, \qquad E_k \to E_{0k} \equiv E_k^0$$

giving the zeroth order terms. It is possible to expand $u_{1k}$ in terms of the unperturbed eigenfunctions $u_k^0$ via a series of coefficients $c_{n'k}^{(1)}$:

$$u_{1k} = \sum_{n'} c_{n'k}^{(1)} u_{n'}^0 \tag{3.11a}$$

or, more generally,

$$u_{nk} = \sum_{n'} c_{n'k}^{(n)} u_{n'}^0\,. \tag{3.11b}$$

Substitution of (3.11a) into (3.10b), multiplication by $u_k^{0*}$ and finally integration over the relevant volume gives

$$E_{1k} = (H')_{kk}\,, \tag{3.12}$$

where $(H')_{kk}$ is the diagonal element of the matrix of the perturbing Hamiltonian $H'$ evaluated with the use of the unperturbed wave function $u_k^0$ ($u_k^{0*}$ = complex conjugate). Hence, this equation enables the first order perturbation of the energy to be obtained.

By multiplying by $u_l^{0*}$ where $l \neq k$, instead of multiplying by $u_k^{0*}$, and integrating, the solution for $c_{lk}^{(1)}$ which is obtained is

$$c_{lk}^{(1)} = \frac{H'_{lk}}{E_l^0 - E_k^0} \,. \tag{3.13}$$

From this it can be seen that for the degenerate case the expansion coefficients become infinite and the method is unsuitable.

### 3.2.3 Application of perturbation theory to a degenerate energy band system

When the system is degenerate it is preferable to use matrices. The method will be outlined here but for further details of the approach the reader is recommended to read a suitable background text on quantum mechanics (e.g. White, 1966).

If for the unperturbed case certain of the eigenfunctions $u_n^0$ are equal, then the energy eigenvalues are also equal. Assume $E_k^0 = E^0 = \ldots = E_r^0$. The Hamiltonian $H_0$ is diagonal in the $u_n^0$ representation but this does not necessarily apply to $H'$ which can have any element of its matrix different from zero. The matrix of $H = H_0 + H'$ in the $u_n^0$ representation is written so as to collect together on the diagonal the equal eigenvalues; i.e. the matrix takes the form:

$$\begin{bmatrix} E_a^0 + H'_{aa} & H'_{ab} & & H'_{al} & H'_{ar} & H'_{as} \\ & E_b^0 + H'_{bb} & & & & \\ & & E_k^0 + H'_{kk} & H'_{kl} & & \\ & & H'_{lk} & E_l^0 + H'_{ll} & & \\ & & H'_{rk} & & E_r^0 + H'_{rr} & \\ H'_{sa} & & & & & H'_{ss} \end{bmatrix}$$

Instead of continuing with the general case, let us consider the application of the method to a particular crystallographic structure. The obvious structure to consider is the diamond or, more generally, the zinc-blende structure because of its common occurrence among narrow-bandgap materials. From group theory the functions for the conduction band are singly degenerate and for the valence band are triply degenerate. The functions describing the conduction band have the symmetry properties of s functions (designated $S$) so that the $\Gamma_1$ representation is used, whereas the valence band is designated by p functions $x$, $y$, $z$ (designated $X$, $Y$, $Z$) with the representation $\Gamma_{15}$.

If $E_c^0$ and $E_v^0$ are eigenvalues for the unperturbed equation (at $k = 0$, $E_c^0 - E_v^0 = E_g$, the energy gap), equation (3.8) takes the following form for the $x$ direction:

$$(H^0 - E_c^0)u_s^0(x) = 0 \tag{3.14a}$$

$$(H^0 - E_v^0)u_x^0(x) = 0 \tag{3.14b}$$

$$(H^0 - E_v^0)u_y^0(x) = 0 \tag{3.14c}$$

$$(H^0 - E_v^0)u_z^0(x) = 0 \tag{3.14d}$$

or generally

$$(H^0 - E_n^0)u_n^0(x) = 0\ . \tag{3.14}$$

For the perturbed case the complete matrix takes the form

$$\begin{bmatrix} E_c^0+\dfrac{\hbar^2k^2}{2m} & k_xP & k_yP & k_zP \\ k_xP & E_v^0+\dfrac{\hbar^2k^2}{2m} & 0 & 0 \\ k_yP & 0 & E_v^0+\dfrac{\hbar^2k^2}{2m} & 0 \\ k_zP & 0 & 0 & E_v^0+\dfrac{\hbar^2k^2}{2m} \end{bmatrix}$$

The contribution from the $\hbar^2k^2/2m$ term is self-evident. The matrix elements from the $\boldsymbol{k}\cdot\boldsymbol{p}$ term must be evaluated individually. $\boldsymbol{p}$ is replaced by the operator $-\hbar\nabla$ and elements of the form $\langle u_m^0|-\mathrm{i}\hbar\boldsymbol{k}\cdot\nabla|u_n^0\rangle$ evaluated. The following are examples of contributions that arise when expressions for angular symmetry of the waveforms as given on page 47 are used:

$$\frac{\hbar}{m}k_x\langle u_s^0|p_x|u_s^0\rangle = -\frac{\mathrm{i}\hbar^2}{m}k_x\int_{\text{all space}} u_s^0\frac{\partial u_s^0}{\partial x}\,\mathrm{d}x = 0\ ,$$

$$\frac{\hbar}{m}k_x\langle u_s^0|p_x|u_x^0\rangle = -\frac{\mathrm{i}\hbar^2}{m}k_x\int_{\text{all space}} u_s^0\frac{\partial u_x^0}{\partial x}\,\mathrm{d}x \equiv k_xP\ ,$$

[Sometimes $P$ is put equal to $(-\mathrm{i}\hbar/m)\langle u_s^0|p_x|u_x^0\rangle$ to make $P$ real.]

$$\frac{\hbar}{m}k_y\langle u_s^0|p_y|u_x^0\rangle = 0 \qquad \frac{\hbar}{m}k_z\langle u_s^0|p_z|u_x^0\rangle = 0\ ,$$

$$\frac{\hbar}{m}k_x\langle u_x^0|p_x|u_x^0\rangle = -\frac{\mathrm{i}\hbar^2}{m}k_x\int_{\text{all space}} u_x^0\frac{\partial u_x^0}{\partial x}\,\mathrm{d}x = 0\ ,$$

as the last integral gives the form $\int x\,\mathrm{d}x$.

Equating the determinant of this matrix for $H$ to energy gives

$$\begin{vmatrix} E_c^0+\dfrac{\hbar^2k^2}{2m}-E & k_xP & k_yP & k_zP \\ k_xP & E_v^0+\dfrac{\hbar^2k^2}{2m}-E & 0 & 0 \\ k_yP & 0 & E_v^0+\dfrac{\hbar^2k^2}{2m}-E & 0 \\ k_zP & 0 & 0 & E_v^0+\dfrac{\hbar^2k^2}{2m}-E \end{vmatrix} = 0\ . \tag{3.15}$$

As some of the elements are zero, this determinant can be expanded easily to give the equation

$$\left(E_{\mathrm{v}}^{0}+\frac{\hbar^2k^2}{2m}-E\right)^2\left[\left(E_{\mathrm{c}}^{0}+\frac{\hbar^2k^2}{2m}-E\right)\left(E_{\mathrm{v}}^{0}+\frac{\hbar^2k^2}{2m}-E\right)-k^2P^2\right]=0\,, \qquad (3.16)$$

(with $k_x^2+k_y^2+k_z^2 = k^2$). There are four solutions for $E$ as given in table 3.4. Solutions A and B correspond to heavy-hole bands which split when spin–orbit coupling is taken into account (see the following section), solution C is the light-hole valence band and solution D the conduction band. The $\hbar^2k^2/2m$ term is usually much smaller than the $k^2P/E_{\mathrm{g}}$ term so that C and D are mirror images of each other about the middle of the energy gap. The effective masses of each band $[m^* = \hbar^2(\partial^2E/\partial k^2)^{-1}]$ are equal at the band edges.

**Table 3.4.** Energy expressions for the bands from the $\boldsymbol{k}\cdot\boldsymbol{p}$ model. Energies are measured from the valence band edge. Perturbations from higher and lower bands are ignored.

| | No spin term | Spin term, $E_{\mathrm{g}} \ll \Delta$ | Spin term, $k \approx 0$ |
|---|---|---|---|
| A Heavy-hole valence band | $\frac{\hbar^2k^2}{2m}$ (A and B) | $\frac{\hbar^2k^2}{2m}$ | $\frac{\hbar^2k^2}{2m}$ |
| B Spin–orbit split valence band | | $-\Delta+\frac{\hbar^2k^2}{2m}-\frac{k^2P^2}{3(E_{\mathrm{g}}+\Delta)}$ | $-\Delta+\frac{\hbar^2k^2}{2m}-\frac{k^2P^2}{3(E_{\mathrm{g}}+\Delta)}$ |
| C Light-hole valence band | $\frac{\hbar^2k^2}{2m}+\frac{1}{2}E_{\mathrm{g}}-(\frac{1}{2}E_{\mathrm{g}}^2+k^2P^2)^{1/2}$ | $\frac{\hbar^2k^2}{2m}+\frac{1}{2}E_{\mathrm{g}}-\frac{1}{2}(E_{\mathrm{g}}^2+\frac{8}{3}k^2P^2)^{1/2}$ | $\frac{\hbar^2k^2}{2m}-\frac{2k^2P^2}{3E_{\mathrm{g}}}$ |
| D Conduction band | $\frac{\hbar^2k^2}{2m}+\frac{1}{2}E_{\mathrm{g}}+(\frac{1}{2}E_{\mathrm{g}}^2+k^2P^2)^{1/2}$ | $\frac{\hbar^2k^2}{2m}+\frac{1}{2}E_{\mathrm{g}}+\frac{1}{2}(E_{\mathrm{g}}^2+\frac{8}{3}k^2P^2)^{1/2}$ | $\frac{\hbar^2k^2}{2m}+\frac{k^2P^2}{3}\left(\frac{2}{E_{\mathrm{g}}}+\frac{1}{E_{\mathrm{g}}+\Delta}\right)$ |

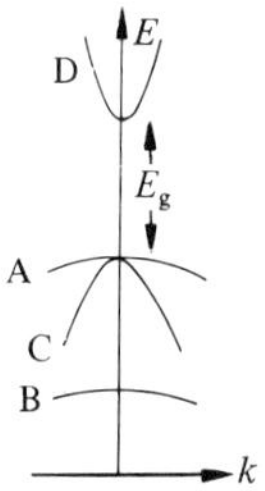

### 3.2.4 Inclusion of spin–orbit coupling

In order to introduce the spin–orbit interaction it is necessary to include an energy term in the Hamiltonian of the form

$$\frac{\hbar}{4m^2c^2}[\nabla V(\boldsymbol{r})\times\boldsymbol{p}]\cdot\boldsymbol{\sigma}\,,$$

where $\sigma_x$, $\sigma_y$, and $\sigma_z$ are the Pauli matrices. This term is introduced as a perturbation acting on the Bloch function $\exp(\mathrm{i}\boldsymbol{k}\cdot\boldsymbol{r})u_{\boldsymbol{k}}(\boldsymbol{r})$, where $u_{\boldsymbol{k}}(\boldsymbol{r})$ is the periodic function as considered earlier. If it is written in the form of an operator acting on the periodic function rather than the complete Bloch function, then the term in the Hamiltonian takes the form

$$\frac{\hbar}{4m^2c^2}\{[\nabla V(\boldsymbol{r})\times \boldsymbol{p}]\cdot\sigma+[\nabla V\times \boldsymbol{k}]\cdot\sigma\}\,.$$

The first term which is independent of $k$ is the dominant term and can be compared to atomic spin–orbit splitting. The second term is the contribution to the spin–orbit energy arising from the crystal momentum and can be neglected here.

Equation (3.5) now takes the form

$$\left\{\frac{p^2}{2m}+\frac{\hbar}{m}\boldsymbol{k}\cdot\boldsymbol{p}+\frac{\hbar}{4m^2c^2}[\nabla V(\boldsymbol{r})\times\boldsymbol{p}]\cdot\sigma+V(\boldsymbol{r})\right\}u_k(\boldsymbol{r})=E'_k u_k(\boldsymbol{r})\,, \tag{3.17}$$

where

$$E'_k=E_k-\left(\frac{\hbar^2}{2m}\right)k^2\,.$$

Applying the model to the symmetry of the zinc-blende structure gives doubly degenerate functions for the conduction band ($S\uparrow$ and $S\downarrow$ where $\uparrow$ means spin up and $\downarrow$ means spin down) and sixfold degeneracy for the valence band ($X\uparrow$, $Y\uparrow$, $Z\uparrow$, $X\downarrow$, $Y\downarrow$, $Z\downarrow$). Solution of the resulting $8\times 8$ interaction matrix gives energy values for the bands as listed in columns 2 and 3 of table 3.4. The term $\Delta$ is the spin–orbit splitting of the valence band, and if $\boldsymbol{k}$ is in the $x$ direction its value is

$$\Delta=\frac{3\hbar\mathrm{i}}{4m^2c^2}\left\langle Y\left|\frac{\partial V}{\partial y}p_z-\frac{\partial V}{\partial z}p_y\right|Z\right\rangle\,. \tag{3.18}$$

Column 2 gives the results when $E_g$ is much smaller than the spin–orbit splitting $\Delta$, and column 3 gives the results for $k\approx 0$ when $E_g$ is larger than the spin–orbit splitting.

### 3.2.5 Brief discussion of the Kane model

For narrow-bandgap materials the important case is, of course, that of $E_g \ll \Delta$. Whereas in classical band theory a parabolic band structure arises (figure 3.6), in this case the band can be very slender (the narrower the gap, the more slender or 'skinny' is the band), with a much smaller effective mass and one which varies according to the filling of the band.

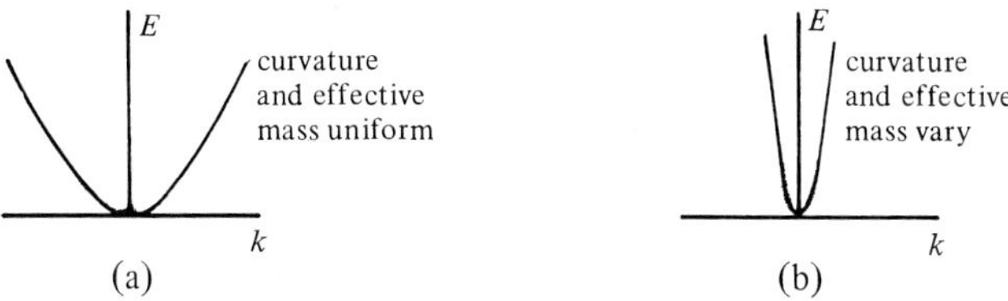


**Figure 3.6.** $E$ versus $k$ curves for (a) parabolic and (b) Kane bands.

Because of the form of the band, a very small carrier concentration will cause filling of the band well away from the band edge. Thus for a small-bandgap material at a finite temperature such that $x_g = E_g/kT$ is also small, the Fermi energy moves well into the band and consequently degenerate rather than classical statistics must be used. The consequences of this for electrical transport properties in semimetal and narrow-bandgap materials will be considered in the next chapter.

The dependence of the effective mass at the band edge on the reduced bandgap $x_g$ can be shown by expanding equation D as a Taylor series. For the conduction band,

$$E = \tfrac{1}{2}E_g + \tfrac{1}{2}E_g\left(1 + \frac{4P^2k^2}{3E_g^2}\right) = E_g + \frac{2P^2k^2}{3E_g} \, .$$

If measured relative to the conduction band edge, this gives $E = 2P^2k^2/3E_g$, and if this is compared with the standard form $E = \hbar^2k^2/2m^*$ then

$$m^* = \frac{3\hbar^2 E_g}{4P^2} = \frac{3\hbar^2 x_g k_B T}{4P^2} \, .$$

### 3.2.6 Cohen's model

Although Kane's band model takes into account interband interaction, it only applies to isotropic bands (particularly those in InSb). An extension to bands with energy extrema possessing no special symmetry within the Brillouin zone has been made by Cohen (1961) with particular application to bismuth, but the model is applicable to other narrow-bandgap materials. He used an analysis similar to that of Kane for a series of energy bands indexed by $n$, where $n = 0$ represents the valence band, $n = 1$ the conduction band and $n > 1$ higher bands. All bandgaps are large except the one between band 0 and band 1. Cohen obtained expressions for the energy surfaces for the bands for three separate symmetry cases, these three cases being considered to have particular application to bismuth (but see section 6.2.3 for a further discussion of bismuth).

The simplest case is that for which rotation plus reflection symmetry applies. The energy surfaces take the form given by the equation

$$\left(E - \frac{\hbar^2k_2^2}{2m_2}\right)\left(E + E_g + \frac{\hbar^2k_2^2}{2m_2'}\right) = E_g\left(\frac{\hbar^2k_1^2}{2m_1} + \frac{\hbar^2k_3^2}{2m_3}\right) . \tag{3.19}$$

Here subscripts 1, 2, and 3 refer to the three orthogonal axes. $m_1$ and $m_3$ are the band-edge effective masses in the transverse directions and $m_2$ is the band-edge effective mass in the longitudinal direction. The wavenumbers $k$ are defined for similar directions. With $m_1$, $m_2$, and $m_3$ referring to the lowest conduction band, $m_2'$ is the longitudinal band-edge effective mass for the valence band. The presence of the terms $E_g$ and $m_2'$ produce the divergence from an ellipsoidal–parabolic model. The energy surfaces are ellipsoidal only for $E$ significantly less than $E_g$; otherwise, although sections normal to axis 2 are ellipses with axes along directions 1 and 3, the areas of the normal sections vary differently with change of $p_2$ as compared with the variation which occurs in the ellipsoidal case. A small value of $m_2'$ leads to a dumbbell shape for the Fermi surface (Allgaier, 1966). Where transverse effective masses can be taken as equal, equation (3.19) simplifies to

$$E = \frac{\hbar^2 k_t^2}{2m_t[1+(E/E_g)+(\hbar^2 k_2^2/2m_2' E_g)]} + \frac{\hbar^2 k_2^2}{2m_2}, \qquad (3.20)$$

where $m_t$ is the transverse effective mass and $k_t$ also refers to the transverse direction.

When reflection symmetry does not apply, equation (3.19) becomes more complex. $p_1$ is replaced by $p_1 + \Delta p_1$ where $\Delta p_1$ depends on many of the other parameters, and there are extra energy terms. The energy surfaces again have elliptical cross sections normal to axis 2 with axes

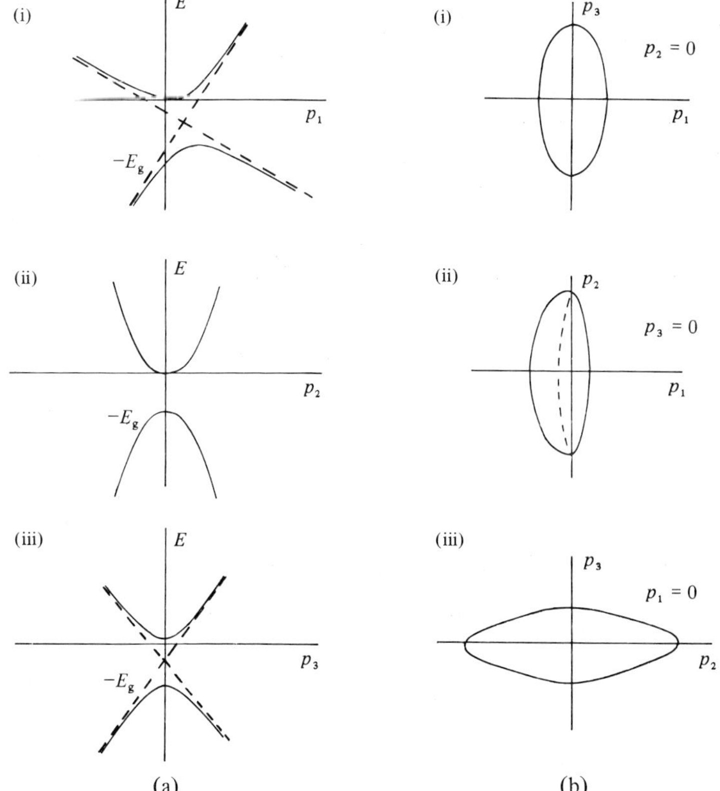


**Figure 3.7.** The Cohen model for the case of rotation symmetry. (a) Sketches of $E$ versus $p$ for (i) $p = p_1$, (ii) $p = p_2$, and (iii) $p = p_3$. (b) Principal sections of the energy surfaces for (i) $p_2 = 0$, (ii) $p_3 = 0$, and (iii) $p_1 = 0$. (After Cohen, 1961.)

along 1 and 3, and the extreme values of $p_2$ are $\pm(2m_2E)^{1/2}$ as they were for the previous case. However, the centres of the ellipses are shifted along axis 1 by $\Delta p_1$, the shift being a function of $p_2$. (It is zero for the extreme value of $p_2$ and increases quadratically with decreasing $p_2$.) The curves showing energy as a function of momentum along the three axes are shown in figure 3.7. It can be seen that the maximum in the valence band is displaced from the minimum in the conduction band. Also shown are principal sections for the energy surfaces. One feature of interest is that a lens-shaped Fermi surface is possible in the planes containing axes 1 and 2.

When rotation symmetry does not apply but reflection symmetry does, there are again further terms in equation (3.19) and $p_3$ is changed in a way analogous to the change of $p_1$ previously. $p_2$ need no longer have extreme values equal to $\pm(2m_2E)^{1/2}$ and quite complicated shapes for the energy surfaces can be set up. The models for all the symmetry cases do in fact have a large number of parameters such as energy gap, Fermi energy, effective masses, and angle of tilt of the energy surfaces.

**References**

Allgaier, R. S., 1966, *Phys. Rev.,* **152**, 808.
Callaway, J., 1964, *Energy Band Theory* (Academic Press, New York).
Cohen, M. H., 1961, *Phys. Rev.,* **121**, 387.
Cracknell, A. P., 1968, *Applied Group Theory* (Pergamon Press, Oxford).
Dresselhaus, G., 1955, *Phys. Rev.,* **100**, 580.
Herring, C., 1940, *Phys. Rev.,* **57**, 1169.
Kane, E. O., 1957, *J. Phys. Chem. Solids,* **1**, 249.
Kane, E. O., 1966, *Semiconductors and Semimetals,* Eds R. F. Willardson, A. C. Beer (Academic Press, New York), **1**, 75.
Kohn, W., Rostocker, N., 1954, *Phys. Rev.,* **94**, 411.
Korringa, J., 1947, *Physica,* **13**, 392.
Parmenter, R. H., 1955, *Phys. Rev.,* **100**, 573.
Quinn, C. M., 1973, *An Introduction to the Quantum Theory of Solids* (Oxford University Press, Oxford).
Slater, J. C., 1937, *Phys. Rev.,* **51**, 846.
White, R. L., 1966, *Basic Quantum Mechanics* (McGraw-Hill, New York).
Wooster, W. A., 1973, *Tensors and Group Theory for the Physical Properties of Crystals* (Oxford University Press, Oxford).
Ziman, J. M., 1972, *Principles of the Theory of Solids* (Cambridge University Press, Cambridge).

# Statistics and transport properties

In a metal the electrons fill the available states up to the Fermi level (for $T = 0$ K) and the density of states is close to that for the free-electron model. The electrons obey Fermi–Dirac statistics and the system is said to be highly degenerate. In a pure semiconductor there are far fewer electrons in the conduction band and classical statistics are applicable. This is a nondegenerate system. Between these two limiting cases lie semimetals and heavily-doped semiconductors, whose electron densities can be such that statistics of the Fermi–Dirac type are applicable, but which may not show certain features which come out of the limiting case of complete degeneracy as applicable to metals. In some respects the statistics for these materials are the most complicated of all. This chapter will discuss these statistics and, particularly, how they are involved in explaining the transport properties of semimetals and narrow-bandgap semiconductors.

## 4.1 Fermi–Dirac and Maxwell–Boltzmann statistics

As is shown by most standard textbooks on statistical mechanics, the probability $f(E)$ of occupation of an energy level $E$ in an electron gas is given by

$$f(E) = \frac{1}{\exp[(E-E_F)/k_B T]+1} \tag{4.1}$$

where $E_F$ is the Fermi energy (or chemical potential). At zero temperature all the electrons pack into the lowest available states subject to the Pauli exclusion principle. As the temperature increases the box-like shape of the occupation versus energy graph disappears and there is a gradual fall-off from $f(E) = 1$ to $f(E) = 0$ (figure 4.1a).

The density of energy states available in the energy range $E$ to $E+dE$ is given by

$$g(E)\,dE = \frac{(2m_0)^{3/2}E^{1/2}}{4\pi^2\hbar^3}\,dE , \tag{4.2}$$

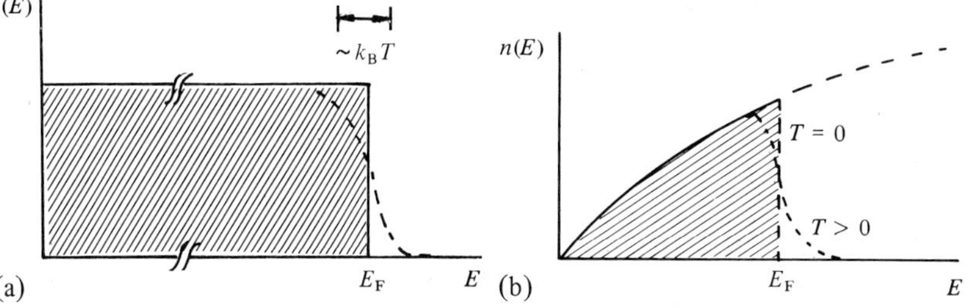


**Figure 4.1.** (a) Probability of occupation of energy states $f(E)$ for Fermi–Dirac statistics and (b) density of electrons $n$ as functions of energy $E$.

where $m_0$ is the mass of the free electron and $\hbar = h/2\pi$, $h$ being Planck's constant. This expression is derived by considering the acceptable wave functions for electrons confined within a finite box such that they satisfy the Schrödinger equation and the boundary conditions, and that $p = \hbar k$, $E = p^2/2m_0$. The density of electrons within a particular energy interval is obtained by multiplying the density of states function by the probability function and by 2 to take account of electrons of opposite spins. This is illustrated in figure 4.1b. The total electron density is

$$n = \int_0^\infty 2\mathrm{f}(E)\mathrm{g}(E)\,\mathrm{d}E \tag{4.3}$$

$$= \frac{(2m_0)^{3/2}}{2\pi^2\hbar^3}\int_0^\infty \frac{E^{1/2}}{\exp[(E-E_\mathrm{F})/k_\mathrm{B}T]+1}\,\mathrm{d}E\,. \tag{4.4}$$

For $T \to 0$ K, the integral becomes $\int_0^{E_\mathrm{F}} E^{1/2}\,\mathrm{d}E$ and gives an expression for the Fermi energy of

$$E_\mathrm{F} = \frac{(3\pi^2 n)^{2/3}}{2m_0}\,. \tag{4.5}$$

For electron densities as found in metals ($n \approx 5 \times 10^{28}$ $\mathrm{m}^{-3}$); this gives a Fermi energy of approximately 6 eV and a characteristic temperature $T_0$ ($= E_\mathrm{F}/k_\mathrm{B}$) of 70000 K. This means that Fermi–Dirac statistics apply rigorously up to the melting points of all metals.

In a classical (Maxwell-Boltzmann) gas the probability of occupation of energy level $E$ is given by

$$\mathrm{P}(E) = A\exp(-E/k_\mathrm{B}T) \tag{4.6}$$

where $A$ is a constant of proportionality. Equations (4.1) and (4.6) take the same form if $\exp(E_\mathrm{F}/k_\mathrm{B}T) \ll 1$, whereupon $A \approx \exp(E_\mathrm{F}/k_\mathrm{B}T)$. This will occur for very large $T$ or very small $E_\mathrm{F}$. In particular, it will apply to pure semiconductors at all temperatures except $T \to 0$ K because of their much lower carrier concentrations and hence small values of $E_\mathrm{F}$. In obtaining the Maxwell–Boltzmann distribution, the Pauli exclusion principle is not involved. However, since the number of states available is much greater than the number of occupying particles, there is little likelihood of two particles being in the same state. There is also the difference that in the derivation of Fermi–Dirac statistics indistinguishability of the particles is assumed whereas in the derivation of Maxwell–Boltzmann statistics the particles are regarded as distinguishable. This affects the number of available microstates (i.e. the number of configurations in phase space) in which the particles can exist, but does not affect the form of the distribution because $A$ in equation (4.6) acts as a normalisation factor. For a discussion of this see, for instance, Jackson (1968).

Plots of the distribution functions in materials possessing carrier concentrations of magnitude between those for pure semiconductors and metals usefully show the intermediate statistical case, and they also, for low and high temperatures, tend towards the limiting cases. These are shown in figure 4.2 for two examples:

(i) a Fermi energy of 0·06 eV at $T = 0$ K corresponding to a carrier concentration of $2{\cdot}3 \times 10^{25}$ m$^{-3}$ [from equation (4.5)],

(ii) a Fermi energy of 0·6 eV at $T = 0$ K corresponding to a carrier concentration of $5{\cdot}4 \times 10^{26}$ m$^{-3}$.

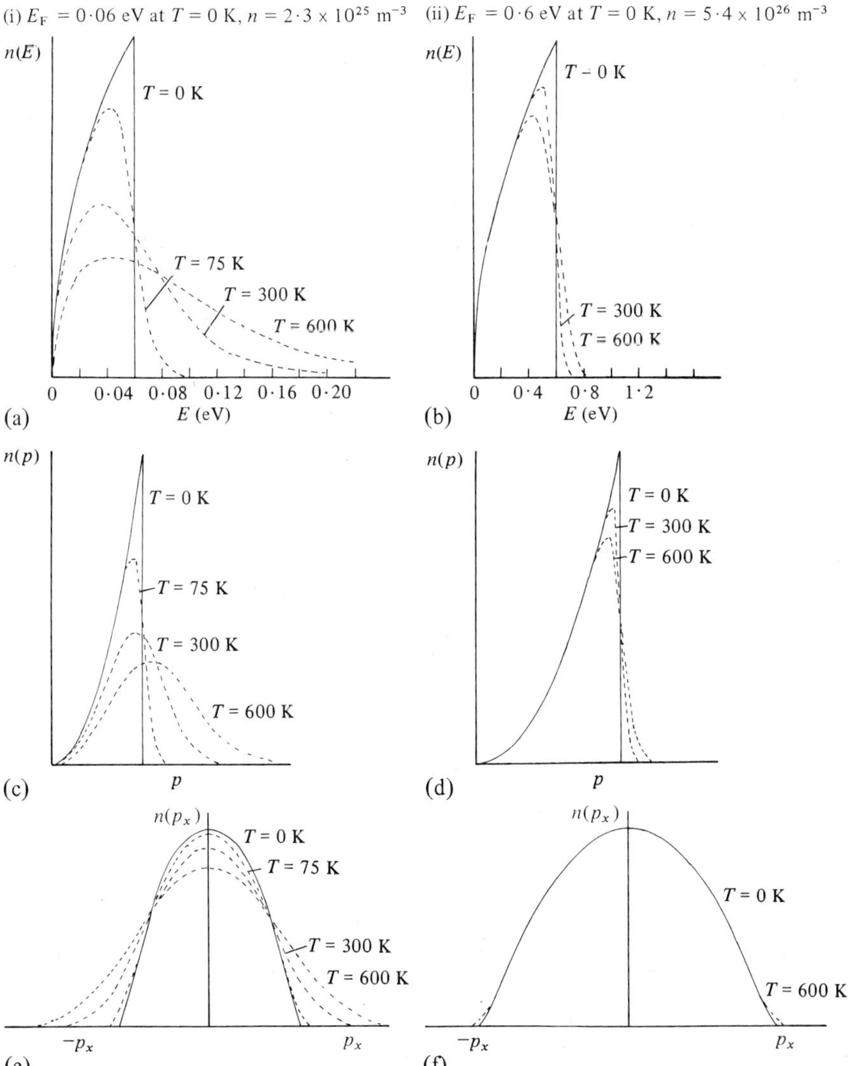


**Figure 4.2.** Carrier distribution functions for cases involving statistics intermediate between Fermi–Dirac and Maxwell–Boltzmann. Energy, momentum, and $x$-component momentum distributions are shown for two different concentrations (see text).

In the derivation, the electron mass of $9{\cdot}1 \times 10^{-31}$ kg is used rather than an effective mass $m^*$, and the form of the curves assumes an exact $E^{1/2}$ energy dependence. The Fermi energy requires to be recalculated at each finite temperature assuming constancy of carrier concentration [although the correction is negligible for case (ii)]. The scales of the graphs in figures 4.2 are arbitrary but have been chosen to normalise the corresponding heights of the zero-temperature curves for the two examples.

Figures 4.2a and 4.2b show plots of $n(E) = 2\mathrm{g}(E)\mathrm{f}(E)$ as a function of energy (equivalent to figure 4.1b). Curves corresponding to $E_\mathrm{F} = 0{\cdot}06$ eV at $T = 0$ K distinctly show the Fermi–Dirac form at $T = 0$ K changing to a classical form at 600 K. It can be seen that with the increase of temperature a large proportion of the electrons receive an increase in energy. In contrast, curves corresponding to $E_\mathrm{F} = 0{\cdot}6$ eV show a very small deviation from the Fermi–Dirac form even at 600 K. Only electrons close to the Fermi energy obtain an increase in energy and it is these electrons that then contribute to charge transfer in transport processes. This effect is, of course, even more enhanced in a metal. Because many fewer electrons are involved, the free path length is much greater than in a classical system. From the graphs it is possible to see that only some electrons will contribute to the specific heat in a metal.

Momentum (or velocity) distributions are illustrated in figures 4.2c and 4.2d. The momentum distribution curve is obtained from the energy distribution curve with the use of the relationships $E = p^2/2m_0$ and $m_0\,\mathrm{d}E = p\,\mathrm{d}p$. This yields

$$n(p)\,\mathrm{d}p = \frac{1}{\pi^2\hbar^3}\frac{p^2\,\mathrm{d}p}{\exp[(E-E_\mathrm{F})/k_\mathrm{B}T]+1}\,. \tag{4.7}$$

Although the form of the curves is different from that for energy, the curves again show the main features of classical, intermediate, and Fermi–Dirac statistics. Finally, curves 4.2e and 4.2f show the momentum components in one direction only. It can be demonstrated that the distribution function is

$$n(p_x)\,\mathrm{d}p_x = \left(\frac{m_0 k_\mathrm{B}T}{2\pi^2\hbar^3}\right)\ln\{\exp[(E_\mathrm{F}-p_x)^2/2m_0]+1\}\,\mathrm{d}p_x\,, \tag{4.8}$$

where $p_x$ is the component of momentum in the $x$ direction. For $T \rightarrow 0$ K the curves show the characteristic cutoff at the Fermi momentum, but as the temperature increases the curves broaden out, this broadening being much more significant in the less-degenerate case.

## 4.2 The transport equation

It is clear from the graphs already shown that calculation of any transport effects requires a knowledge of the distribution of the carriers over the various states of the system when the system is acted upon by external forces such as an electric field or a thermal gradient. It is necessary to use the carrier distribution function $f(\boldsymbol{r}, \boldsymbol{p}, t)$ where $\boldsymbol{r}$ is the position vector, $\boldsymbol{p}$ the momentum and $t$ is the time. From Liouville's theorem for an isolated system (see Landau and Lifshitz, 1959),

$$\frac{\mathrm{d}f}{\mathrm{d}t} = \frac{\partial f}{\partial t} + \left(\frac{\partial f}{\partial t}\right)_{\text{drift}} + \left(\frac{\partial f}{\partial t}\right)_{\text{carriers}} = 0 \, . \tag{4.9}$$

For stationary processes or ones in which the relaxation time is much greater than the carrier relaxation time,

$$\frac{\partial f}{\partial t} = 0$$

and

$$\left(\frac{\partial f}{\partial t}\right)_{\text{drift}} = -\left(\frac{\partial f}{\partial t}\right)_{\text{collisions}} \, . \tag{4.10}$$

Application of the external fields means that the average momentum (see figures 4.2e and 4.2f) is no longer zero. The carrier distribution is changed with respect to both momentum and position. Hence,

$$\frac{\partial f(\boldsymbol{r} \cdot \boldsymbol{p})}{\partial t} = \boldsymbol{r}\frac{\partial f}{\partial \boldsymbol{r}} + \boldsymbol{p}\frac{\partial f}{\partial \boldsymbol{p}} \, . \tag{4.11}$$

To obtain a suitable expression for the quantities in equation (4.11), it is necessary to make a number of assumptions. Let $f_0(\boldsymbol{r}, \boldsymbol{p})$ be the equilibrium distribution function and assume that

(i) $f_0(\boldsymbol{r}, \boldsymbol{p})$ is replaced by $f(\boldsymbol{r}, \boldsymbol{p})$ when a small field is applied, the field being such that $f - f_0 \ll f_0$;

(ii) the change in the distribution takes the form (Wilson, 1953)

$$f - f_0 = \boldsymbol{v} \cdot \chi(\boldsymbol{r}, \boldsymbol{p}) \frac{\partial f_0}{\partial E} = \boldsymbol{v} \cdot \delta(\boldsymbol{r}, \boldsymbol{p}) \tag{4.12}$$

(i.e. expanding as a Taylor expansion), where $\boldsymbol{v}$ is velocity and $\delta(\boldsymbol{r}, \boldsymbol{p})$ contains the $\partial f_0/\partial E$ term and gives the departure of the particle system from equilibrium;

(iii) the change of carrier energy on collision is small ($\Delta E \ll E$);

(iv) the scattering process can be represented by a relaxation time $\tau(\boldsymbol{p})$ such that

$$\left(\frac{\partial f}{\partial t}\right)_{\text{collisions}} = -\frac{f - f_0}{\tau(\boldsymbol{p})} \, . \tag{4.13}$$

Assumption (iii) is necessary for this and scattering must produce a random distribution of velocities with the probabilities of transition to $\boldsymbol{v}$ and $-\boldsymbol{v}$ being equal. Expressing the relaxation time in this form implies a return to equilibrium at a rate which decreases exponentially with time. The magnitude of the relaxation time will depend on the nature of the scattering.

Equations (4.11) and (4.13) give

$$v\frac{\partial f}{\partial r}+\frac{\mathrm{d}\boldsymbol{p}}{\mathrm{d}t}\frac{\partial f}{\partial \boldsymbol{p}} = -\frac{f-f_0}{\tau(\boldsymbol{p})} . \tag{4.14}$$

The force on the carriers when they are in electric and magnetic fields is given by

$$\frac{\mathrm{d}\boldsymbol{p}}{\mathrm{d}t} = e(\boldsymbol{E}+v\times \boldsymbol{B}) , \tag{4.15}$$

where $e$ is the charge on the carriers (this is an absolute value and so negative for electrons), $\boldsymbol{E}$ is the electric field and $\boldsymbol{B}$ the magnetic induction. Substitution of equation (4.15) in (4.14) gives $f-f_0$ and hence $\delta(\boldsymbol{r},\boldsymbol{p})$ (see Tsidil'kovskii, 1962):

$$\delta = \frac{\delta_0+e\tau(m^{-1}\delta_0\times \boldsymbol{B})+(e\tau)^2|m^{-1}|\delta_0\cdot \boldsymbol{B}m\boldsymbol{B}}{1+(e\tau)^2|m^{-1}|(m\boldsymbol{B}\cdot \boldsymbol{B})} , \tag{4.16}$$

where $|m^{-1}|$ is the determinant of the reciprocal mass tensor $m^{-1}$ and

$$\delta_0 = \left\{-eE+\left[T\frac{\partial}{\partial T}\left(\frac{E_\mathrm{F}}{T}\right)+\frac{E}{T}\right]\frac{\partial T}{\partial r}\right\}\tau\frac{\partial f_0}{\partial E} \tag{4.17}$$

(but see page 69 for comments on the use of the relaxation time $\tau$). For spherical energy surfaces corresponding to a minimum at the centre of the zone in $\boldsymbol{k}$-space, $m^{-1} = 1/m^*$. More generally, band anisotropy must be considered. Transport of electrical charge and energy is given by

$$\boldsymbol{j} = e\int_0^\infty v\,\mathrm{d}n , \tag{4.18a}$$

$$\boldsymbol{W} = \int_0^\infty Ev\,\mathrm{d}n , \tag{4.18b}$$

where $\boldsymbol{j}$ is the electrical current density and $\boldsymbol{W}$ is the thermal current density excluding heat flow due to the lattice. Detailed solution of the transport equation is not possible here, but an indication will be given of the forms of solution for different types of band structure and, in particular, that for the parabolic band.

As the number of carriers per unit volume in an element of momentum space is $(2f/h^3)\,\mathrm{d}\boldsymbol{p}$,

$$\boldsymbol{j} = \frac{2e}{h^3}\int_0^\infty vf\,\mathrm{d}\boldsymbol{p} = \frac{2e}{h^3}\int_0^\infty v(v\cdot\delta)\,\mathrm{d}\boldsymbol{p} . \tag{4.19}$$

If $\mathrm{d}p$ is used instead of $\mathrm{d}\boldsymbol{p}$ and a factor of $\frac{4}{3}\pi$ incorporated for integrating over angles in spherical coordinates, equation (4.19) becomes

$$\boldsymbol{j} = \frac{8\pi e}{3m^{*2}h^3}\int_0^\infty \delta p^4\,\mathrm{d}p . \tag{4.20a}$$

Similarly, equation (4.18b) becomes

$$W = \frac{8\pi}{3m^{*2}h^3}\int_0^\infty E\delta p^4\,\mathrm{d}p\,. \tag{4.20b}$$

If the magnetic field is orthogonal to the applied magnetic field or to the temperature gradient, then the following equations are obtained:

$$j = e^2(I_{10}E' + I_{20}E' \times h) - e\left(I_{11}\frac{\partial \ln T}{\partial r} + I_{21}\frac{\partial \ln T}{\partial r} \times h\right) \tag{4.21a}$$

and

$$W = e(I_{11}E' + I_{21}E' \times h) - \left(I_{12}\frac{\partial \ln T}{\partial r} + I_{22}\frac{\partial \ln T}{\partial r} \times h\right) \tag{4.21b}$$

where

$$E' = E - \frac{T}{e}\frac{\partial}{\partial r}\left(\frac{E_F}{T}\right), \qquad h = \frac{B}{B}\,.$$

Other geometries can be considered. For instance, Harman and Honig (1967) have analysed the case of the Corbino disc where electrical connections are attached to the inner and outer peripheries of a circular disk and the current flows radially.

It is possible to partially transform the equations for $j$ and $W$ so that $j$ is the dependent variable (Mazur and Prigogine, 1951; Callen, 1952). This gives the electric field in the presence of a thermal gradient:

$$E'' = E - \frac{1}{e}\frac{\partial E_F}{\partial r} = \boldsymbol{\rho}(B)j - \boldsymbol{\alpha}(B)\frac{\partial T}{\partial r}\,, \tag{4.22a}$$

and the heat flow in the presence of a thermal gradient and an electric current:

$$W' = W - \frac{E_F}{e}j = -\boldsymbol{\pi}(B)j - \mathbf{K}(B)\frac{\partial T}{\partial r}\,. \tag{4.22b}$$

For the total thermal conductivity, the lattice contribution $K_L$ gives a contribution to $W'$ of $-K_L\,\partial T/\partial r$. The elements of the tensors $\boldsymbol{\rho}(B)$, $\boldsymbol{\alpha}(B)$, $\boldsymbol{\pi}(B)$, and $\mathbf{K}(B)$ are given in table 4.1 (Putley, 1960).

To see how the tensors are related to the usual terminology for the galvanomagnetic and thermomagnetic effects, it is necessary to express them fully in subscript notation. Rewriting equations (4.22a) and (4.22b) gives

$$E_i'' = \rho_{ik}(B)j_k - \alpha_{ik}(B)\frac{\partial T_k}{\partial r}\,, \tag{4.23a}$$

$$W_i' = -\pi_{ik}(B)j_k - K_{ik}(B)\frac{\partial T_k}{\partial r}\,. \tag{4.23b}$$

The coefficient tensors can be written (Beer, 1963; Bhagavantam and Pantulu, 1964) as a power series in magnetic field provided the series is converging (i.e. for small magnetic field):

$$\rho_{ik}(B) = \underset{\substack{\text{electrical}\\ \text{resistivity}}}{\rho_{ik}} + \underset{\substack{\text{Hall}\\ \text{effect}}}{\rho_{ikl}B_l} + \underset{\text{magnetoresistivity}}{\rho_{iklm}B_lB_m} + \ldots \tag{4.24a}$$

$$\alpha_{ik}(B) = \underset{\substack{\text{thermoelectric}\\ \text{power}}}{\alpha_{ik}} + \underset{\substack{\text{Nernst}\\ \text{effect}}}{\alpha_{ikl}B_l} + \underset{\substack{\text{longitudinal Nernst-}\\ \text{Ettingshausen effect}}}{\alpha_{iklm}B_lB_m} + \ldots \tag{4.24b}$$

$$\pi_{ik}(B) = \pi_{ik} + \underset{\substack{\text{Ettingshausen}\\ \text{effect}}}{\pi_{ikl}B_l} + \pi_{iklm}B_lB_m + \ldots \tag{4.24c}$$

$$K_{ik}(B) = \underset{\substack{\text{thermal}\\ \text{conductivity}}}{K_{ik}} + \underset{\substack{\text{Righi-Leduc}\\ \text{effect}}}{K_{ikl}B_l} + \underset{\substack{\text{Maggi-Righi-Leduc}\\ \text{effect}}}{K_{iklm}B_lB_m} + \ldots \tag{4.24d}$$

The longitudinal effects correspond to the even powers in $B$ and the transverse effects to the odd powers. Also, by application of Onsager's principle (Landau and Lifshitz, 1960) we have

$$\rho_{ik}(B) = \rho_{ki}(-B)\,, \qquad K_{ik}(B) = K_{ki}(-B)\,, \qquad \pi_{ik}(B) = T\alpha_{ki}(-B)\,.$$

Note that the expressions refer to a *temperature gradient* along a particular axis of the crystal; this is called the isothermal case. An alternative situation is one in which the *heat flow* $h$ is in a particular direction; this is called the adiabatic case. Figure 4.3 depicts thermal conductivity as measured in the isothermal case and thermal resistivity as measured in the adiabatic case. Included in the diagrams are the conductivity ($K$) and resistivity ($R$) ellipsoids used to represent the magnitude of conductivity and resistivity in any particular direction (Nye, 1957). For anisotropic materials resistivity will be the reciprocal of conductivity along the principal axes only. Relationships exist between the isothermal and adiabatic coefficients (Heurlinger, 1915; 1916; Bridgman, 1924).

**Table 4.1.** Elements of the galvanomagnetic and thermomagnetic tensors.

| Composite tensor | Longitudinal effect | Transverse effect |
|---|---|---|
| **Galvanomagnetic effects** | | |
| | Resistance/magnetoresistance | Hall effect |
| $\boldsymbol{\rho}(B)$ | $\rho_{xx} = \rho_{yy} = \dfrac{I_{10}}{e^2(I_{10}^2+I_{20}^2)}$ | $\rho_{xy} = -\rho_{yx} = \dfrac{I_{20}}{e^2(I_{10}^2+I_{20}^2)}$ |
| | Peltier effect | Ettingshausen effect |
| $\boldsymbol{\pi}(B)$ | $\pi_{xx}(B) = T\alpha_{xx}(-B)$ | $\pi_{xy}(B) = T\alpha_{yx}(-B)$ |
| **Thermomagnetic effects** | | |
| | Thermoelectric effect | Nernst effect |
| $\boldsymbol{\alpha}(B)$ | $\alpha_{xx} = \alpha_{yy} = E_{\mathrm{F}}^{*} - \dfrac{1}{Te}\dfrac{I_{10}I_{11}+I_{20}I_{21}}{I_{10}^2+I_{20}^2}$ | $\alpha_{xy} = -\alpha_{yx} = \dfrac{1}{Te}\dfrac{I_{10}I_{21}-I_{20}I_{11}}{I_{10}^2+I_{20}^2}$ |
| | Total thermal conductivity | Righi-Leduc effect |
| $\mathbf{K}(B)$ | $K_{xx} = K_{yy} = \dfrac{1}{T}\left(I_{12}+\dfrac{I_{21}^2I_{10}-2I_{11}I_{20}I_{21}-I_{11}^2I_{10}}{I_{10}^2+I_{20}^2}\right)+K_{\mathrm{L}}$ | $K_{xy} = -K_{yx} = -\dfrac{1}{TK_{xx}}\left(I_{22}-\dfrac{I_{21}^2I_{20}+2I_{10}I_{11}I_{21}-I_{11}^2I_{10}}{I_{10}^2+I_{20}^2}\right)$ |

In practice, transport effects are usually measured on long samples so that adiabatic coefficients are obtained, and isothermal coefficients calculated from these. In addition, the coefficients are also related by the Kelvin equation $TQ_{\mathrm{i}} = PK_{\mathrm{i}}$, where $Q$ is the Nernst coefficient, $P$ is the Ettingshausen coefficient and subscript i refers to isothermal conditions (see Jan, 1957), and by the Wiedemann–Franz equation $R_{\mathrm{H_i}} = S\rho_{\mathrm{i}}$, in which $R_{\mathrm{H}}$ is the Hall coefficient and $S$ the Righi–Leduc coefficient given by $S = \partial T_y/(B_z \partial T_x)$, and which can be derived either from free-electron theory (Sommerfeld and Frank, 1931) or from the generalised law of Wiedemann–Franz (Kohler, 1941).

For solution of the transport integrals $I_{ij}$, it is necessary to substitute suitably for the dependence of relaxation time on energy. The relaxation time is put in the form

$$\tau = \tau_0 \left(\frac{E}{k_{\mathrm{B}}T}\right)^{r-\frac{1}{2}}, \tag{4.25}$$

where $\tau_0$ is a function independent of energy $E$ and $r$ is the scattering-mechanism parameter. $r$ can vary from 0 for acoustic scattering to 2 for ionised impurity scattering and can take values of $\frac{1}{2}$ and 1 for other types of scattering (see for instance, Tsidil'kovskii, 1962). For some scattering mechanisms the concept of relaxation time may not hold, in which case other methods of solving the equations for the transport properties become necessary. If more than one scattering mechanism is involved then Matthiessen's rule applies,

$$\frac{1}{\tau} = \frac{1}{\tau_1} + \frac{1}{\tau_2} + \frac{1}{\tau_3} + \ldots \tag{4.26}$$

where $\tau_i$ refers to the relaxation time associated with the $i$th scattering mechanism.

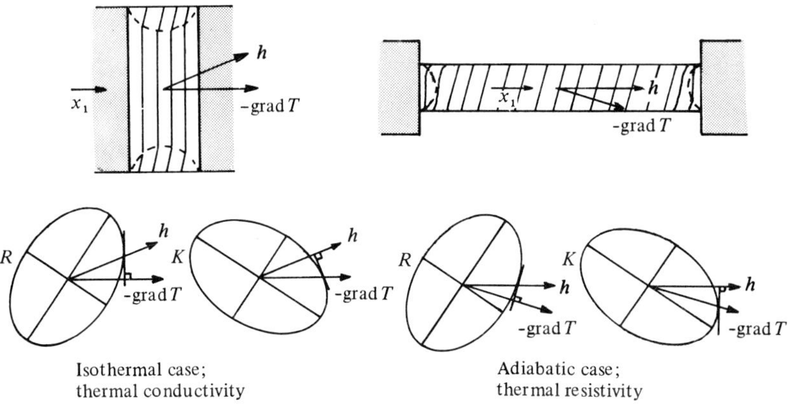


**Figure 4.3.** Isothermal and adiabatic heat flow. (After Nye, 1957.)

**4.3 Solution of the transport equation for a single parabolic band**
For spherical bands and $E = p^2/2m^*$ such that

$$\frac{\partial f_0}{\partial E} = \left(\frac{m^*}{p}\right)\frac{\partial f_0}{\partial p}$$

we have

$$I_{ij} = \frac{8\pi}{3h^3}(eB)^{i-1}\int_0^\infty p^3\left(\frac{\tau}{m^*}\right)^i \frac{\mathrm{d}f_0}{\mathrm{d}p}\frac{E^j}{1+\beta^2}\mathrm{d}p\,, \tag{4.27}$$

where $\beta = (e\tau B)/m^*$.

For low and high magnetic field conditions those effects which depend on the strength of the magnetic field can be calculated by expanding the integrals as power series in $\beta$, by a standard expansion of $1+\beta$ where $\beta$ is small or $(1+1/\beta)$ where $\beta$ is large. For intermediate field values numerical evaluation of the integrals is necessary except for the limiting cases of complete degeneracy or nondegeneracy. One possible way of expressing the integrals is as functions of the measurable parameters of carrier concentration $n$ and conductivity mobility $\mu$:

$$I_{ij} = -\frac{n(\mu B)^i}{eB}\left[\int_0^\infty p^3\left(\frac{\tau}{m^*}\right)^i E^j\frac{\mathrm{d}f_0}{\mathrm{d}p}\frac{1}{1+\beta^2}\mathrm{d}p \Bigg/ \int_0^\infty p^3\frac{\mathrm{d}f_0}{\mathrm{d}p}\mathrm{d}p\right] \times\left(\int_0^\infty p^3\frac{\mathrm{d}f_0}{\mathrm{d}p}\mathrm{d}p \Bigg/ \int_0^\infty p^3\frac{\tau}{m^*}\frac{\mathrm{d}f_0}{\mathrm{d}p}\mathrm{d}p\right)^i\,, \tag{4.28}$$

as

$$n = \frac{8\pi}{h^3}\int_0^\infty p^2 f_0\mathrm{d}p\,, \qquad \sigma = -\frac{8\pi}{h^3}\int_0^\infty p^3\frac{\mathrm{d}f_0}{\mathrm{d}p}\mathrm{d}p \tag{4.29}$$

(shown by integrating by parts) and

$$\mu = \int_0^\infty p^3\frac{\tau}{m^*}\frac{\mathrm{d}f_0}{\mathrm{d}p}\mathrm{d}p \Bigg/ e\int_0^\infty p^3\frac{\mathrm{d}f_0}{\mathrm{d}p}\mathrm{d}p \tag{4.30}$$

from electrical conductivity $\sigma = e\mu n$. Substituting for $\tau$, $f_0$, and $E\ (= p^2/2m^*)$ gives

$$\int_0^\infty p^3\left(\frac{\tau}{m^*}\right)^i E^j\frac{\mathrm{d}f_0}{\mathrm{d}p}\frac{1}{1+\beta^2}\mathrm{d}p = -\frac{\tau_0^i m^{*\,3/2-i}}{k_\mathrm{B}T}\int_0^\infty\frac{E^{3/2+j+ir-i/2}\exp[(E-E_\mathrm{F})/k_\mathrm{B}T]}{[1+W^2(E/k_\mathrm{B}T)^{2r-1}]\{\exp[(E-E_\mathrm{F})/k_\mathrm{B}T]+1\}^2}\mathrm{d}E \tag{4.31}$$

where

$$W = \frac{e\tau_0}{m^*}B = \mu_0 B\,.$$

The use here of mobility $\mu_0$, which is different slightly from the conductivity mobility $\mu$, follows that of Tsidil'kovskii. The alternative approach (see for instance Harman and Honig, 1967) retains a $\partial E/\partial p$ factor obtained when substituting for the effective mass $m^*$.

If

$$\mathcal{J}(P, W, Q, E_F^*) = \int_0^\infty \frac{x^P \exp(x - E_F^*)}{(1 + W^2 x^Q)[\exp(x - E_F^*) + 1]^2} dx \quad (4.32)$$

where $x = E/k_B T$, $P = \frac{3}{2} + j + ir - \frac{1}{2}i$, and $Q = 2r - 1$, then

$$I_{ij} = (\mu B)^i (k_B T)^j \frac{n}{eB} \frac{\mathcal{J}(P, W, Q, E_F^*)}{\mathcal{J}(\frac{3}{2}, 0, 0, E_F^*)} \left[ \frac{\mathcal{J}(\frac{3}{2}, 0, 0, E_F^*)}{\mathcal{J}(1 + r, 0, 0, E_F^*)} \right]^i, \quad (4.33)$$

the terms in $\tau_0$ and $m^*$ having cancelled. Integrals of the form

$$\int_0^\infty \frac{x^m}{1 + \exp(x - E_F^*)} dx = F_m(E_F^*) \quad (4.34)$$

are called Fermi–Dirac integrals, and listings for these at half-integer intervals are given by Tauc (1962) for $-4 < E_F^* < 20$, half-integer values having been calculated by McDougall and Stoner (1938) and integral values by Rhodes (1950). As $\mathcal{J}(\frac{3}{2}, 0, 0, E_F^*)$ is $\frac{3}{2} F_{½}(E_F^*)$ and $\mathcal{J}(1 + r, 0, 0, E_F^*)$ is $F_r(E_F^*)/(1 + r)$ (by integrating by parts), equation (4.33) becomes

$$I_{ij} = (\mu B)^i (k_B T)^j \frac{n}{eB} \frac{\mathcal{J}(P, W, Q, E_F^*)}{\frac{3}{2} F_r(E_F^*)} \left[ \frac{3F_{½}(E_F^*)}{2(1 + r) F_r(E_F^*)} \right]^i. \quad (4.35)$$

Hence the transport integrals can be calculated as a function of $\mu B$ for different reduced Fermi levels as required.

As has been discussed already, classical statistics are suitable for semiconductors provided they do not have particularly large carrier concentrations (in practice, $n < \sim 10^{23}$ m$^{-3}$). In these cases, where $E_F^* \ll 1$,

$$F_m(E_F^*) \approx \int_0^\infty x^m \exp(-x)\, dx = \exp(E_F^*)\Gamma(m + 1). \quad (4.36)$$

The gamma function $\Gamma(m)$ has the following properties;

$$\Gamma(m) = (m - 1)\Gamma(m - 1),$$
$$\Gamma(m) = (m - 1)! \text{ for } m \text{ an integer},$$
$$\Gamma(\tfrac{1}{2}) = \pi^{½},$$

and hence is calculated rather more easily than the Fermi–Dirac integrals. The range of $E_F^*$ values where approximations are suitable is indicated by figure 4.4 (Wilson, 1953) showing $F_{½}(E_F^*)$ and the possible approximations $\frac{2}{3} F_{3/2}(E_F^*)$ and $\frac{1}{2}\pi^{½} \exp E_F^*$.

In those cases where $E_F^* \gg 1$, the Fermi–Dirac integral can be written in the form of a rapidly converging series

$$F_m(E_F^*) = \frac{E_F^{*m+1}}{2m} + \tfrac{1}{6}\pi^2 m E_F^{*m-1} + \tfrac{7}{360}\pi^2 m(m-1)(m-2) E_F^{*m-3} + \ldots. \quad (4.37)$$

Table 4.2 shows expressions for some of the limiting cases where one type of carrier is present.

**Table 4.2.** Selected transport coefficients (single carrier case).

| Coefficient | Arbitrary relaxation time $\tau = \tau_0\left(\frac{E}{k_B T}\right)^{r-\frac{1}{2}}$ | | $r = 0$ (equivalent to mean free path independent of energy; $l = l_0 E^r$) | |
|---|---|---|---|---|
| | classical limit | degenerate case | classical limit | degenerate case |
| Electrical conductivity, $\sigma$ | $\frac{4ne^2\tau_0\Gamma(r+2)}{3\pi^{\frac{1}{2}}m^*}$ <br> $n = \frac{2\pi^{\frac{3}{2}}(2m^*k_B T)^{\frac{3}{2}}}{h^3}\exp E_F^*$ | $\frac{16\pi m^* e^2 l}{3h^3} k_B T F_0(E_F^*)$ <br> $n = \frac{4\pi(2m^*k_B T)^{\frac{3}{2}}}{h^3} F_{\frac{1}{2}}(E_F^*)$ | $ne\mu$ | as for arbitrary relaxation time |
| Hall coefficient, $R_H$ | $-\frac{3\pi^{\frac{1}{2}}\,\Gamma(2r+\frac{3}{2})}{4ne\,\Gamma^2(r+2)}$ | $-\frac{3(2r+\frac{1}{2})}{2(r+1)^2 ne}\frac{F_{\frac{1}{2}}F_{2r-\frac{1}{2}}}{F_r^2}$ | $-\frac{3\pi}{8ne}$ | $-\frac{3}{4ne}\frac{F_{\frac{1}{2}}F_{-\frac{1}{2}}}{ln^2(1+\exp E_F^*)}$ <br> $\rightarrow -\frac{1}{ne}$ for $E_F^* \rightarrow \infty$ |
| Wiedemann–Franz–Lorenz ratio, $K_{el}/\sigma T$ | $(r+2)\left(\frac{k_B}{e}\right)^2$ | $\frac{k_B^2}{e}\left[\frac{(r+3)F_{r+2}}{(r+1)F_r} - \frac{(r+2)^2F_{r+1}^2}{(r+1)^2F_{r,}^2}\right]$ | $2\left(\frac{k_B}{e}\right)^2$ | $\frac{k_B^2}{e}\left(\frac{3F_2}{F_0} - \frac{4F_1^2}{F_0^2}\right)$ <br> $\rightarrow \frac{\pi^2}{3}\left(\frac{k_B}{e}\right)^2$ for $E_F^* \rightarrow \infty$ |
| Thermoelectric power, $\alpha$ | $-\frac{k_B}{e}(r+2-E_F^*)$ | $-\frac{k_B}{e}\left[\frac{(r+2)F_{r+1}}{(r+1)F_r} - E_F^*\right]$ | $-\frac{k_B}{e}(2-E_F^*)$ | $-\frac{k_B}{e}\left[\frac{2F_1}{\ln(1+\exp E_F^*)} - E_F^*\right]$ |
| Nernst coefficient, $Q$ | $\frac{(r-\frac{1}{2})k_B\tau_0\Gamma(2r+\frac{3}{2})}{m^*\Gamma^2(r+2)}$ | | $-\frac{3\pi}{16}\left(\frac{k_B}{e}\right)\mu \quad \left(\mu = \left\|\frac{\sigma}{ne}\right\|\right)$ | |

More detailed expressions for two-band and multi-band conductors are given respectively by Putley (1960) and Harman and Honig (1967). The expressions as obtained by Harman and Honig for the Righi–Leduc effect (this effect is included in table 4.1 but not in table 4.2) do not take fully into account the lattice thermal conductivity. Although this is unimportant for metals where thermal conductance is primarily electronic, it is necessary to include the lattice conductivity for semiconductors as has been done by Putley. In addition there is some variation of sign convention. Putley, and Harman and Honig give a positive sign to the Seebeck coefficient when the carriers are positive holes, this being the convention largely used by experimentalists, but the conventions of Putley, and Harman and Honig give opposite signs for the transverse thermomagnetic effects. These points are discussed by Putley (1975) who also extends his two-band treatment to a tabulation of the transverse galvano- and thermomagnetic coefficients for an isotropic multi-band conductor.

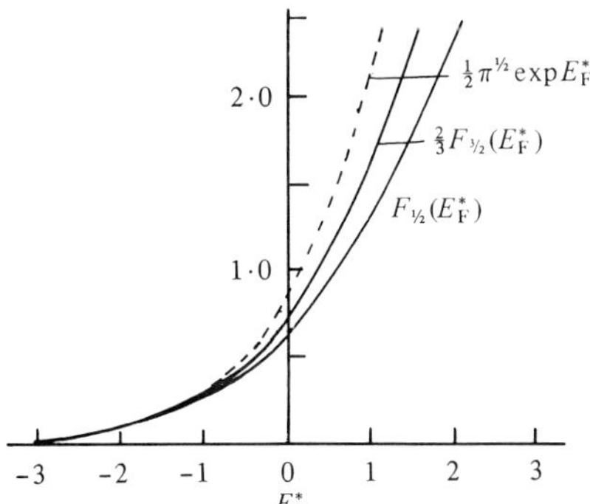


**Figure 4.4.** Fermi–Dirac functions for $F_{1/2}(E_F^*)$ and $\frac{2}{3}F_{3/2}(E_F^*)$ and also $\frac{1}{2}\pi^{1/2}\exp E_F^*$.

**4.4 Solution of the transport equation for a single Kane-type band**

The Kane band form for $E_g \ll \Delta$ (where $\Delta$ is the spin–orbit splitting) as given by equation D in table 3.4 is,

$$E = \frac{\hbar^2 k^2}{2m_0} - \tfrac{1}{2}E_g + \tfrac{1}{2}(E_g^2 + \tfrac{8}{3}P^2k^2)^{1/2} ,$$

where the energy is now measured from the *bottom of the conduction band*. This applies to the narrow-bandgap intermetallic semiconductors of type III–V and to some II–VI semiconductors. The $\hbar^2k^2/2m_0$ term ($m_0$ is the free electron mass) can usually be ignored.

$$E \approx -\tfrac{1}{2}E_g + \tfrac{1}{2}(E_g^2 + \tfrac{8}{3}P^2k^2)^{1/2} . \qquad (4.38)$$

To evaluate the transport integrals for this case it is necessary to solve for $k$ and hence $p$ in terms of $x$ ($= E/k_B T$);

$$k = \gamma^{1/2} x^{1/2} (x + x_g)^{1/2} = \frac{p}{\hbar} \tag{4.39}$$

where

$$\gamma = \frac{3}{2}\left(\frac{k_B T}{P}\right)^2 , \quad x_g = \frac{E_g}{k_B T} .$$

In particular $k$ or $p$ must be used rather than $E$ in equation (4.3). From equation (4.29) and $\partial x/\partial p = 2x^{1/2}(x+x_g)^{1/2}/(2x+x_g)$ we obtain

$$n = \frac{\gamma^{3/2}}{2\pi^2}\int_0^\infty \frac{x^{1/2}(x+x_g)^{1/2}(2x+x_g)}{\exp(x-E_F^*)+1}\,dx . \tag{4.40}$$

More generally, the transport integrals $I_{ij}$ can be obtained as for the parabolic case but with correct substitution for $E$ etc., including correct substitution for $\partial f_0/\partial E$ to give the equation corresponding to (4.27). Generalised Fermi functions of the form

$$F(x_g, E_F^*) = \frac{2}{\pi^{1/2}}\int_0^\infty \frac{x^{1/2}(1+x/x_g)^{1/2}(1+2x/x_g)}{\exp(x-E_F^*)+1}\,dx ,$$

which reduce to ordinary Fermi–Dirac integrals of the form $F_{1/2}(E_F^*)$ when $x_g \to \infty$, have been calculated by Bebb and Ratliff (1971).

The density-of-states function in the case of the Kane model is (Kołodziejczak, 1961)

$$g(E) = \frac{(2m_d)^{3/2}}{4\pi^2\hbar^3}\left[E\left(1+\frac{E}{E_g}\right)\right]^{1/2}\left(1+\frac{2E}{E_g}\right) , \tag{4.41}$$

where $m_d$ is the band edge effective density. This expression should be compared with equation (4.2) for a parabolic band and like equation (4.2) it does not include the factor of 2 for electrons of opposite spin. The solutions for expressions for the other limiting case of the Kane band form where $E_g \gg \Delta$ differ only trivially.

The expression for the density-of-states function for the Cohen model (see page 58) may also be compared. For the case of rotation plus reflection symmetry, a density-of-states function of the form (Strel'chenko, 1966; Allgaier, 1966)

$$g(E) = \frac{(2m_d)^{3/2}}{4\pi^2\hbar^3}E^{1/2}\left[1+\frac{5E}{3E_g}\left(1+\frac{m_2}{5m_2'}\right)\right] \tag{4.42}$$

is obtained where again the factor 2 for spin is excluded. $m_2$ and $m_2'$ are the longitudinal band-edge effective masses for the conduction and valence bands respectively when the expression applies to the conduction band.

## 4.5 Solution of the transport equation for anisotropic materials

A complete discussion of the necessary treatment for anisotropic materials is beyond the scope of this book and yet is particularly important for many of the materials to be considered. It is intended here to compromise by indicating the sort of difficulties involved in determining the transport properties, without including a full mathematical analysis.

The treatment so far has been applicable to materials whose energy surfaces in momentum- or $k$-space can be represented by spheres centred at the origin of the coordinate system. Many materials such as bismuth and bismuth telluride do not exhibit the single minimum but have band edges at two or more positions in $k$-space. The energy surfaces are no longer spherical; they must be represented by ellipsoids. The model required to discuss this form of band structure is called the many-valley band model. These cases are illustrated in figure 4.5 by two-dimensional projections of the energy contours near to the band extrema (maxima or minima) on to the $k_z = 0$ plane in $k$-space. The outer octagonal boundary indicates the first Brillouin edge as in the case of f.c.c. crystals. Figure 4.5a illustrates spherical energy surfaces centred at $k = 0$, and figure 4.5b elliptical surfaces centred about six equivalent points in $k$-space. These two cases can be combined to give the situation shown in figure 4.5c. Figure 4.5d illustrates the case for two band edges which are degenerate but which are spherically symmetric, and combined with 4.5a gives figure 4.5e. Finally, figure 4.5f shows a generalisation of figure 4.5d, in the same way that 4.5b is a generalisation of 4.5a, but is a model that has not proved particularly useful when applied to real materials.

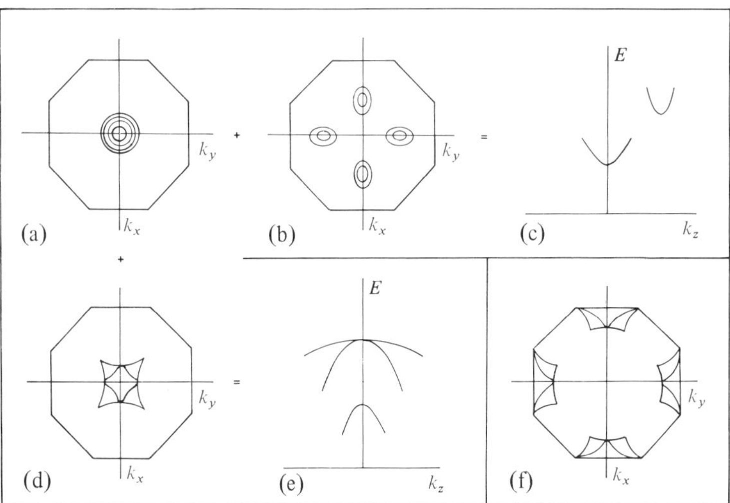


**Figure 4.5.** Possible band structure models. (After Harman and Honig, 1967.) For description see text.

Figure 4.5.e should be compared with the figure given with table 3.4; it is the typical situation when spin-orbit interaction is occurring and one of three valence bands is separated from the other two at $\boldsymbol{k} = 0$.

One of the problems of dealing with anisotropic cases is that usually a number of coordinate systems are used. Firstly there are the symmetry axes of the particular crystal involved. These may not be rectangular Cartesian coordinates, in which case it may be desirable to introduce pseudo-symmetry axes which are orthogonal to each other. Secondly there are the axes used to describe the position and orientation of the energy ellipsoids. The axes of this coordinate system are directed along the principal axes of the ellipsoids and can be related, of course, to the crystal symmetry axes. Finally there is the laboratory system of coordinates used to describe the sample configuration and the directions of the applied magnetic, electric, or thermal fields.

In the ellipsoidal many-valley model, where the energy-momentum dependence is parabolic, the variation in energy near a minimum is given by

$$E = \frac{\hbar^2}{2}\left(\frac{k_x^2}{m_x^*} + \frac{k_y^2}{m_y^*} + \frac{k_z^2}{m_z^*}\right) , \tag{4.43}$$

where $k_x$, $k_y$, $k_z$ are referred to the axes of the ellipsoid, and $m_x^*$, $m_y^*$, and $m_z^*$ are the effective masses corresponding to the three principal axes. For the density-of-states equation (4.2) to be applicable it is necessary to use a density-of-states effective mass $m_\mathrm{d}^*$:

$$m_\mathrm{d}^* = (m_x^* m_y^* m_z^*)^{1/3} . \tag{4.44}$$

This takes account of the fact that the volume of the energy ellipsoid depends on the product of the semi-major axes $[k_x^*/(2m_x^*)^{1/2}$, $k_y^*/(2m_y^*)^{1/2}$, $k_z^*/(2m_z^*)^{1/2}]$. This density-of-states effective mass is different from the conductivity (or inertial) effective mass which is defined as an averaged effective mass given by

$$\frac{1}{m_\mathrm{c}^*} = \frac{1}{3}\left(\frac{1}{m_x^*} + \frac{1}{m_y^*} + \frac{1}{m_z^*}\right) . \tag{4.45}$$

Equation (4.16) was given in a form which is applicable to anisotropic materials only if the relaxation time $\tau$ does not depend on direction in $k$-space. This is not generally true, and Harman and Honig (1967) treat the anisotropic case with both relaxation time and the effective mass as tensors. If it is assumed that energy is conserved during the scattering processes or that all the carrier velocities are randomised in the scattering processes, then three relaxation times can be used which are the principal components of the relaxation time tensor (Herring and Vogt, 1956). For some crystal classes the principal axes of the $\boldsymbol{\tau}$ tensor coincide with the principal axes of the reciprocal mass tensor.

Mackay and Sybert (1969) have considered the isothermal conductivity for an arbitrarily-oriented magnetic field for both the specific case of an isotropic relaxation time and the generalised case of an anisotropic relaxation time. They introduced a $\boldsymbol{\tau}$ which satisfies the symmetry elements of the crystal as a whole rather than the local symmetry at an energy extremum: this has the disadvantage that $\boldsymbol{\tau}$ is not simultaneously diagonal with $\boldsymbol{m}^{-1}$. Aubrey (1971) gives a method of obtaining the magnetoconductivity tensor for the group V semimetals that is valid over the classical range of magnetic fields. The method involves considering the mobility tensors associated with the individual band extrema. The partial conductivities are found in this way and the total magnetoconductivity tensor obtained by summing the partial contributions. Expressions are cumbersome for the arbitrary directions of magnetic field, but take on simplified forms when $\boldsymbol{B}$ is along the crystallographic axes. The explicit expressions applicable to antimony and arsenic are given for $\boldsymbol{B}$ along the binary, bisectrix, and trigonal directions (i.e. axes 1, 2, and 3 respectively).

In adding contributions to conductivity from different bands it is usual to assume that the contribution to transport from one ellipsoid is independent of the contributions from the others.

## 4.6 Tensor components of the transport coefficients

The numbers of components involved in equations 4.24a–4.24d can be obtained from group theory and are given in table 4.3 (from Bhagavantam and Pantulu, 1964) in order to show the increasing complexity of the

**Table 4.3.** Number of tensor coefficients in the 32 crystal classes.

| Property | Crystal classes | | | | | | | | | | |
|---|---|---|---|---|---|---|---|---|---|---|---|
| | 1<br>$\bar{1}$ | 2<br>$m$<br>$2/m$ | 222<br>$2mm$<br>$mmm$ | 4<br>$\bar{4}$<br>$4/m$ | $4mm$<br>$\bar{4}2m$<br>422<br>$4/mmm$ | 3<br>$\bar{3}$ | $3m$<br>32<br>$\bar{3}m$ | $3/m$<br>6<br>$6/m$ | $\bar{6}2m$<br>$6mm$<br>622<br>$6/mmm$ | 23<br>$m3$ | $\bar{4}3m$<br>432<br>$m3m$ |
| Electrical resistivity, thermal conductivity | 6 | 4 | 3 | 2 | 2 | 2 | 2 | 2 | 2 | 1 | 1 |
| Hall effect, Righi-Leduc effect, thermoelectric power | 9 | 5 | 3 | 3 | 2 | 3 | 2 | 3 | 2 | 1 | 1 |
| Nernst effect, Ettingshausen effect | 27 | 13 | 6 | 7 | 3 | 9 | 4 | 7 | 3 | 2 | 1 |
| Magnetoresistivity, Maggi-Righi-Leduc effect | 36 | 20 | 12 | 10 | 7 | 12 | 8 | 8 | 6 | 4 | 3 |
| Application (examples) | | $AuTe_2$ | | | $Cd_3As_2$ | | Sb<br>Bi<br>As<br>$Bi_2Te_3$ | C | | | α-Sn<br>HgTe<br>InSb<br>$Mg_2Pb$ |

transport effects for decreasing crystal symmetry. Electrical resistivity and thermal conductivity are second-rank tensors relating two vectors and have the smallest number of independent coefficients. The thermoelectric power tensor is also second rank, relating two vectors, but is not symmetrical and has more independent coefficients in certain of the crystal classes. The Hall effect and Righi–Leduc effect are third-rank tensors relating the axial vector $\boldsymbol{B}_l$ to an antisymmetric second-rank tensor (although because of their intrinsic symmetry the tensors are equivalent to second rank polar tensors[(2)]). The Nernst and Ettingshausen effects are also third rank and relate the axial vector $\boldsymbol{B}_l$ to a general second-rank tensor. Finally in the list are magnetoresistance and the Maggi–Righi–Leduc effect (variation of thermal conductivity in a magnetic field) and these are fourth-rank tensors relating two second-rank symmetric tensors.

## 4.7 Two-band Hall effect and magnetoresistance

Whereas a full anisotropic treatment of transport effects is required for many materials, much simpler situations are often dealt with. One such is the case of n-type semiconductors having two conduction band minima and being doped sufficiently for electron carriers to be present in both. Determination of the magnetoresistance and Hall coefficient as functions of (magnetic field)$^{-2}$ enables the band parameters to be obtained (Kwan *et al.*, 1971) and the method is outlined here.

For two isotropic independent bands, the variation of Hall coefficient $R_{\mathrm{H}}$ and resistivity $\rho$ with magnetic induction $B$ can be expressed as functions of the Hall coefficients $R_1$ and $R_2$ for the separate bands and the conductivities $\sigma_1$ and $\sigma_2$:

$$R_{\mathrm{H}}(B) = \frac{(R_1\sigma_1^2+R_2\sigma_2^2)+R_1R_2\sigma_1^2\sigma_2^2(R_1+R_2)B^2}{(\sigma_1+\sigma_2)^2+\sigma_1^2\sigma_2^2(R_1+R_2)^2B^2}\ ; \tag{4.46}$$

$$\rho(B) = \frac{(\sigma_1+\sigma_2)+\sigma_1\sigma_2(\sigma_1R_1^2+\sigma_2R_2^2)B^2}{(\sigma_1+\sigma_2)^2+\sigma_1^2\sigma_2^2(R_1+R_2)^2B^2}\ . \tag{4.47}$$

In cases where there is high degeneracy in both bands, the expression for the Hall coefficient can be approximated (Willardson *et al.*, 1954) to:

$$R_{\mathrm{H}} \approx -\frac{1}{en_1}\frac{1+(n_2/n_1)(\mu_2/\mu_1)^2}{[1+(n_2/n_1)(\mu_2/\mu_1)]^2} \tag{4.48}$$

and

$$R_{\mathrm{H}} \approx -\frac{1}{e(n_1+n_2)} \tag{4.49}$$

for small and strong magnetic fields respectively. The latter case is useful in that it gives the total extrinsic carrier concentration without any

(2) If a quantity transforms according to $T_{ijk...} = +l_{ip}l_{jq}l_{kr} \ldots T_{pqr...}$, where $l_{ip}$ etc. are direction cosines, it is called an axial tensor; if it transforms according to $T_{ijk...} = -l_{ip}l_{jq}l_{kr} \ldots T_{pqr...}$, it is called a polar tensor; a similar terminology is used for vectors.

weighting factors involving mobilities but usually requires rather a high field in practice. The above expressions do not include a numerical factor because it is usually close to unity. Expressions for magnetoresistance in low and high magnetic fields can be obtained similarly.

Various methods have been used for solving equations (4.46) and (4.47) for $R_H(B)$ and $\rho(B)$. Generally it is assumed that the variation of $R_1$, $R_2$, $\sigma_1$ and $\sigma_2$ with $B$ is negligible compared with the variation of $R_H$ and $\sigma$. If $R_0$, $\rho_0$, and $\sigma_0$ are zero field values of the Hall coefficient, resistivity, and conductivity, and $\Delta R_H$ and $\Delta\rho$ are the changes of the Hall coefficient and resistivity when the magnetic field is applied, then the equations can be plotted in the form

$$\frac{\rho_0}{\Delta\rho} = tB^{-2}+u \; , \tag{4.50}$$

$$-\frac{R_0}{\Delta R_H} = sB^{-2}+a \; , \tag{4.51}$$

where

$$t = \frac{\sigma_0-\sigma_1}{\sigma_1(R_0\sigma_0-R_1\sigma_1)^2} \; , \tag{4.52}$$

$$u = \frac{\sigma_1(R_0\sigma_0+R_1\sigma_0-2R_1\sigma_1)^2}{(\sigma_0-\sigma_1)(R_0\sigma_0-R_1\sigma_1)^2} \; , \tag{4.53}$$

$$s = \frac{R_0\sigma_0(\sigma_0-\sigma_1)^2}{\sigma_1^2(R_0\sigma_0+R_1\sigma_0-2R_1\sigma_1)(R_0\sigma_0-R_1\sigma_1)^2} \; , \tag{4.54}$$

$$a = \frac{R_0\sigma_0(R_0\sigma_0+R_1\sigma_0-2R_1\sigma_1)}{(R_0\sigma_0-R_1\sigma_1)^2} \; . \tag{4.55}$$

Two of the four equations are required to solve for $R_1$ and $\sigma_1$ and Kwan *et al.* consider solutions when taking different pairs. Because the intercepts may be uncertain, the use of $s$ and $t$ may be preferable. Values of $R_2$ and $\sigma_2$ are found from equations (4.46) and (4.47) with $B \to 0$. Carrier concentrations and mobilities should then be calculable, although some knowledge of the type of scattering and/or the degeneracy of the material is necessary. Even then there may be difficulty in specifically obtaining $\mu_2$ and $n_2$ rather than the product $\mu_2 n_2$.

Expressions for other transport coefficients where two types of carriers are involved are given, for example, by Putley, 1960.

## 4.8 Quantisation effects

Investigation of semimetals in particular has involved such phenomena as the de Haas–van Alphen and the Shubnikov–de Haas effects which arise from quantisation of the electron orbits in a magnetic field. A detailed analysis of these effects cannot be attempted here but an indication will be given of how these effects are useful in determining data concerning the shape of actual Fermi surfaces.

The de Haas–van Alphen effect is an oscillatory variation of magnetic susceptibility $\chi$ and was first observed in bismuth. In the absence of a magnetic field, the electrons have a continuous range of energies up to the Fermi energy and in momentum space they occupy all the available $k$ values. When a magnetic field is applied in the $z$ direction, although $k_z$ is still unquantised, only discrete energy values $E_n$ are possible and spacing of these energy levels increases with increase of the magnetic field. Expressions have been obtained for the values of $E_n$ for spherical and ellipsoidal energy surfaces but the calculation becomes more difficult for more complex surfaces.

Taking the case of a spherical energy surface in zero magnetic field, consider the contours which are described by the intersection of planes of $k_z = 0$ with this energy surface. These contours will be closed orbits containing an area $A$ which is a function of $E$ and $k_z$. When a magnetic field is applied, the quantised values of $E$ can be shown to be such that

$$A = 2\pi(n+\gamma)\frac{eB}{\hbar}, \tag{4.56}$$

where $n$ is an integer and $\gamma$ is a constant which in the present case of a quadratic surface is $\frac{1}{2}$. This can be considered as equivalent to the spherical surface becoming instead a series of coaxial cylinders about the $k_z$ axis (figure 4.6a). The possible energy levels are highly degenerate so that electrons which would otherwise occupy the continuous range of energies between one level and the next can now occupy the single energy level. The maximum energy level remains little changed, but as the magnetic field is increased and the energy surface alters the mean energy $\bar{E}$ shows considerable oscillation in its value. The magnetic moment of the system of electrons is given by $\mu = -\partial E/\partial B$, and the susceptibility by $\chi = -\partial^2 E/\partial B^2$; the magnetic moment is an oscillatory function of $1/B$

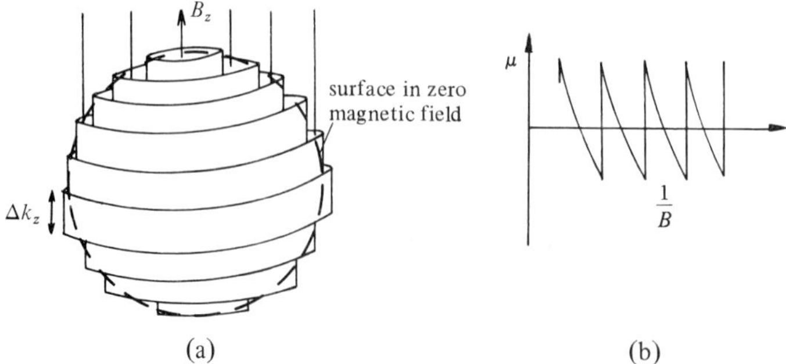


**Figure 4.6.** Quantisation effects: (a) tubes of quantised energy; (b) variation of magnetic moment as a function of $1/B$ arising as a result.

(figure 4.6b) and the oscillations occur at equal intervals such that

$$\Delta(1/B) = \frac{2\pi e}{\hbar A} , \qquad (4.57)$$

where $A$ is the extremal Fermi surface area normal to the direction of the magnetic field. Hence, by measurement of $\Delta(1/B)$ as a function of $B$ in different directions, considerable information is obtained regarding the Fermi surface. In semimetals where the Fermi surfaces are small, the periods are well-spaced and much lower magnetic fields are required than in the case of metals. The method can detect very small changes of area; theoretically it has a sensitivity of about 1 part in $10^4$, whereas in practice it is difficult to obtain the magnetic field calibration to better than 1 part in $10^3$. The amplitude of the effect is reduced away from absolute zero temperature by the fact that the Fermi surface is no longer perfectly sharp. It is also reduced by electron collisions which produce collision broadening of the quantised energy levels.

An effective mass can be defined from

$$m^* = \frac{\hbar^2}{2\pi}\left(\frac{\mathrm{d}A}{\mathrm{d}E}\right)_{k_z} , \qquad (4.58)$$

and the values for this for the extremal cross sections can be determined as a function of angle from the temperature dependence and field dependence of the effect. The reciprocal of the carrier relaxation time can be obtained as a function of angle also, and provided the Fermi surface is centrosymmetric the shape of the energy surfaces (e.g. prolateness of the ellipsoids) and the electron velocities can be derived.

The various transport effects such as electrical and thermal conductivity, thermoelectric effect, and Hall effect can all show a periodic dependence on $(1/B)$ (for a review see Kahn and Frederikse, 1959). One which has been studied particularly is the Shubnikov–de Haas effect which is the oscillatory dependence of electrical resistance on magnetic field, this oscillatory term being superimposed on a monotonic variation of resistance in a magnetic field, which may, in fact, be considerably larger than the oscillatory contribution. Although the coupling of electrons to scattering agents is independent of the magnetic field, the probability of scattering is proportional to the number of states into which the scattering can occur. The latter is proportional to the density of states, which oscillates at the de Haas–van Alphen frequency where the relaxation time oscillates, and this effect is the main contribution to the Shubnikov–de Haas effect. It can be shown that the relative amplitude depends on $(2n)^{-1/2}$ where $n$ is the carrier concentration, so that the effect is strong in semimetals where the carrier concentration is small compared with metals and it is particularly useful in semiconductors where the density of states is too small for the de Haas–van Alphen effect to be observed.

**References**

Allgaier, R. S., 1966, *Phys. Rev.*, **152**, 808.
Aubrey, J. E., 1971, *J. Phys. F.*, **1**, 493.
Bebb, H. B., Ratliff, C. R., 1971, *J. Appl. Phys.*, **42**, 3189.
Beer, A. C., 1963, *Galvanomagnetic Effects in Semiconductors*, Supplement 4 to *Solid State Physics* (Academic Press, New York).
Bhagavantam, S., Pantulu, P. V., 1964, *Proc. Indian Acad. Sci.*, **60**, 1.
Bridgman, P. W., 1924, *Solid State Phys.*, **4**, 199.
Callen, H. B., 1952, *Phys. Rev.*, **85**, 16.
Harman, T. C., Honig, J. M., 1967, *Thermoelectric and Thermomagnetic Effects and Applications* (McGraw-Hill, New York).
Herring, C., Vogt, E., 1956, *Phys. Rev.*, **101**, 944.
Heurlinger, T., 1915, *Ann. Phys.*, **84**, 48.
Heurlinger, T., 1916, *Phys. Z.*, **17**, 221.
Jackson, E. A., 1968, *Equilibrium Statistical Mechanics* (Prentice-Hall, Englewood Cliffs, NJ).
Jan, J.-P., 1957, *Solid State Phys.*, **5**, 1.
Kahn, A. H., Frederikse, H. P. R., 1959, *Solid State Phys.*, **9**, 257.
Kohler, M., 1941, *Ann. Phys. (Leipzig)*, **40**, 601.
Kołodziejczak, J., 1961, *Acta Phys. Pol.*, **20**, 289, 379.
Kwan, C. C. Y., Basinski, J., Woolley, J. C., 1971, *Phys. Status Solidi*, **B48**, 699.
Landau, L. D., Lifshitz, E. M., 1959, *Course in Theoretical Physics: 5. Statistical Physics* (Pergamon Press, Oxford).
Landau, L. D., Lifshitz, E. M., 1960, *Course in Theoretical Physics: 6. Electrodynamics of Continuous Media* (Pergamon Press, Oxford).
McDougall, J., Stoner, E. C., 1938, *Philos. Trans. R. Soc. London, Ser. A.*, **237**, 67.
Mackay, H. J., Sybert, J. R., 1969, *Phys. Rev.*, **180**, 678.
Mazur, P., Prigogine, I., 1951, *J. Phys. Radium*, **12**, 616.
Nye, J. F., 1957, *Physical Properties of Crystals* (Oxford University Press, Oxford).
Putley, E. H., 1960, *The Hall Effect and Related Phenomena* (Butterworths, London).
Putley, E. H., 1975, *J. Phys. C.*, **8**, 1937.
Rhodes, P., 1950, *Proc. R. Soc. London, Ser. A.*, **204**, 396.
Sommerfeld, A., Frank, N. H., 1931, *Rev. Mod. Phys.*, **3**, 1.
Strel'chenko, E. G., 1966, *Sov. Phys. - Solid State*, **8**, 772.
Tauc, J., 1962, *Photo- and Thermoelectric Effects in Semiconductors* (Pergamon Press, Oxford).
Tsidil'kovskii, I. M., 1962, *Thermomagnetic Effects in Semiconductors* (Infosearch, London).
Willardson, R. K., Harman, T. C., Beer, A. C., 1954, *Phys. Rev.*, **96**, 1512.
Wilson, A. H., 1953, *The Theory of Metals*, 2nd edition (Cambridge University Press, Cambridge).

# 5

# Device applications

Much of the work on the narrow-bandgap semiconductors and particularly that on the mixed crystal systems has been directed towards applications in devices. This chapter will discuss the background to the types of applications mentioned in part II. However, another volume in the series, *Physics of Solid State Devices,* by T. H. Beeforth and H. J. Goldsmid, has already dealt at length with device applications. In the chapter here, the main concern will be with those properties which are required to make a material suitable for a particular device application.

## 5.1 Photovoltaic detectors

These usually consist of a $p$–$n$ junction whose electrical properties are sensitive to infrared radiation which is incident normal to the plane of the junction (figure 5.1). The chalcogenide mixed crystal systems are particularly attractive for use in these detectors as their bandgaps can be tailored to suit the radiation wavelength of interest. The radiation is absorbed over the surface of the detector, typically within 1 $\mu$m of the surface, producing electrons and holes which diffuse towards the $p$–$n$ junction where they produce an increase in the minority carrier concentration. This leads to a change in the junction potential. Thus, if the junction is short-circuited externally, there is a current flow to return minority carrier concentrations at the junction back to equilibrium. If a reverse bias is applied, the photocurrent adds to the normal reverse diode current; in effect the application of the bias voltage moves the operating point to some desired position on the current–voltage characteristic curve away from the origin. The signal radiation is usually chopped so as to produce a fluctuating voltage which can be distinguished from DC voltages produced by background radiation sources. Photovoltaic detectors have the advantages over photoconductive detectors (see next section), in that they are faster and need lower power for driving.

It is common practice to have a thin n-type layer on top of a slab of p-type material; similar arguments to those that follow would apply to a p-type layer on an n-type slab. The active part of the device is the depletion layer at the junction plus regions on either side of approximately

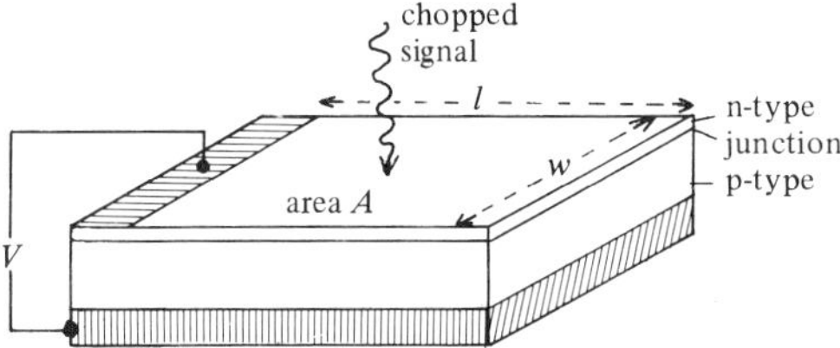


**Figure 5.1.** Configuration for a photovoltaic detector.

the width of the diffusion length. The sensitivity of a photovoltaic detector can be specified completely by two parameters, the quantum efficiency $\zeta$, defined as the number of light-generated carriers reaching the junction per incident photon, and the incremental resistance $R$ [$\equiv (\partial V/\partial I)_{V=0}$]. Analysis of the detector performance is slightly different according to whether or not background radiation is assumed. The analysis for this mode of operation will follow that of Melngailis and Harman (1970) and assume no background radiation.

The total current $I_t$ in the diode consists of the photocurrent $I_p$ plus the dark current $I(V)$ which is given in the ideal case by

$$I(V) = I_0\left(\exp\frac{eV}{k_B T} - 1\right) , \tag{5.1}$$

where $I_0$ is the diode saturation current. Hence,

$$I_t = -I_p + I(V) , \tag{5.2}$$

as the photocurrent flows in a reverse direction to the forward diode current. The photocurrent can be expressed as

$$I_p = e\zeta N , \tag{5.3}$$

where $N$ is the number of incident photons per second. For small voltages the diode current $I(V)$ is assumed to depend linearly on the voltage. Hence,

$$I_t = -I_p + \frac{V}{R} . \tag{5.4}$$

For a detector in the photovoltaic mode the total current flowing is zero. Thus,

$$V = RI_p = e\zeta NR . \tag{5.5}$$

The power of the incident radiation $P_\lambda$ is given by the number of photons $N$ multiplied by the energy per photon $E_\lambda$ so that

$$V = \frac{e\zeta RP_\lambda}{E_\lambda} \tag{5.6}$$

and the voltage responsivity $\mathcal{R}_{V\lambda}$ is given by

$$\mathcal{R}_{V\lambda} = \frac{V}{P_\lambda} = \frac{e\zeta R}{E_\lambda} . \tag{5.7}$$

For calculating detectivity it is usual to use noise-equivalent power, $P_{N\lambda}$. For this the signal voltage is equated to the r.m.s. noise voltage for 1 Hz bandwidth. The Johnson noise voltage is given by

$$\overline{V^2} = 4k_B TR\Delta f \tag{5.8}$$

so that

$$\frac{e\zeta R P_{N\lambda}}{E_\lambda} = (\overline{V^2})^{1/2} = (4k_B T R)^{1/2} ,$$

whence

$$P_{N\lambda} = \frac{2E_\lambda (k_B T)^{1/2}}{e\zeta R^{1/2}} . \tag{5.9}$$

If the area of the detector is $A$, the detectivity is defined as

$$D_\lambda^* \equiv \frac{A^{1/2}}{P_{N\lambda}} . \tag{5.10}$$

The expression is based on the fact that detectivity should depend on the square root of the detector area. This is not always so, particularly in the far-infrared. Hence

$$D_\lambda^* = \frac{e\zeta (AR)^{1/2}}{2E_\lambda (k_B T)^{1/2}} . \tag{5.11}$$

It can be seen that detectivity is proportional to efficiency and to the square root of the diode resistance. From equation (5.1) the incremental resistance is $k_B T/eI_0$.

$$D_\lambda^* = \frac{e^{1/2}\zeta A^{1/2}}{2E_\lambda I_0^{1/2}} = \frac{e^{1/2}\zeta}{2E_\lambda j_0^{1/2}} \tag{5.12}$$

where $j_0$ is the saturation current density. If desired, $j_0^{1/2}$ can be expressed as a function of the equilibrium densities of holes and electrons and their mobilities and half-lives. In order to make the saturation current small, majority carrier densities must be made large to keep the minority carrier densities small; however, they must not be made so large that conduction is no longer by minority carrier injection.

If the calculation is repeated for the case of reverse bias it is necessary to involve the shot noise (which in the special case of zero bias reduces to the Johnson noise) and it is found that detectivity is increased by a factor of $2^{1/2}$.

To maximise efficiency $\zeta$ it is found necessary to have a small reflection coefficient for the radiation at the surface of the diode, a small surface-recombination velocity and a junction depth which is small compared with the hole diffusion length. The response speed is limited either by the lifetimes of the excited carriers or by circuit parameters such as the junction capacitance.

## 5.2 Photoconductive detectors

In the absence of any signal radiation, the voltage across the photoconductive diode is

$$V_0 = \frac{I_0 l}{(\sigma_d + \sigma_b) wt} , \tag{5.13}$$

where $l$ is the length of the diode, $w$ the width, and $t$ the thickness. $\sigma_d$ is the dark conductivity and $\sigma_b$ is the conductivity arising from any background radiation. The chopped signal radiation gives rise to an AC voltage of

$$V_s = \frac{\sigma_s}{\sigma} V_0 , \tag{5.14}$$

where $\sigma$ is the total conductivity with background and signal radiation present. The signal radiation is usually small such that $\sigma_s \ll \sigma_d + \sigma_b$ so that $\sigma \approx \sigma_d + \sigma_b$. If the electron and hole concentrations at thermal equilibrium are $n_0$ and $p_0$ respectively and their mobilities are $\mu_n$ and $\mu_p$, then the dark conductivity is

$$\sigma_d = e(n_0\mu_n + p_0\mu_p) . \tag{5.15}$$

Also, if $n_b$ is the excess electron–hole concentration produced by the background radiation, then

$$\sigma_b = en_b(\mu_n + \mu_p) . \tag{5.16}$$

When the chopped signal radiation is incident, it is necessary to take into account the lifetimes $\tau_n$ and $\tau_p$ of the excess electrons and holes which are excited. If $\zeta$ is the quantum efficiency as in the case of the photovoltaic detector and $J_s$ is the number of signal photons arising per unit area in unit time, the signal photoconductivity is given by

$$\sigma_s = \frac{\zeta J_s e}{t}[\mu_n\tau_n + \mu_p\tau_p] . \tag{5.17}$$

Hence,

$$V_s \approx \frac{\sigma_s}{\sigma_d + \sigma_b} V_0 = \frac{\zeta J_s(\mu_n\tau_n + \mu_p\tau_p)V_0}{t[(n_0\mu_n + p_0\mu_p) + n_b(\mu_n + \mu_p)]} . \tag{5.18}$$

Relating the signal current to the energy per photon $E_\lambda$ ($= h\nu = hc/\lambda$) we obtain

$$J_s = \frac{P_{N\lambda}}{lwE_\lambda} = \frac{P_{N\lambda}}{AE_\lambda} . \tag{5.19}$$

From equations (5.18) and (5.19), the voltage responsivity in the photoconductive mode is

$$\mathcal{R}_{V\lambda} = \frac{V_s}{P_{N\lambda}} = \frac{\zeta(\mu_n\tau_n + \mu_p\tau_p)V_0}{AtE_\lambda[(n_0\mu_n + p_0\mu_p) + n_b(\mu_n + \mu_p)]} . \tag{5.20}$$

This expression is analogous to equation (5.7) for the photovoltaic mode. Approximations can be made. Typically in n-type material, $n_0 \gg p_0$; also $\sigma_b \ll \sigma_d$ and $\tau_n \approx \tau_p$ in many practical cases. These conditions then give

$$\mathcal{R}_{V\lambda} \approx \frac{\zeta\tau V_0}{AtE_\lambda n_0} . \tag{5.21}$$

Similarly an expression for detectivity can be obtained, analogous to the expression in equation (5.12) for the photovoltaic mode. Considering the Johnson noise-limited case we have

$$D^*_\lambda = \frac{A^{1/2}}{P_{N\lambda}} = \frac{A^{1/2}\mathcal{R}_{V\lambda}}{V_s}$$

$$\approx \frac{\zeta\tau V_0}{A^{1/2}tE_\lambda n_0 V_s} \ . \tag{5.22}$$

Putting $V_s = (\overline{V^2})^{1/2} = (4k_B TR\Delta f)^{1/2}$ where $\Delta f = 1$ Hz and

$$R = \frac{l}{(\sigma_d + \sigma_b)wt} \approx \frac{l}{en_0\mu_n wt}$$

gives

$$D^*_\lambda = \frac{\zeta\tau V_0}{2E_\lambda l}\left(\frac{e\mu_n}{n_0 tk_B T}\right)^{1/2} . \tag{5.23}$$

From this equation, the conditions for maximising detectivity can be seen. A large bias field $V_0/l$ is required applied to a thin sample operating at a low temperature and $\tau$ and $\mu_n$ require maximising whilst $n_0$ needs to be minimised. However, as the applied field is increased, Joule heating and generation–recombination noise become important. Under the limiting condition of generation–recombination noise the detectivity becomes independent of the material parameters of the photoconductor except for the effect these parameters have on the quantum efficiency.

### 5.3 Narrow-bandgap semiconductor lasers

Semiconductor laser emission was first reported by Hall *et al.* (1962) and laser action has been observed in many semiconductor materials including narrow-bandgap semiconductors such as $Pb_{1-x}Sn_xTe$, $Pb_{1-x}Sn_xSe$, and $Hg_{1-x}Cd_xTe$. The highest performance has been obtained from the first of these systems and, although several pumping systems have been used (electrical injection of the carriers either at the p–n junction or in the vicinity of metal–semiconductor contacts, optical pumping or electron-beam pumping), p–n diodes have been found best. The geometry of the laser diode must be such as to provide the required cavity for the internal reflection of the emitted radiation. Narrow-bandgap semiconductor lasers have been reviewed by Harman (1971).

Figure 5.2 shows a typical arrangement for such a laser device. With no bias applied to the junction the Fermi energy is constant across the device, as is shown in figure 5.3a. Hatched areas represent conduction electrons and holes. Between the p-type and n-type regions is the depletion layer where a lower concentration of conduction electrons and holes exists. Once a voltage is applied to the junction, the equilibrium Fermi energy is replaced by quasi-Fermi levels, $E^n_{qF}$ being the level applicable to the electrons and $E^p_{qF}$ to the holes. While the applied voltage is low and the difference between the quasi-Fermi levels $E^{n}_{qF} - E^{p}_{qF}$ is small, there is a small amount of spontaneous radiation of frequency $\nu$ where $h\nu \approx E_g$. However, once the applied voltage has been increased such that $E^{n}_{qF} - E^{p}_{qF} > h\nu$, then there are electrons and holes existing in the

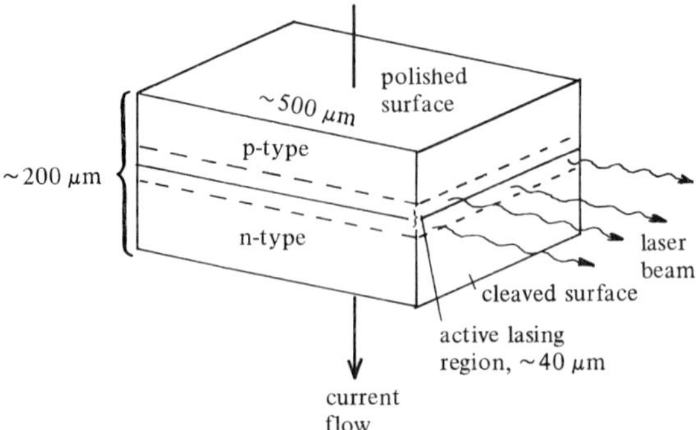


**Figure 5.2.** Configuration for a semiconductor laser.

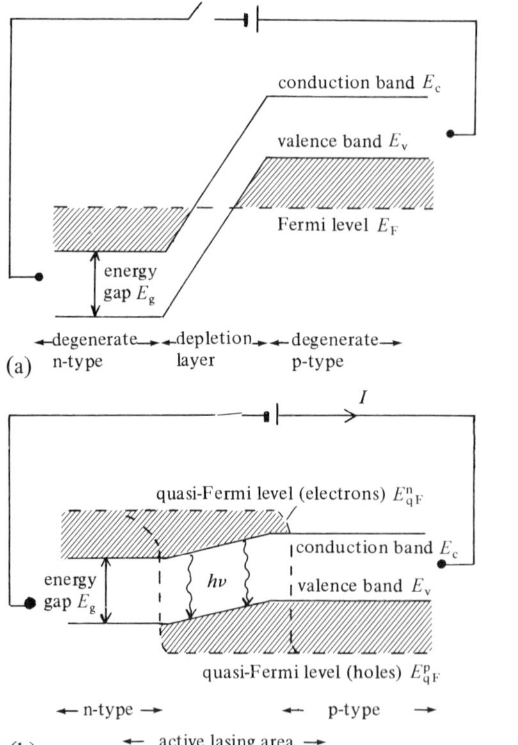


**Figure 5.3.** Energy levels across the junction of the semiconductor laser: (a) open-circuit, zero bias; and (b) short-circuit, forward bias.

same region, and upon recombination these give out stimulated radiation of energy $h\nu$. At even higher applied voltages (this is the case shown in figure 5.3b) carriers can diffuse into regions of the opposite type and nonequilibrium populations of carriers are established. An inverted population can be set up extending beyond the depletion layer. Provided diffusion lengths are sufficiently large and energy losses are small, the n–p diode will lase over most of this region.

## 5.4 Hall effect and magnetoresistance devices

The origin of the Hall and magnetoresistance coefficients can be seen from equation (4.24a) in chapter 4. The Hall effect is given by

$$R_H = \frac{E_y}{I_x B_z} , \tag{5.24}$$

where $I_x$ is the total current and $B_z$ is the magnetic induction. For a material of known Hall coefficient, measurement of the transverse voltage $E_y$ gives a measure of $I_x B_z$ and hence a measure of $B_z$ for fixed $I_x$. For good sensitivity it is usually necessary to obtain a maximum Hall voltage for a given magnetic field, $B_z$. In a degenerate semiconductor, the Hall coefficient is given by

$$R_H = \pm \frac{1}{ne} , \tag{5.25}$$

where the negative sign applies to electron carriers and the positive sign to holes. For a nondegenerate semiconductor, where the carriers are subject to acoustic-mode lattice scattering, the Hall coefficient is given by a similar expression:

$$R_H = \pm \frac{3}{8ne} . \tag{5.26}$$

In each case decrease in the carrier concentration $n$ increases the Hall coefficient. However, it is also important for the carriers to have a high mobility. This is because the Hall element must be incorporated into an output circuit. For maximum power transfer the load resistance must equal the transverse resistance of the Hall element (i.e. the resistance in the direction in which the Hall voltage is developed). It can be shown that the transfer efficiency in this case depends on $R_H \sigma$ which equals the Hall mobility $\mu_H$. However, when the Hall element is not delivering into a matched circuit, but into a high impedance amplifier, the product $\mu_H R_H$ should be large. Hence the requirement then is for a material having both low carrier concentration and high carrier mobility.

Indium antimonide with its electron mobility at room temperature as high as 8 $m^2\ V^{-1}\ s^{-1}$ is suitable and much used. However, the fact that it has a small energy gap means that the intrinsic material has a rather low electrical resistivity so that it is usual to dope it with acceptor impurity such that the hole concentration becomes an order of magnitude greater than the electron concentration. The Hall coefficient for this two-carrier case is now given by

$$R_H \approx \frac{p\mu_p^2 - n\mu_n^2}{e(p\mu_p + n\mu_n)^2} \tag{5.27}$$

and the Hall coefficient remains negative provided $p/n < (\mu_n/\mu_p)^2$. The electron mobility is halved as compared with the undoped samples, but because this is still much larger than the hole mobility the material still appears to be n-type. To provide samples of sufficiently high resistivity for matching to conventional amplifiers, the indium antimonide samples need to be of thin-film form. The electron mobility tends to be much less than in the bulk samples but this can be largely prevented by using recrystallisation techniques (Carroll and Spivak, 1966; Wieder, 1966). Other high-mobility materials have also been investigated, for instance cadmium arsenide (see section 8.4.4).

The magnetoresistance effect is the change of resistance of a sample in a magnetic field; as used in devices it is usually the change of longitudinal resistance when a transverse magnetic field is applied to the sample (i.e. the same configuration as for the Hall effect). Whereas the Hall effect is proportional to the magnetic field, the magnetoresistance usually depends on the second power of the magnetic field. In general, Hall effect devices tend to be superior in low magnetic fields but magnetoresistance devices become better in high magnetic fields, the change-over point being $\mu B = 1$. For maximum magnetoresistance it is necessary to eliminate the electric field in the transverse direction. This can be achieved by the so-called Corbino disc in which the electrical leads are connected to the inner and outer peripheries of the disc and the current and energy flow is radial. The magnetic field is applied in a direction perpendicular to the main surface of the disc. The galvanomagnetic effects for the Corbino disc geometry have been discussed extensively by Harman and Honig (1967). The magnetoresistance is also large in samples which are short and wide because the end electrodes tend to short out the Hall field. The geometry can be taken one step further by using a number of samples and metal strips in series. This enables the Hall element to have a suitable overall length to provide the desired total resistance. Weiss (1966) has obtained an equivalent arrangement by using a eutectic of InSb + 1·8% NiSb where the NiSb takes the form of fine needles and provides the necessary strips to reduce the Hall effect. A similar type of eutectic between $Cd_3As_2$ and NiAs has been grown by Hiscocks (1969) and is discussed in section 8.4.3.

## 5.5 Thermoelectric devices

Thermoelectric devices either convert heat energy into electric energy or the reverse and in the absence of a magnetic field depend on the Seebeck and Peltier effects and to a lesser extent the Thomson effect. These effects are as follows.

*Seebeck effect.* If an electric circuit is composed of two dissimilar metals and one of the junctions between the metals is maintained at a different temperature to the other, then a current flows around the circuit. An absolute measure of the effect for one metal (corresponding to using a superconductor as the other metal) gives the thermoelectric power (measured as voltage established for unit temperature difference when open-circuit).

*Peltier effect.* When an electric current passes across the junction between two dissimilar metals, heat is in general liberated or absorbed at the junction.

*Thomson effect.* When an electric current passes through a single homogeneous but unequally heated conductor, heat is absorbed or liberated. (This is additional to Joule heating.)

The thermoelectric figure of merit for a given material is defined by

$$z = \frac{\alpha^2 \sigma}{K}, \tag{5.28}$$

where $\alpha$ is the absolute thermoelectric power, $\sigma$ the electrical conductivity and $K$ the thermal conductivity. Obviously it is necessary to have a good electrical conductivity to minimise Joule heating and a low thermal conductivity so that a temperature differential can be maintained. For a couple it can be shown (see for instance Goldsmid, 1960) that the equivalent expression is

$$z = \frac{\alpha_{pn}^2}{[(K_p/\mu_p)^{1/2} + (K_n/\mu_n)^{1/2}]^2}, \tag{5.29}$$

where $\alpha_{pn}$ is the thermoelectric power for the particular pair of materials and p and n refer to the p-type and n-type arms respectively. For thermoelectric couples of practical use the figure of merit expressions for the arms are generally similar in magnitude and hence expression (5.28) is more usually employed.

In a metal $K/\sigma$ is fixed (equal to $\frac{1}{3}\pi^2 k_B^2 T/e^2$ from the Wiedemann–Franz–Lorenz Law) so that maximising $z$ involves maximising $\alpha$. In fact the Seebeck coefficients of metals are too low. Semiconductors however, can have very large Seebeck coefficients but these are usually accompanied by very small electrical conductivities. Figure 5.4 shows in a schematic way how the electrical conductivity and the reciprocal of thermal conductivity may be expected to change with carrier concentrations in different types of materials, also how the square of the thermoelectric power will vary, and hence shows how the thermoelectric figure of merit can be expected to maximise for semiconductors having fairly high carrier concentrations. In practice use is made of semiconductors to which impurities have been added. Although the presence of the impurities tends to decrease the Seebeck coefficient, electrical conductivity is increased and the lattice contribution to the thermal conductivity is reduced. A high carrier mobility is required along with a large effective mass. These two conditions tend to be contradictory, but if a multi-valley semiconductor is chosen it is the inertial effective mass which determines the carrier mobility, whereas it is the overall density-of-states effective mass which determines $z$. To keep the lattice thermal conductivity low, materials of high atomic weight are chosen as these have a low Debye temperature, low melting point, and hence large lattice vibrations at any given temperature. Thus, many-valley semiconductors of high atomic weight doped and/or in an alloy prove to be best, for instance alloys involving $Bi_2Te_3$. In fact, it is necessary for the bandgap to be sufficiently large for a Seebeck coefficient of approximately $\pm 200\ \mu V\ K^{-1}$ to be possible without the production of a significant number of minority carriers. Hence, the energy gap must be much greater than $k_B T$. The materials of interest at approximately room temperature (for instance $Bi_2Te_3$ alloyed with other tellurides and selenides) have energy gaps near the upper limiting value of $10 k_B T$ ($T$ being here room temperature) established for the purposes of this review (for $Bi_2Te_3$, $E_g \approx 5 k_B T$); other materials significant particularly at higher temperatures and having larger bandgaps are not considered.

Figure 5.5 shows diagrammatically the three possible modes of operation of a thermoelectric device: as a generator, as a heat pump, and as a refrigerator. It is usual, but not essential, for one arm of the device to be of n-type material and the other of p-type. Figures of merit have not been achieved such as to make thermoelectric devices suitable for general applications, except in special cases such as for generators for

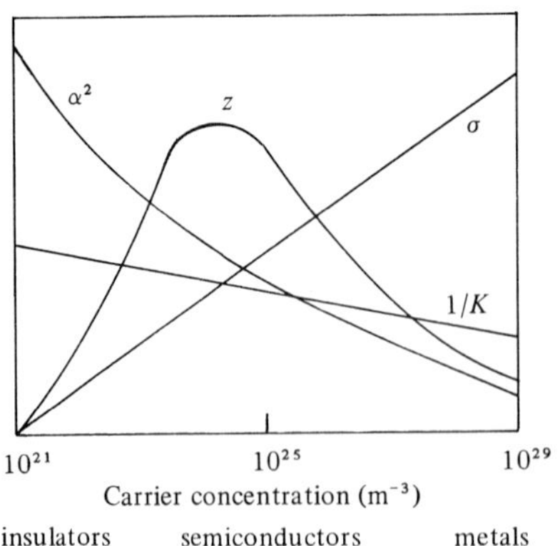


**Figure 5.4.** Variation of $\sigma$, $1/K$, $\alpha^2$, and $z$ with carrier concentration (schematic).

inaccessible locations or as refrigerators where only a small amount of heat extraction is required. In actual operation of the devices the quantity of interest becomes in the case of a generator the efficiency and in the case of a refrigerator or a heat pump the coefficient of performance. The efficiency of a thermoelectric generator is given by the ratio of the electric power output to the nett rate of heat transfer from the hot to the cold junction. This efficiency can be shown to be equal to the product of the Carnot efficiency [$= (T_0 - T_L)/T_0$, where $T_0$ is the junction temperature and $T_L$ the terminal temperature] and the device efficiency which depends on $z$, $T_0$, $\Delta T = T_0 - T_L$ and $R_L/R$, where $R_L$ is the load resistance and $R$ is the internal resistance of the device. The efficiency of the thermoelectric generator increases with increase of the figure of merit $z$ but not linearly (see Ioffe, 1957, for graphs showing the dependence). The coefficient of performance for a refrigerator (or heat pump) is given by the ratio of the cooling (or heating) capacity to the electrical power consumption. It depends on similar factors to the efficiency for a thermoelectric generator, although the overall expression is almost the inverse of that for the efficiency of a generator. The maximum coefficient of performance also increases with increase of $z$ (see Ure and Heikes, 1961).

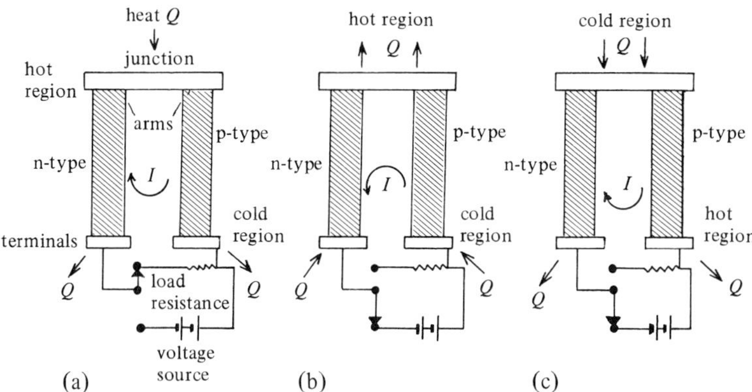


**Figure 5.5.** Modes of operation of thermoelectric devices: (a) generator, (b) heat pump, and (c) refrigerator.

## 5.6 Thermomagnetic devices

For many materials, the thermoelectric effect can be enhanced in a magnetic field. This will not be considered further here however; what will be considered are transverse devices which become possible when a magnetic field is applied orthogonal to either a temperature gradient (hence establishing a voltage transverse both to magnetic field and temperature gradient—the Nernst effect) or to an electric current (hence establishing a temperature gradient transverse to both magnetic field and electric current—the Nernst–Ettingshausen effect). How these effects arise was discussed in section 4.2.

A figure of merit for the Nernst–Ettingshausen effect can be defined in an analogous way to that for the thermoelectric effect:

$$z_{xy}^{\mathrm{i}} = \frac{B_z^2 Q_{\mathrm{i}}^2}{\rho_x^{\mathrm{i}} K_y^{\mathrm{i}}} = \frac{(\alpha_{xy}^{\mathrm{i}})^2}{\rho_x^{\mathrm{i}} K_y^{\mathrm{i}}} \; . \tag{5.30}$$

Here the index i refers to isothermal conditions as discussed on page 68. $\rho_x^{\mathrm{i}}$ is the isothermal electrical resistivity of the material in the direction of current flow, $K_y^{\mathrm{i}}$ is the isothermal thermal conductivity in the transverse direction (the direction in which the temperature gradient is established) and $B_z$ is the magnetic field. Although the expressions for the figures of merit for the thermoelectric and thermomagnetic effects are analogous, the energy processes involved are rather different and conditions of optimisation are not the same for each. Although $z$ (and $zT$) values for the thermomagnetic mode tend to be lower than for the thermoelectric mode, nevertheless it is possible to have comparable cooling because of a higher coefficient of performance for a Nernst–Ettingshausen-type refrigerator. Also, cooling improves very much more rapidly with increase of $z_{xy}T$ than it does for $zT$ so that although material of $zT \sim 0{\cdot}9$ (for bismuth telluride alloys) are available, a $z_{xy}T$ value of $0{\cdot}5$ would provide a Nernst–Ettingshausen cooler of superior performance to the thermoelectric cooler. In addition, because it depends on the transverse effect, the Nernst–Ettingshausen device has the advantage in that it can be cascaded into a tapered block as shown in figure 5.6 (O'Brien and Wallace, 1958), instead of having repeated stages as is necessary for a Peltier–Seebeck device. In fact, if rectangular geometry is used, then an arrangement such as a spiral screw becomes possible (Norwood, 1963; Guthrie, 1965). It is also an advantage of the Nernst–Ettingshausen devices that the heat is transferred across junctions which are separated from those through which the electric current passes.

It can be shown that there are a number of requirements for a material to be suitable for use in a Nernst–Ettingshausen-type device. The material should have equal concentrations of electrons and holes (i.e. it should be intrinsic) so that the Ettingshausen effect is enhanced by generation and recombination of electrons and holes. The mobilities of the electrons and

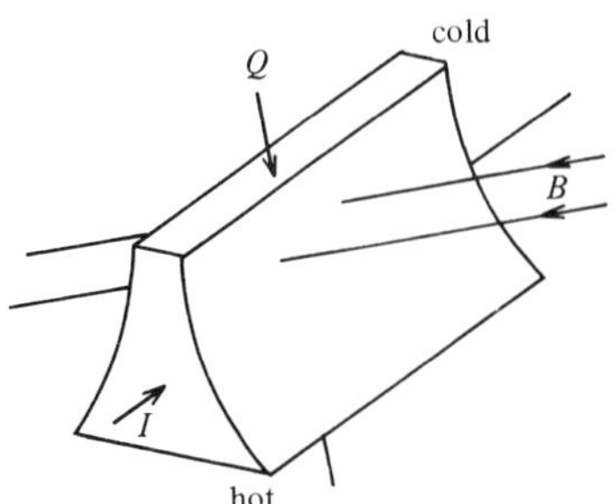


**Figure 5.6.** Cascaded Nernst–Ettingshausen device.

holes ($\mu_n$ and $\mu_p$) are required to be high and if possible greater than 1 $m^2$ $V^{-1}$ $s^{-1}$. Also, it can be shown that $\mu_n \mu_p/(\mu_n + \mu_p)$ should be large, so it is desirable that the mobilities be approximately equal. Carrier concentrations should be high, and it is unfortunate in this respect that carrier mobilities tend to decrease as carrier concentrations increase. The bandgap should be small or preferably the material should be semimetallic with the conduction and valence bands overlapping by $\sim k_B T$. As with thermoelectric materials, the electrical resistivity should be low in order to keep down the Joule heating, and, so that a large transverse temperature gradient can be established, the lattice thermal conductivity should be low (although not necessarily the total thermal conductivity as the device should be operated in a high magnetic field where the effects of the electronic conductivity are largely removed). The latter condition makes desirable the use of compounds containing elements of high atomic weight. In addition, materials with a high magnetoresistance and high electron and hole density-of-states effective masses give better results. If the material is anisotropic, then it is possible to maximise the $[\mu_n \mu_p/(\mu_n + \mu_p)]$ factor, whereas the lattice thermal conductivity tends to be insensitive to crystal direction.

On the basis of these requirements Harman and Honig (1963) listed materials considered worth investigating in this context, including Bi, Bi–Sb alloys, HgTe, HgSe, grey Sn, $Mg_2Pb$, single-crystal graphite, and $Cd_3As_2$. All are relevant to this book, but in various ways they have tended to be disappointing and device application still largely awaits the right material being found and being produced by a suitable growth technique. With the materials available, fields of 1·5 T or higher are necessary. Fields of 10 T are readily available from superconducting magnets but, although these do not require electric power, they do require cryogenic cooling.

Actual calculation of efficiency or coefficient of performance depends on the exact configuration used and whether the devices are operated in the isothermal or adiabatic modes. In addition, hybrid coolers have been examined in which a Peltier cascade acts as a variable temperature heat sink for a Nernst–Ettingshausen type of cooler (Madigan, 1964). Thermoelectric and thermomagnetic devices are discussed in detail by Harman and Honig (1967).

Thermal radiation detectors based on the Nernst–Ettingshausen effect have also been investigated. A thin slab of material, thickness $t$, is used such that the radiation is incident on one of the main surfaces, so setting up a temperature gradient through the thickness. The sample is placed in a magnetic field such that the direction of the field is across the width $w$ of the sample, and a voltage is established along the length $L$ orthogonal to magnetic field and established temperature gradient. A feature of the Nernst detector is that the thermal time constant and the responsivity are decoupled (Washwell *et al.*, 1970). The responsivity for a thermal detector is flat to a frequency given by

$$\tau = a\frac{C}{K}t^2 , \tag{5.31}$$

where $a \approx 0{\cdot}4$, $K$ is the thermal conductivity, $C$ is the heat capacity per unit volume, $t$ is the thickness, and $\tau$ is in seconds (Smith *et al.*, 1957). This expression assumes complete absorption at the surface; if this is not the case, then a more complicated expression is necessary. The output voltage is given by

$$V = \frac{QBIL}{K} , \tag{5.32}$$

where $Q$ is the Nernst coefficient and $I$ is the intensity of the thermal radiation. Using equations (5.31) and (5.32), and also $R = L/tw =$ the electrical resistance, $A = lw =$ the area of the detector surface, we find (when $t$ is eliminated);

$$\frac{V}{I} = \frac{A^{1/2}R^{1/2}\tau^{1/4}}{a^{1/4}}\frac{QB}{K^{3/4}\rho^{1/2}C^{1/4}} = \frac{A^{1/2}R^{1/2}\tau^{1/4}}{a^{1/4}}F \tag{5.33}$$

where

$$F = \frac{QB}{K^{3/4}\rho^{1/2}C^{1/4}} \tag{5.34}$$

is taken as the merit factor for thermomagnetic detector material (Goldsmid and Sidney, 1971).

By reducing the thickness of the sample, the response time $\tau$ can be decreased, and hence the frequency of operation increased. In ordinary thermal radiation detectors the responsivity depends on the temperature difference and hence also on the thickness. Consequently in these detectors the responsivity is decreased as the speed of response is increased. In the Nernst detector the responsivity is given by

$$R_s = \frac{BQ}{Kw} \tag{5.35}$$

for low frequencies (i.e. it is independent of $t$). Another factor used for describing the performance of the Nernst detectors is the detectivity

$$D^* = \frac{R_s}{V_n}(wL\Delta f)^{1/2} \tag{5.36}$$

where $V_n$ is the noise voltage and $\Delta f$ is the electrical bandwidth. Sources of noise are mainly Johnson noise and noise due to thermal fluctuations.

Washwell *et al.* (1970) compared the performance of Bi and $Bi_{97}Sb_3$ Nernst detectors with that of other detectors such as thermocouples and bolometers and showed that at 300 and 200 K the Nernst detector has larger $D^*$ values for $\tau < 0{\cdot}4$ ms and $\tau < 100$ ms respectively, and that at

77 K the Nernst detector has a higher value of $D^*$ for all values of $\tau$. Eutectics consisting of semiconductor plus good-conducting needles have been investigated, the heat flow being in the direction of the needles (which short-circuit any electric field in this direction). Paul and Weiss (1968) investigated InSb–NiSb and Goldsmid and Sidney (1971) investigated $Cd_3As_2$–NiAs. In the latter case no improvement was found over $Cd_3As_2$ (see also section 8.4.4).

## 5.7 Gunn effect devices

The Gunn effect (Gunn, 1963) is the occurrence of high-frequency (in the microwave region) current variations on the application of high DC electric fields to certain semiconductors. For a semiconductor to exhibit the effect, it must possess a negative change of electron mobility with electric field (referred to as NDM, negative differential mobility) once the electric field has reached a threshold value. Usually this decrease in mobility arises from the transfer of electrons from a high-mobility central valley in the conduction band to satellite valleys which are at a higher energy but where the mobility is lower. Under these conditions, the charge distribution in the semiconductor sample becomes unstable and there is established a domain of high electric field, and this domain passes down the sample to the anode where it collapses. On formation of the domain the current in the sample falls, whereas on collapse of the domain the current increases again, allowing the nucleation of a new domain at the cathode. Hence a regular series of pulses is observed.

In fact, this mode of operation, the so-called 'transit' mode, is not the only one possible. It has the disadvantage that the frequency of operation, $\nu_t$, is fixed by the length of the sample. It is possible for the sample to be made part of a resonant circuit, in which case the voltage across the sample will tend to be modulated at the resonant frequency. Although the domain reaches the anode as in the previous mode of operation, this arrangement can advance or delay the time at which the electric field returns to a sufficient value for the next domain to start. Thus the frequency of operation can be varied but only from $\sim\frac{1}{2}\nu_t$ to $2\nu_t$. Alternatively, there is the limited space-charge accumulation (LSA) mode of operation where the time for a domain to form is of the same order as the dielectric relaxation time $\tau_r$ ($= \epsilon/\sigma$ where $\epsilon$ is the electric permittivity and $\sigma$ the electrical conductivity). By modulating the field at a frequency $\gg 1/\tau_r$, the domains are prevented from forming properly but are produced at the frequency of modulation. By using large DC fields and also a large AC field superimposed, it is possible to have high frequency of operation, but the frequencies must be above a certain value (the ratio of carrier concentration to frequency must lie between $2 \times 10^{10}$ and $10^{11}$ s m$^{-3}$) and frequencies up to $10^{11}$ Hz have been used (Beeforth and Goldsmid, 1970).

As a consequence the Gunn effect has application in tunable oscillators and in voltage-controlled frequency switches. It is usually necessary for the bandgap of the semiconductor to be greater than the energy difference between the main conduction band minimum and satellite minima. However, the Gunn effect has been observed in indium antimonide without this restriction applying (see section 8.1.4) and this opens the possibility of the effect being practicable in many other materials.

**References**

Beeforth, T. H., Goldsmid, H. J., 1970, *Physics of Solid State Devices* (Pion, London).
Carroll, J. A., Spivak, J. F., 1966, *Solid-State Electron.*, **9**, 383.
Goldsmid, H. J., 1960, *Applications of Thermoelectricity* (Methuen, London).
Goldsmid, H. J., Sidney, K. R., 1971, *J. Phys. D.*, **4**, 869.
Gunn, J. B., 1963, *Solid State Commun.*, **1**, 88.
Guthrie, G. L., 1965, *J. Appl. Phys.*, **36**, 3118.
Hall, R. N., Fenner, G. E., Kingsley, J. D., Soltys, T. J., Carlson, R. O., 1962, *Phys. Rev. Lett.*, **9**, 366.
Harman, T. C., 1971, *Proceedings of the International Conference on Semimetals and Narrow Gap Semiconductors, Dallas (1970), J. Phys. Chem. Solids*, **32**, suppl. 1, 363.
Harman, T. C., Honig, J. M., 1963, *Semicond. Prod. Solid State Technol.*, **6**, 19.
Harman, T. C., Honig, J. M., 1967, *Thermoelectric and Thermomagnetic Effects and Applications* (McGraw-Hill, New York).
Hiscocks, S. E. R., 1969, *J. Mater. Sci.*, **4**, 773.
Ioffe, A. F., 1957, *Semiconductor Thermoelements and Thermoelectric Cooling* (Infosearch, London), p.42.
Madigan, J. R., 1964, *Solid-State Electron.*, **7**, 643.
Melngailis, I., Harman, T. C., 1970, in *Semiconductors and Semimetals*, **5**, Eds R. K. Willardson, A. C. Beer (Academic Press, New York), p.111.
Norwood, M. H., 1963, *J. Appl. Phys.*, **34**, 594.
O'Brien, B. J., Wallace, C. S., 1958, *J. Appl. Phys.*, **29**, 1010.
Paul, B., Weiss, H., 1968, *Solid-State Electron.*, **11**, 979.
Smith, R. A., Jones, F. E., Chasmar, R. P., 1957, *The Detection and Measurement of Infrared Radiation* (Oxford University Press, Oxford).
Ure, R. W., Jr., Heikes, R. R., 1961, in *Thermoelectricity: Science and Engineering*, Eds R. R. Heikes, J. W. Ure, Jr. (Interscience, New York), chapter 15.
Washwell, E. R., Hawkins, S. R., Cuff, K. F., 1970, *Appl. Phys. Lett.*, **17**, 164.
Weiss, H., 1966, *Solid-State Electron.*, **9**, 443.
Wieder, H. H., 1966, *Solid-State Electron.*, **9**, 373.

# Part 2

# Particular systems

# 6

# Elements

## 6.1 Grey tin (α-tin)

In this, the first chapter dealing with specific materials which are either semimetallic or are semiconductors with very narrow bandgaps, the few elements which fulfil this classification will be considered. It is perhaps fitting to begin with grey tin which has a classic zero bandgap structure that is used as a comparison for those of other semimetallic materials.

### 6.1.1 Crystal growth

α-Sn is the low temperature allotrope of tin and transforms back to the ordinary white metallic form at 13°C. This transformation to the white form is relatively fast, whereas if white tin is retained below this temperature it takes days or weeks to become α-Sn. Grey tin has the diamond structure (*Fd*3*m*) with $a_0 = 6{\cdot}49$ Å. This compares with $a_0 = 5{\cdot}83$ Å and $c_0 = 3{\cdot}18$ Å for the tetragonal structure of white tin. The transformation to grey tin involves an increase in volume and hence, unless special preparative methods are used, the resulting α-Sn sample is cracked and even powdery.

Single crystals are best grown by recrystallisation from a solution of tin in mercury according to the method of Ewald and Tufte (1958). Mercury has a low solubility in grey tin and hence the crystals grown in this way have a very low mercury content. The apparatus used for this technique is shown in figure 6.1. It is retained within a refrigerator operated at −55°C and the mercury solution is supersaturated locally at −30°C. The tin content is maintained from a supply of white tin kept at −20°C in the lower portion of the apparatus. There is thus a temperature gradient between this region and the higher region kept at −30°C, and so

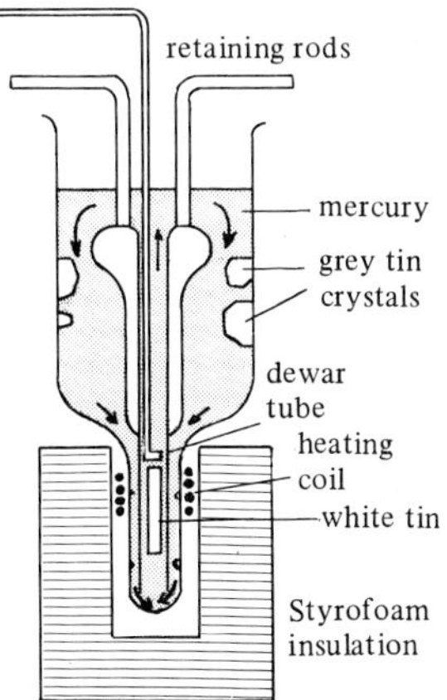


**Figure 6.1.** Growth of grey tin single crystals (Ewald and Tufte, 1958).

there is convection within the solution, with a radial flow outwards at the top allowing crystal growth on the cooler walls just below the surface of the mercury. As they form, the small crystals are removed repeatedly. This reduces the number of nuclei present until eventually there are sufficiently few to allow a small number of large crystals to grow. It is probably necessary to initiate nucleation and growth by the presence of a small amount of grey tin powder, but once crystal growth of grey tin has occurred in the apparatus, growth of grey tin crystals can continue. Crystals having dimensions of ~1 cm are grown by the method. The original crystals contained about $10^{-3}$ atomic percent mercury and were p-type but, with improved growth conditions and purification of the mercury, n-type crystals have been obtained.

Growing grey tin from a mercury solution is laborious; for this reason crystals grown by transforming cast single-crystal wires of white tin are sometimes used for electrical measurements. Growth by this method tends to be much quicker; for instance, Lavine and Ewald (1971) quote a transformation rate of 0·5 cm per week. They started the transformation at −35°C and then maintained the temperature at a few degrees below 0°C. Maintaining the temperature at a slightly higher value would improve the crystal quality but decrease the rate of growth. However, because the method does not produce such good crystals as the recrystallisation method, it is necessary to select lengths of the transformed wire which are free of cracks and other large-scale imperfections.

### 6.1.2 Band structure

Attempts to deduce the band structure from transport effects led to the conclusion that grey tin is a semiconductor having a small bandgap of 0·08 to 0·09 eV, but more recent theoretical work in conjunction with experiment has led to the acceptance of a band structure model of the type put forward by Groves and Paul (1963). In this model there is a small indirect bandgap between the upper valence band and a second conduction band, thus explaining the earlier experimental results, but the gap between the main valence and conduction bands at $\boldsymbol{k} = 0$ is zero. Figure 6.2 shows this structure. The double group representations for the bands are shown with the single group representations in brackets. Deduction of the band form was a typical application of the $\boldsymbol{k} \cdot \boldsymbol{p}$ method discussed in chapter 3. Figure 6.2 may be compared with figure 3.4 discussed for zinc blende; $E(k)\uparrow = E(-k)\downarrow$ and $E(k)\uparrow = E(k)\downarrow$ leading to double spin degeneracy of the bands. Because low temperature (4 K) measurements of oscillatory magnetoresistance show an electron effective mass of $0{\cdot}02m_0$ independent of carrier concentration, the conduction band must be parabolic. This is compatible on the $\boldsymbol{k} \cdot \boldsymbol{p}$ model with a zero bandgap but not with a finite one of 0·1 eV. The gap between the $\Gamma_7^-$ and $\Gamma_8^+$ fits the rule proposed by Herman (1955) that this gap should vary systematically for compounds across the periodic table and depend on $\lambda^2$ where $\lambda$ represents the strength of the lattice potential (see figure 6.3).

It is interesting to note that the $\Gamma_7^-$ valence band in grey tin is the conduction band in germanium and the $\Gamma_8^+$ conduction band is a valence band because of a reversal of sign of the energy difference in the $\boldsymbol{k} \cdot \boldsymbol{p}$ model.

With warping neglected, the two bands can be represented by $E_{k^\pm} = (A \pm B)k^2$ where $A$ and $B$ are inverse effective mass parameters. As the degeneracy of the bands is symmetry-induced, it can be removed with a uniaxial stress. If a compressional stress is applied along the (001) direction, an energy gap is opened out and the isotropic bands become anisotropic (Pikus and Bir, 1960). The degeneracy can be removed and an energy gap induced also by application of a magnetic field (Kowalski and Zawadski, 1973) and this should be observable in transport effects at low temperatures by a decrease of intrinsic carrier concentration with increasing magnetic field due to a freeze-out effect. With the small effective masses involved, magnetic field quantisation occurs and Kowalski and Zawadski have indicated the magnetic sub-bands which should occur.

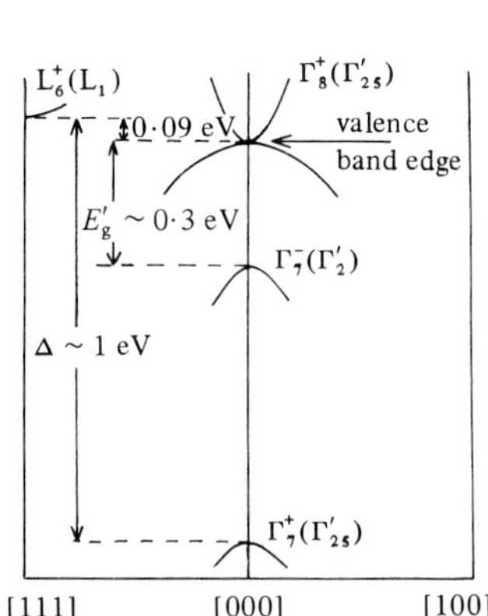


Figure 6.2. Band structure model of grey tin.

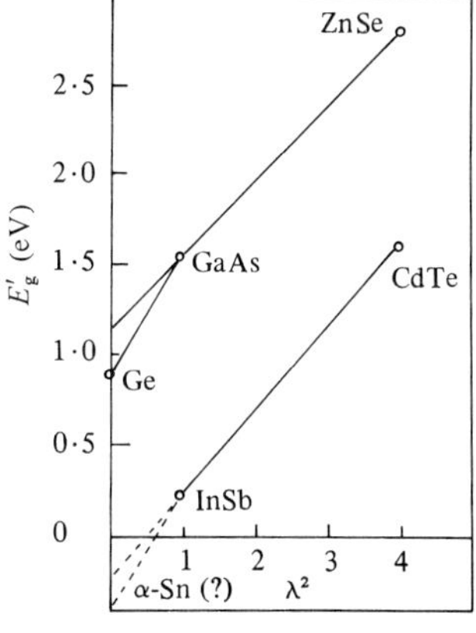


**Figure 6.3.** Variation of $E_g' = E(\Gamma_7^-) - E(\Gamma_8^+)$ with $\lambda^2$ for the Ge and α-Sn isoelectronic sequence (from Groves and Paul, 1963).

### 6.1.3 Properties

Lavine and Ewald (1971) have measured the electrical conductivity, Hall coefficient, and magnetoresistance of grey tin as functions of temperature, magnetic field, and doping, and have interpreted their results in terms of the band model of Groves and Paul. Figure 6.4 shows Hall data for antimony-doped samples of various donor concentrations at 4·2 K. For samples possessing carrier concentrations less than $5 \times 10^{23}$ m$^{-3}$ conduction is confined to the $\Gamma_8^+$ conduction band and the Hall coefficient is field-dependent. Because it was not possible with the fields available to obtain a high field value of the Hall coefficient ($R_\infty$), the electron concentrations were calculated by replotting the curves in the form $\Delta R_H/R_H(0)B^2$ against $B$, where $R_H(0)$ is the Hall coefficient extrapolated to zero field $B$ and $\Delta R_H$ is the change of Hall coefficient on application of field $B$. This method is suitable, as degenerate statistics apply at this temperature. Typical data obtained from the electrical conductivity and Hall measurements are shown in table 6.1. $n_0$ is the carrier concentration in the central conduction minimum and $n_1$ is the carrier concentration in the $\langle 111 \rangle$ minima ($n_t = n_0 + n_1$). $n_0$ could be measured also from the period of the oscillatory magnetoresistance. (The much heavier mass of the $\Gamma_6^+$ electrons means that quantisation of their energy would not occur until much larger values of magnetic field than for the case of the $\Gamma_8^+$ electrons.) Using data of Booth and Ewald (1968) who related the Fermi energy $E_F$ to the carrier concentration $n_0$, Lavine and Ewald then deduced $E_F$ and the density-of-states effective mass $m_{1d}^*$ from the standard expression

$$n_1 = \frac{\pi^2}{3}\left(\frac{2m^* k_B T}{\hbar^2}\right)^{3/2}\left(\frac{E_F - E_g}{k_B T}\right)^{3/2}$$

which applies for extreme degeneracy. This was assuming parabolic bands (not strictly true for the $\Gamma_6^+$ band edge) and also $E_g$ independent of the total donor concentration. However, values of $E_g = \left(0 \cdot 092 \begin{smallmatrix} +0 \cdot 005 \\ -0 \cdot 001 \end{smallmatrix}\right)$ eV and

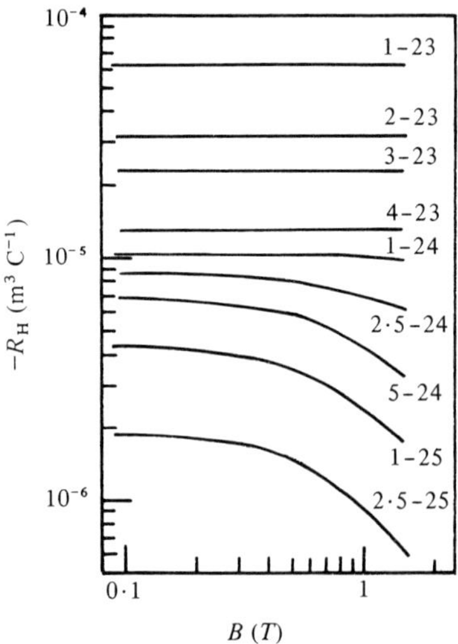


**Figure 6.4.** The field dependence of the Hall coefficient for antimony-doped samples of grey tin at 4·2 K. The curves are denoted by their donor concentrations; e.g. 1-23 = 1 × $10^{23}$ donors m$^{-3}$ (from Lavine and Ewald, 1971).

$m_{1d}^* \leqslant 0{\cdot}21 m_0$ were deduced, although the presence of an impurity band-tail consisting of states below the ellipsoid band edge was noted. Electron mobilities in the two types of conduction minima were calculated and the results were found to be consistent with screened ionised-impurity scattering at 4·2 K. Table 6.2 gives mobility values as deduced from the Hall and conductivity data for heavily doped samples. For low doping concentrations it is necessary to assume that the dielectric constant depends on the amount of doping.

When the temperature was increased from 4·2 K, samples of donor concentration $2 \times 10^{23} \leqslant N_D \leqslant 2{\cdot}5 \times 10^{24}$ m$^{-3}$ showed an increase in Hall coefficient due to thermal transfer of electrons from the ⟨000⟩ to the ⟨111⟩ minima (figure 6.5), the effect being consistent with the deduced values of energy gap and effective mass. The change in energy gap with temperature was shown to be $dE_g/dT = -4 \times 10^{-5}$ eV K$^{-1}$. For intrinsic material, the Hall coefficient showed a $T^{-3/2}$ dependence, this dependence being characteristic of a zero bandgap material (figure 6.6). For the pure sample, magnetoresistance was found to be very large, indicating two-band effects right down to 4·2 K. Highly doped samples showed a sharp

**Table 6.1.** Summary of experimental parameters of antimony-doped α-tin samples at 4·2 K. $R_H(0)$ and $\sigma(0)$ are the Hall coefficient and electrical conductivity in zero magnetic field, $N_D$ is the donor concentration, $n_H$ the Hall number [$= -1/R_H(0)e$], $n_0$ the electron concentration in the central band, $n_1$ the electron concentration in the ⟨111⟩ valleys and $n_t$ the total electron concentration (Lavine and Ewald, 1971).

| $10^{-24}N_D$ (m$^{-3}$) | $10^{-24}n_H$ (m$^{-3}$) | $10^{-24}n_0$ (m$^{-3}$) | $10^{-24}n_t$ (m$^{-3}$) | $10^6 R_H(0)$ (m$^3$ C$^{-1}$) | $\sigma(0)$ ($\Omega^{-1}$ m$^{-1}$) | $\mu_H = R_H\sigma$ (m$^2$ V$^{-1}$ s$^{-1}$) | $10^{-24}n_1$ (m$^{-3}$) |
|---|---|---|---|---|---|---|---|
| 0·10 | 0·10 | 0·10 | - | 62·5 | 114·5 | 7·16 | - |
| 0·40 | 0·37 | 0·40 | - | 17·0 | 286·0 | 4·85 | - |
| 0·50 | 0·44 | 0·48 | - | 14·3 | 452·0 | 6·45 | - |
| 1·0 | 0·61 | 0·61 | - | 10·3 | 788·0 | 8·08 | 0·39 |
| 2·5 | 0·73 | 0·68 | 2·3 | 8·6 | 842·0 | 7·20 | 1·8 |
| 5·0 | 0·92 | 0·79 | 4·6 | 6·8 | 904·0 | 6·17 | 4·2 |
| 10 | 1·5 | 0·96 | 9·4 | 4·3 | 1030 | 4·43 | 9·0 |
| 25 | 3·5 | 1·3 | 28 | 1·8 | 1220 | 2·20 | 24 |

**Table 6.2.** Electron concentration ratios and electron mobilities in heavily doped α-tin samples at 4·2 K (Lavine and Ewald, 1971).

| $10^{-24}N_D$ (m$^{-3}$) | $\frac{n_1}{n_0}$ | $\frac{\mu_1}{\mu_0}$ | $\mu_0$ (m$^2$ V$^{-1}$ s$^{-1}$) | $\mu_1$ (m$^2$ V$^{-1}$ s$^{-1}$) |
|---|---|---|---|---|
| 2·5 | 2·7 | 0·021 | 7·6 | 0·16 |
| 5·0 | 5·3 | 0·022 | 7·2 | 0·15 |
| 10 | 9·4 | 0·022 | 5·4 | 0·12 |
| 25 | 18·2 | 0·028 | 3·2 | 0·09 |

decrease in magnetoresistance as the temperature was decreased and the samples became extrinsic. The hole mobility showed a $T^{-3/2}$ dependence in the intrinsic range and is probably dominated by phonon scattering down to 30 K.

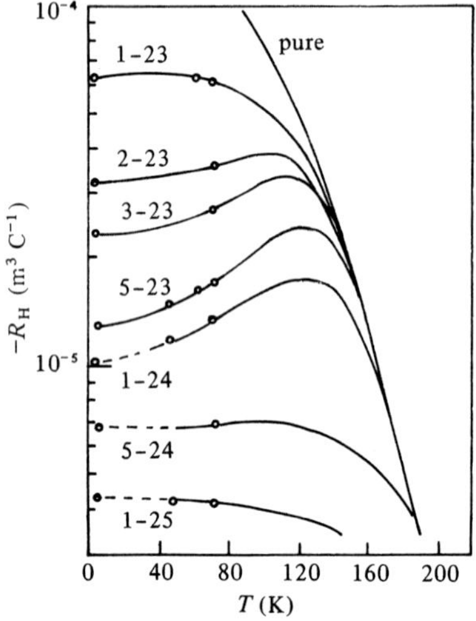


**Figure 6.5.** The temperature dependence of the low-field Hall coefficient for antimony-doped samples. Circles represent DC measurements at fixed temperatures as used to calibrate the AC temperature sweep measurements shown by the continuous lines (Lavine and Ewald, 1971).

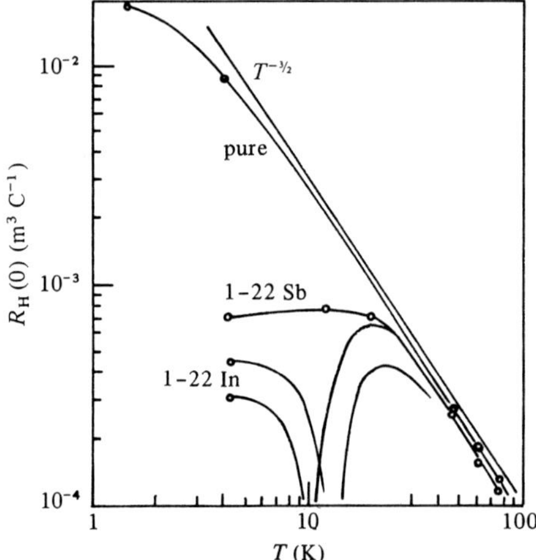


**Figure 6.6.** Temperature dependence of the low-field Hall coefficient for pure and lightly-doped n-type and p-type samples. Circles and continuous lines represent DC and AC measurements as in figure 6.5 (Lavine and Ewald, 1971).

## 6.2 Bismuth, antimony, and arsenic

### 6.2.1 Introduction

These three elements are semimetallic and all have a similar rhombohedral crystal structure (often called the arsenic structure) of point group $R\overline{3}m(D_{3d}^5)$ with two atoms per unit cell, one at $(u, u, u)$ and the other at $(-u, -u, -u)$ (figure 6.7a). $u$ takes the value of 0·237 for Bi, 0·233 for Sb, and 0·226 for As. Table 6.3 (Wyckoff, 1963) gives the lattice parameters for the three elements and also for the corresponding hexagonal cell containing six atoms at $[000;\ \frac{2}{3}\frac{1}{3}\frac{1}{3};\frac{1}{3}\frac{2}{3}\frac{2}{3}] \pm 00u$ where $u$ is as before. The lattice can be obtained from a simple cubic lattice by separating it into two face-centred cubic lattices as shown in figure 6.7b (Abrikosov and Fal'kovskii, 1963). The two sublattices are separated by a translation along the body diagonal so that the corner of one is at the centre of the other. Then by causing a slight trigonal distortion that slightly alters the value of $\alpha$ (which is 60° prior to the distortion), the structures of the three elements can be obtained. The Bravais lattice being only slightly distorted from that for the face-centred cubic sublattice, the Brillouin zones for the structures are as for the face-centred case with just a slight distortion. The usual notation is shown in figure 6.8 (Cohen, 1961) and should be compared with that in figure 3.2, the additional notation being necessary because of the distortion, the effect of which is to squash the truncated octahedron. The main difference is that the **L** points have lower and different symmetry than the **L** points for the f.c.c. structure.

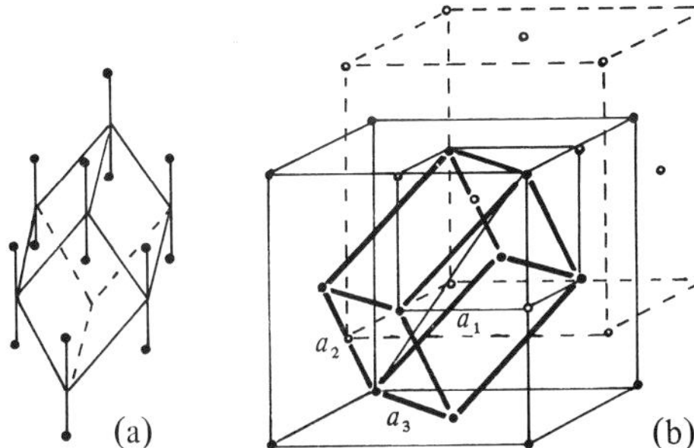


**Figure 6.7.** (a) The structure of arsenic; (b) the relationship of the As (Sb, Bi) structure to the simple cubic lattice. (After Abrikosov and Fal'kovskii, 1963.)

**Table 6.3.** Lattice parameters for bismuth, antimony and arsenic (Wyckoff, 1963).

| | Rhombohedral | | Hexagonal | | |
|---|---|---|---|---|---|
| | $a_0$ (Å) | $\alpha$ | $a_0'$ (Å) | $c_0'$ (Å) | $c_0'/a_0'$ |
| Bi | 4·746 | 57°14·2′ | 4·549 | 11·862 | 2·61 |
| Sb | 4·507 | 57°6·5′ | 4·308 | 11·274 | 2·62 |
| As | 4·131 | 54°10′ | 3·760 | 10·548 | 2·81 |
| Cubic | | 60° | | | 2·45 |

The L designation is given to the lower symmetry points while the two points of higher symmetry are called T points. The ΓT axis is a trigonal axis and there are three binary axes TW which are perpendicular to the mirror plane and three bisectrix axes TU which lie in the mirror plane. Small differences in the shape of the zones occur between Bi, Sb and As but the general shape is the same.

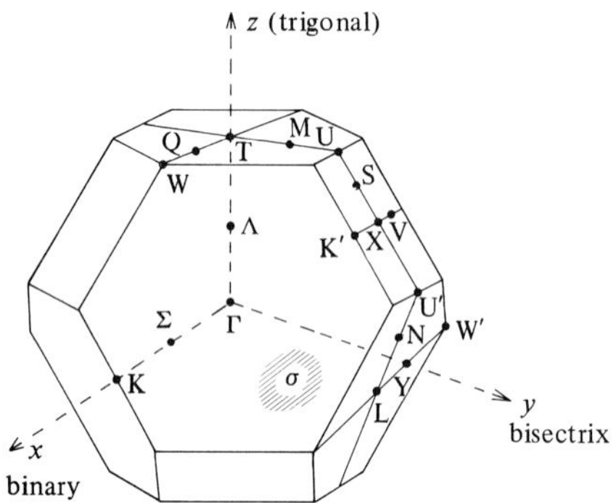


**Figure 6.8.** Brillouin zone for As, Sb, and Bi.

### 6.2.2 Preparation

Of the three semimetals, bismuth is the easier to grow as large single crystals; for instance crystals have been up to 170 mm diameter and 47 kg in weight (Bednarski, 1970). It is usual to grow the crystals by some form of Bridgman technique, and both movement of the charge through a temperature gradient, and lowering of the furnace temperature have been successfully employed. As there is a 3·3% increase in volume on solidification, crystals are grown in soft moulds, usually graphite containers lined with $Al_2O_3$ powder (or other suitable powder such as bismuth or zirconium oxides) to avoid strain which would otherwise be produced in the crystals. A suitable growth rate in a dynamic Bridgman setup is a descent speed of 5 cm $h^{-1}$ and a furnace gradient at the melting point of 15 K $cm^{-1}$ (Van der Planken, 1970). Using a stationary charge and furnace, Bednarski grew large crystals by cooling over a period of 40 to 90 h depending on the size of the crystals. A slow rate of cooling is important because of the low (and anisotropic) thermal conductivity and the fact that the thermal conductivity of molten bismuth is 2·5 times that for solid bismuth.

High-purity antimony is normally prepared by zone-refining; as first reported by Tanenbaum *et al.* (1954) this was carried out in a nitrogen atmosphere. A more recent feature of the method has been to add 0·1% by weight of aluminium to remove traces of arsenic in the molten zone. However, Huntley and Shah (1970) have reported the preparation of particularly high-purity antimony, such that the resistance ratio for resistances at 300 K and 4·2 K was $6 \times 10^3$ for nearly 75% of the zone-refined ingot. A vitreous carbon boat and a dry hydrogen atmosphere at 400 Torr were used (both these features probably helped purity by removing sulphur and oxygen) and the molten zone passed through the ingot six times at 1 cm $h^{-1}$. The ratio of the total ingot length to molten zone length was 15:1. Most of the final ingot was single crystal.

Various methods have been used for arsenic. The main problem is that arsenic sublimes and can only be melted at 817°C under a pressure of 35 atm. Large crystals have been grown by the Bridgman–Stockbarger method in sealed heavy-walled tubes but because arsenic expands on cooling the crystals tend to be strained. Also, crystals of one orientation tend to be produced, the (111) plane being parallel to the crystal length. Using the method Saunders and Lawson (1965) have grown crystals of over 3 cm length and 1 cm diameter. The temperatures of the two zones of the furnace were held such that a 30 K temperature difference existed between the bottom conical tip and the top end of the quartz tube holding the charge, and once the bottom of the tube had reached 820°C, the tube was lowered at 1 cm $h^{-1}$. To remove the effects of slight deviation from cylindrical symmetry in the furnace, the tube was rotated at 10 cycles $min^{-1}$ during lowering. To remove the strain produced in the crystals, they were annealed for a few hours at 300, 400, and finally 500°C.

A rather more elaborate (and hence expensive) arrangement was used by Ketterson and Eckstein (1965) to reduce the strain produced during growth. The crystals were grown in a tapered quartz bomb such that as the material expanded on cooling the component of force parallel to the axis tended to push the sample into a region of larger radius. The charge of arsenic was transferred to this bomb by distillation before the bomb was sealed off. The temperature was raised to 850°C and then lowered over a 48 h period. Again, growth was followed by an anneal (12 h at 375°C and 15 min at 425°C).

Growth from the vapour reduces the need for thick-walled tubes and can reduce strain in the crystals. Jeavons and Saunders (1968) have grown crystals of up to 5 cm length and 2 cm diameter over a period of 7 days in a quartz growth tube of 30 cm length and 2 cm diameter. The furnace was Kanthal-wound with a mullite tube and it was found necessary to smooth out any temperature undulations between the heater turns by lining the tube with stainless steel. The overall temperature profile in the tube was adjusted so that the crystal grew by progressive condensation. Crystals of higher perfection were obtained from the vapour by Shetty *et al.* (1969) in a furnace whose absolute temperature in the range 500–900°C was controlled to 0·1 K. The gradient was adjusted by means of multiply-tapped heaters and also controlled to 0·1 K.

Hydrothermal growth of arsenic has been carried out by Rau and Rabenau (1968) who used hydrogen iodide as solvent. Growth was at a temperature of 450°C over a period of ten days and crystals up to 8 mm long were grown.

### 6.2.3 Band structure

As the group V semimetals have two atoms per unit cell and five electrons per atom, there are sufficient valence electrons to fill completely five bands. From consideration of crystal binding, the important electron states are bonding and antibonding s and p states. The order in which the states occur are: bonding s state lowest, antibonding s state next, and bonding and then antibonding p states following upwards. Both semiconductors and semimetals are found in group V, semimetallic behaviour being the more common. The fact that the group V semimetals have similar crystal structure means that any energy band model based on the rhombohedral distortion is applicable to all. Various types of band calculation have been carried out, particularly for bismuth; most commonly pseudopotential techniques are used. The effect of the distortion is to lower the bands at the L points and to raise them at T points. Compared with the case of a simple cubic lattice, the displacement of the atoms causes a small overlap of the fifth and sixth bands and an equal density of electrons and holes of $\sim 3 \times 10^{26}$ $m^{-3}$. Hence electron pockets should exist at the L points and hole pockets at the T points (although for a considerable time there was controversy as to the location of the electrons and holes). The variation of $u$ from 0·25 and $\alpha$ from 60° is a measure of the distortion and the band structure should exhibit this distortion, and hence give increase of bandgap from bismuth through antimony to arsenic.

Much of the experimental data concerns the energy bands at the L and T points and for many cases a satisfactory representation of the band structure over a restricted region of $\boldsymbol{k}$-space is achieved in terms of two coupled bands (Dresselhaus, 1971), usually on the basis of the $\boldsymbol{k} \cdot \boldsymbol{p}$ method. As discussed in section 3.2.6, Cohen (1961) proposed a nonellipsoidal nonparabolic model for bismuth and this has been widely used for the interpretation of experimental data, although an ellipsoidal nonparabolic model proposed by Lax (Brown *et al.*, 1963) may be adequate.

The total number of electrons can be measured both from the Hall effect and from infrared measurements (near the plasma edge) and has been found to be three times the number of electrons per ellipsoid as determined from Fermi surface dimensions with the aid of de Haas–van Alphen data. This indicates six half-ellipsoids located at the L points. Study of the angular dependence of the cross-sectional areas of the ellipsoids shows them to be highly elongated and prolate with a positive tilt angle to the main symmetry axis of approximately 4° (Brown *et al.*, 1968) or 7° (Michenaud and Issi, 1972). [A positive angle is one measured by a rotation around the binary axis ($+x$) so as to rotate the bisectrix axis ($+y$) toward the trigonal axis ($+z$); this is the convention of Brown *et al.* and is not the one always used for bismuth.] Analysis of cyclotron resonance data (Jain and Koenig, 1962) for holes (on the basis of a parabolic model) showed that they exist in one spheroid.

It indicated a T or Γ location, and support for the T location has come from arguments regarding the phonon frequencies involved in phonon-assisted recombination of electrons and holes (Koenig *et al.*, 1958).

Band parameters for the electron and hole bands are given in table 6.4. There is a band overlap of between 0·036 and 0·038 eV (Isaacson and Williams, 1969). The electron band is somewhat nonparabolic owing to a lower-lying hole band at the L point at a separation of 13 meV at 4 K increasing to 35 meV at 280 K (Vechi and Dresselhaus, 1974). The Fermi energy for the electrons is 28 meV and for the holes 11 meV. Nonparabolic effects for the holes are smaller than for the electrons but are more complicated because there is interaction with more than one other band.

Early measurements indicated the presence of further pockets of carriers other than those at the L and T points, but de Haas–Shubnikov measurements (Brown, 1970) show these to have arisen from the effects of crystal twinning and sample misalignment.

The Fermi surfaces of antimony and arsenic are rather more complicated than those of bismuth, but have been interpreted by the use of pseudo-potential calculations in conjunction with experimental results. It is more difficult to count the number of ellipsoids for the carriers in the cases of antimony and arsenic as there are more marked departures from an exactly ellipsoidal shape. Although it was clearly determined that there were twice as many carrier pocket sites of one carrier sign as of the other, the actual signs were not confirmed fully for antimony and arsenic until the effect of doping on the size of the de Haas–van Alphen periods had been studied (Ishizawa and Tanuma, 1965). In fact the situation is similar to that in bismuth with electrons contained in three ellipsoids around the L points and holes located near the T points. In the case of arsenic the energy surfaces for the holes are multiply-connected in such a way as to

**Table 6.4.** Band parameters for group V semimetals. (After Dresselhaus, 1971.)

| Quantity | Bi | | Sb | | As | | |
|---|---|---|---|---|---|---|---|
| | electrons | holes | electrons | holes | electrons | α holes | γ holes |
| $m_{1'}$[a] | 0·00521[d] | 0·064[d] | 0·093[e] | 0·068[e] | 0·135[g] | 0·106[h] | 0·046[h] |
| $m_{2'}$[a] | 1·21[d] | 0·064[d] | 1·14[e] | 0·92[e] | 1·52[g] | 1·56[h] | 0·016[h] |
| $m_{3'}$[a] | 0·014[d] | 0·69[d] | 0·088[e] | 0·050[e] | 0·127[g] | 0·089[h] | −1·82[h] |
| Tilt-angle[b] | +4°[d] | 0°[d] | −4°[e] | −36°[e] | −4°[g] | −53°[h] | −10°[h] |
| $E_F$ (eV)[c] | 0·0276[d] | 0·0109[d] | 0·096[f] | 0·104[f] | 0·1905[h] | 0·177[h] | 0·0106[h] |
| $n$ ($10^{24}$ $m^{-3}$) | 0·275[d] | 0·275[d] | 55·4[f] | 54·9[f] | 212[h] | 234[h] | ~300[i] |

[a] Effective mass parameters evaluated at the Fermi surface; [b] tilt angle convention of Brown *et al.* (1968) used; [c] Fermi level measured with respect to band extrema; [d] Smith *et al.* (1964); [e] Datars and Vanderkooy (1964); [f] Windmiller (1966); [g] Vanderkooy and Datars (1968); Datars and Vanderkooy (1966); [h] Priestley *et al.* (1967); [i] Fukase (1969).

give rise to a hole Fermi surface as shown in figure 6.9. This is the classic Lin–Falicov Fermi surface as determined in 1966 for the holes in arsenic. Arsenic has four times as many carriers as does antimony and so has a larger Fermi surface than antimony. As the Fermi surface contracts the $\gamma$ necks connecting the $\alpha$ pockets break off and disappear and the shape of the pockets becomes increasingly ellipsoidal. The hole pockets in antimony are approximately ellipsoidal but not as closely so as the Fermi surfaces for electrons. (The holes are at a low symmetry point whereas the electrons are at a point having inversion symmetry.)

In table 6.4 (Dresselhaus, 1971), summarising data for Bi, Sb and As, the band parameters are obtained by means of ellipsoidal approximations. Semi-major axes have been determined from the maximum Fermi-surface cross-sectional areas; these extremal values are not necessarily orthogonal to each other. The $\gamma$ necks in arsenic are treated as consisting of minority carriers and are approximated by hyperboloids truncated where they intersect the $\alpha$ pockets. Because energy gaps in antimony and arsenic are larger than in bismuth, therefore nonparabolic corrections are less important. Nevertheless nonparabolic effects have been detected in antimony for both electrons and holes by de Haas–van Alphen measurements (Brandt *et al.*, 1967).

Although the Fermi surfaces are now well-known for the three elements, the band structure away from the Fermi level is less well determined. Pseudopotential calculations have been made for bismuth (Golin, 1968), antimony (Falicov and Lin, 1966) and arsenic (Falicov and Golin, 1965) and figure 6.10 shows the band-structure calculated for bismuth. Similar figures are not given here for antimony and arsenic because, although there are not direct comparisons between the three band diagrams, nevertheless the three do show certain similarities such as the ordering of

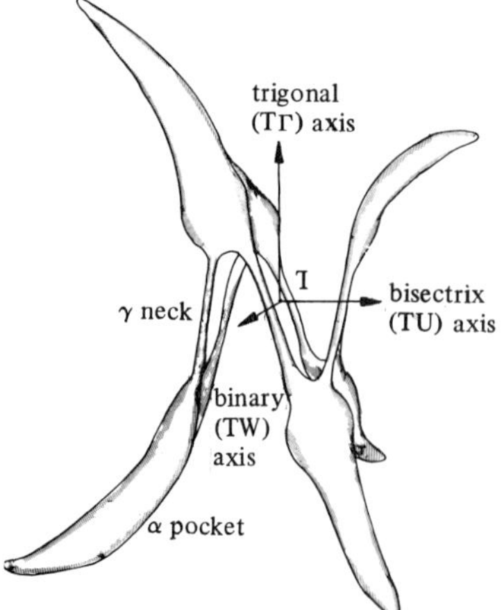


**Figure 6.9.** Hole Fermi surface for arsenic as determined by Lin and Falicov (1966).

the bands and their overall shape, and also, as has already been noted, similar positioning in the Brillouin zone of the carrier pockets. Spin-orbit interaction has been included in the band calculations as it is of major importance for bismuth with its large atomic number. In antimony and arsenic it is large in terms of the magnitude of the band overlaps and bandgaps but not in terms of overall characteristic band widths and so was not included in these calculations.

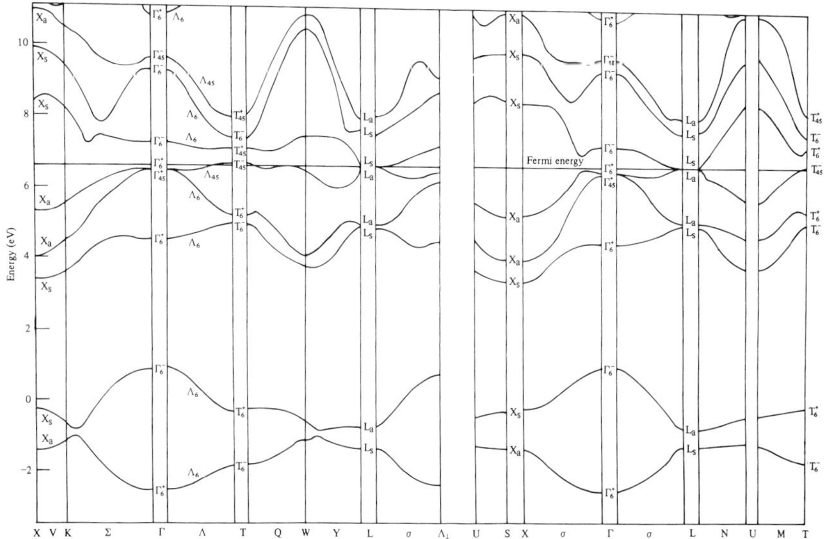


**Figure 6.10.** Pseudopotential calculation of energy bands in bismuth (Golin, 1968).

### 6.2.4 Properties

There has been extensive interest in the transport properties of bismuth because of its very large magnetoresistance and Hall effect—and the galvanomagnetic and thermomagnetic effects have been largely discovered in the material. Bismuth has particular application when alloyed with antimony as a thermoelectric material and the properties of these alloys will be considered in some detail in section 6.3.2.

Theoretical expressions for the galvanomagnetic coefficients in bismuth based on simple transport theory were obtained by Abeles and Meiboom (1956) by taking into account two kinds of charge carrier and anisotropy of carrier mobility and expanding as a power series of $B$ [as in equation (4.24)]. A number of galvanomagnetic tensor components need to be measured for each of the different transport effects (see table 4.3). The two components of the low-field isothermal resistivity $\rho_{11}$ and $\rho_{33}$ and of the low field Hall effect as measured by Michenaud and Issi (1972) and compared with earlier results are shown in figure 6.11. At very low temperatures, as shown by the broken line in figure 6.11a (Friedman, 1967), the resistivity obeys a $T^2$ law, but at higher temperatures the dependence on temperature is linear. This linear dependence is 'accidental' and depends on the detailed temperature dependence of the carrier density and mobility. Figure 6.11c shows the variation of carrier density with temperature from 4·2 K [the value taken at this temperature is from Bhargava (1967) and is $3 \times 10^{23}$ $m^{-3}$ as obtained from the de Haas–van Alphen effect] to 300 K. The variation of carrier density by a factor of 10 is not found in arsenic or antimony. According to a model of Lopez (1968), carrier generation from the main valence band to the main conduction band requires a phonon energy of 43 ± 4 K. Scattering mechanisms governing the variation of mobility with temperature (figure 6.11d) would also be affected by this carrier excitation. Figure 6.11d incorporates results of Hartman (1969) for $\mu_1^n$, $\mu_2^n$, $\mu_3^n$, and $\mu_1^p$, where $\mu^n$ and

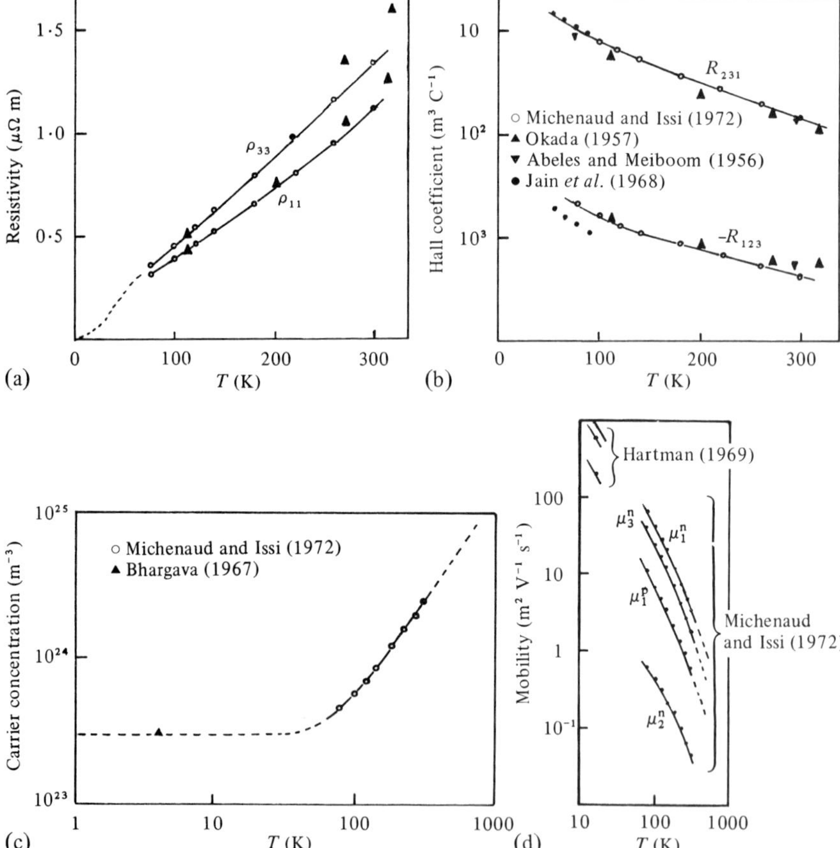


**Figure 6.11.** Properties of bismuth as functions of temperature: (a) the two components of zero-field isothermal resistivity $\rho_{11}$ and $\rho_{33}$; (b) low-field Hall coefficient; (c) carrier density; (d) electron and hole mobilities.

$\mu^p$ are mobilities of the electrons and holes respectively and subscripts 1, 2, and 3 refer to the principal axes of the ellipsoids. The graph shows a $T^{-2}$ dependence at low temperatures (scattering is between electrons in separate valleys), an almost $T^{-2}$ dependence between 77 K and 120 K, and a $T^{-2\cdot5}$ dependence above 140 K. The $\mu_2^n$ dependence is somewhat different as it shows a dependence close to the metallic $T^{-2\cdot5}$ at very low temperatures. Michenaud and Issi also measured the eight magnetoresistance coefficients and found that they all varied as $T^{-2}$ up to nearly 120 K. They calculated the electron ellipsoid tilt angle for the temperature range 77 to 300 K and obtained a variation between 6°40′ and 8°30′, values which are somewhat larger than that given in table 6.4. The angular dependence of magneto-resistance as plotted on polar diagrams and compared with theory based on expressions of Aubrey (1971) is given by Sümengen *et al.* (1974). A value of the tilt angle of the electron ellipsoid of between 6° and 7° is obtained at 20·4 K. An anisotropy of the electron relaxation time of approximately 2·5 was deduced.

Once bismuth is doped, impurity scattering becomes important, and at 4·2 K impurity scattering alone determines the mobilities in the doped samples (Noothoven van Goor, 1971). The thermoelectric power of bismuth has been measured including the variation in a magnetic field. Uher and Goldsmid (1974) have measured the magneto-Seebeck coefficient in the temperature range 20 to 300 K in magnetic fields up to 5·5 T. They obtained a phonon-drag component which persisted in very high fields to well above liquid-nitrogen temperature. The phonon-drag effect (arising from anisotropy in the propagation of lattice vibrations in the presence of a temperature gradient, giving rise to preferential scattering of carriers in the direction of the temperature gradient) had been expected to become negligible above 20 K because anharmonicity of the lattice vibrations above this temperature had been expected to restore equilibrium. According to the theory of Korenblit (1967), $\alpha_{11}(B)$ should saturate at high fields whereas $\alpha_{33}(B)$ should be a linear function of the magnetic field reversing with reversal of the magnetic field and follow a $T^{-2}$ temperature dependence. In fact Uher and Goldsmid found $\alpha_{33}(B)$ to be always positive, although they obtained a temperature dependence of $T^{1\cdot8}$ which is in good agreement with the theory. Nor did they obtain the saturation of $\alpha_{11}(B)$ although checking this at room temperature would have required higher magnetic fields than were available. Jacobson and Ertl (1972) have discussed the application of Korenblit's theory at low temperatures and point out that the intrinsic assumption of $n = p$ cannot be made even for the purest samples. Sümengen and Saunders (1972) have obtained analytic expressions for the thermomagnetic tensor $\alpha(B)$ and show how the umkehr effect and the thermomagnetic power are related to the tilt angle and nature of the Fermi surface.

The effect of uniaxial compression has been studied by Ertl and Rahman (1971) on thermomagnetic effects in bismuth at 80 K. The effects were found to be very sensitive to pressure and indicated not only a decrease in the energy overlap but also increasing participation from a light hole band (separated from the conduction band at the L point by 0·015 eV with no compression). This affects the hole mobility to which the thermomagnetic effects are highly sensitive.

The galvanomagnetic properties of single-crystal antimony have been investigated by Öktü and Saunders (1967) between 77 and 273 K and interpreted in terms of the band model discussed. Carrier mobilities were deduced using two carriers and a multivalley band structure. The temperature dependences of the mobilities of the holes and electrons were found to be $T^{-1\cdot 48}$ and $T^{-1\cdot 42}$ respectively indicating similar scattering mechanisms for both types of carriers. For a degenerate semimetal, the relaxation time for intravalley lattice scattering takes the form (Wilson, 1958) $\tau = \text{const} \times T^{-1}E^{-1/2}$ and, as discussed by Öktü and Saunders, the value of ~1·5 for the exponent compares with −2·1 for bismuth and −1·2 for single crystal graphite. On taking into account phonon densities and carrier density of states, the dependence for the acoustic mode should be −1·0 so that there must be additional contributions such as from intervalley scattering, electron-hole collisions, or two-phonon processes. Values for the mobilities of the electrons and holes and the tilt angles as deduced by Öktü and Saunders are given in table 6.6 where they are compared with values deduced for As and As–Sb by the same group of workers.

Measurements of the Hall coefficients and electrical resistivity have been made below 77 K by Tanaka *et al.* (1968) and Bansal and Duggal (1973), who deduced that intervalley scattering is playing a significant role. Below 20 K the resistivity shows a $T^3$ dependence compared with a $T^2$ dependence for Bi. The Hall coefficient $R_{12,3}$ (magnetic field along the trigonal axis) increases linearly as the temperature increases from 4·2 to 20 K, but $R_{23,1}$ (magnetic field along the bisectrix axis) decreases as the temperature falls. This has not been fully explained, but the explanation may lie in the possible presence of a second set of holes lying in a band with its edge close to the Fermi level. The behaviour is in sharp contrast to the variation above 55 K where the dependence is reversed. Thermomagnetic measurements have been made in the temperature region 2 to 100 K by Bresler and Red'ko (1972) for single crystals oriented along the main crystallographic directions. Phonon drag plays a large role at low temperatures. The partial thermoelectric powers were measured for the electrons and holes and found to be in agreement with theory on the assumption that electron and hole concentrations at 4 K are unequal ($\Delta N = p - n$).

Arsenic has been less extensively investigated than bismuth and antimony. Jeavons and Saunders (1969) obtained the twelve magnetoresistivity tensor components for the low-field case for the temperature region 305 to 77 K. They found the electrons to be sited in pockets tilted at $-8°$ and holes in pockets tilted at $-50°$ (these angles having been converted to the convention of Brown *et al.*). The equal carrier densities were found to be essentially independent of temperature ranging from $1 \cdot 9 \times 10^{26}$ m$^{-3}$ at 77 K to $2 \cdot 1 \times 10^{26}$ m$^{-3}$ at 305 K. Because of the intervalley scattering, the carrier dependences were found to be close to $T^{-1 \cdot 7}$, i.e. considerably greater than $T^{-1 \cdot 0}$.

Effective masses have been measured from cyclotron resonance by Ih and Langenberg (1970, 1971) and by Cooper and Lawson (1971). Ih and Langenberg, who made their measurements at a microwave frequency of 134 GHz, found by comparing results for an ellipsoidal model with de Haas–van Alphen data that at the Fermi surface there is considerable deviation of the energy bands from a quadratic behaviour. Cooper and Lawson, on the basis of their measurements at a frequency of 24 GHz and a temperature of $1 \cdot 15$ K, report the measurement of several mass series, this being possible because of the high $\omega\tau$ value in the crystals. In particular, they measured hole-mass series in the binary-bisectrix plane and found a low-mass series ($0 \cdot 03 m_0$) corresponding to the necks of the hole surface. For the electron ellipsoids, they measured a tilt angle of $-5 \cdot 5° \pm 1 \cdot 5$ with $m_1 = 0 \cdot 134 m_0$, $m_2 = 0 \cdot 140 m_0$, $m_3 = 1 \cdot 35 m_0$, and for the hole ellipsoids $m_1 = 0 \cdot 146 m_0$, $m_2 - 0 \cdot 068 m_0$, $m_3 = 0 \cdot 0946 m_0$, these values being claimed as more accurate than preceding measurements because of the high-quality crystals used.

## 6.3 Alloys of bismuth, antimony, and arsenic

### 6.3.1 Preparation and band structure of Bi–Sb alloys

It has proved very difficult to produce good homogeneous single crystals of Bi–Sb mainly because of the problem of constitutional supercooling (see section 2.3.3). For the alloys there is a large temperature difference between the liquidus and the solidus (50 K for 10% Sb composition) and yet the melting temperature is low (below 300°C). Crystals are usually grown by the zone melting technique (see sections 2.3.2 and 2.6.5), and to prevent the supercooling the maximum rate of growth turns out as a consequence to be very low indeed. Even with a temperature gradient of 60 K cm$^{-1}$, an upper growth rate of $0 \cdot 4$ mm h$^{-1}$ is set. Therefore many early results taken from measurements on crystals grown at faster rates are questionable because of sample inhomogeneity. Further, it is not possible to produce homogeneity by annealing the samples after growth because of the low melting temperature.

The modifications to the band structure which occur when antimony is added to bismuth are of major importance for explaining property changes. Jain (1959) showed conclusively that a semiconductor range occurs within the solid solution range, and the presence of the bandgap was also shown by Esaki tunnelling (Esaki, 1966). The bands involved are those containing electrons and holes at the L points and holes at the T point. It is thought that the direct gap at the L point remains unchanged as antimony is added but that the valence band at the T point moves down so that when it reaches a level lower than the L-point valence band a fixed-bandgap semiconductor is achieved (figure 6.12). However, the effective mass of the T-band holes is significantly larger than that of the L-band holes so that the plot of activation energy against antimony concentration obtained by Jain does not show a flat region until this effect has been eliminated by a further fall in the level of the T-band. At the maximum on the activation curve there are approximately equal contributions to the electrical conductivity from the two sets of holes. More recent work (Yim and Amith, 1972) indicates a maximum bandgap of 0·02 eV occurring at a composition between 12 and 15% Sb, as against the value of 0·014 eV at 12% Sb obtained by Jain.

The bandgap is zero at the point where the L conduction and T valence bands become uncrossed and this corresponds to about 5% Sb. These bands recross when antimony content is increased to 40% and the structure returns to the semimetallic form of antimony. It is thought that the electron and hole concentrations remain equal throughout the composition range. If the bands are taken as parabolic then it is found that the mobility changes little with alloying. In fact evidence from cyclotron resonance would suggest some nonparabolicity of the bands.

If pressure is applied to bismuth, it becomes a semiconductor and it is significant that the rate of change of energy gap with pressure is approximately the same for bismuth and the alloys. Whereas addition of antimony increases the energy gap at the T point only, application of pressure increases the energy gap at both T and L points (Goldsmid, 1970).

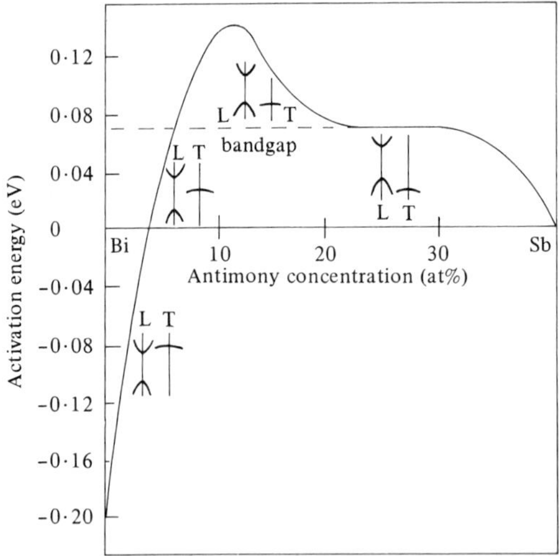


**Figure 6.12.** Variation of band structure as antimony is added to bismuth.

Considerable galvanomagnetic data have been obtained for the alloys, particularly in the region of zero bandgap. For alloys with a small but finite bandgap it is possible to induce a semiconductor–metal transition by application of a magnetic field. In addition, it is possible to transfer electrons from the L-point valence band to the T-point valence band by application of the magnetic field, and the exact sequence of events and the dependence of magnetoresistance, for instance, on magnetic field depends on the precise composition.

### 6.3.2 Properties of Bi–Sb alloys

When antimony is added to bismuth, there is a decrease in the $a_0$ and $c_0$ parameters, but Jain (1959) has shown that there is slight deviation from Vegard's law as shown in figure 6.13. The electrical and thermal properties of Bi–Sb alloys have been measured in some detail in view of the actual and possible applications of the alloys for thermoelectric and thermomagnetic energy conversion. The properties have been reviewed by Goldsmid (1970).

The fact that Bi–Sb alloys show a maximum electrical resistivity when about 15 at% Sb is added was discovered very early (Smith, 1911). It was also found that the negative Seebeck coefficient peaked close to this composition and that the positive temperature dependence of resistivity had a minimum at the same antimony content. However, the difficulty already mentioned of obtaining homogeneous samples means that only recent results are reliable in detail.

Jain (1959) measured electrical resistivity and Hall coefficients on samples prepared by zone levelling; his electrical resistivity results as a function of temperature are shown in figure 6.14. (The resistivity values are normalised with respect to the room temperature values.) It is from these curves that Jain deduced the diagram for energy gap as a function of composition shown in figure 6.12. From curves expressing the number of carriers as a function of temperature for samples of various compositions,

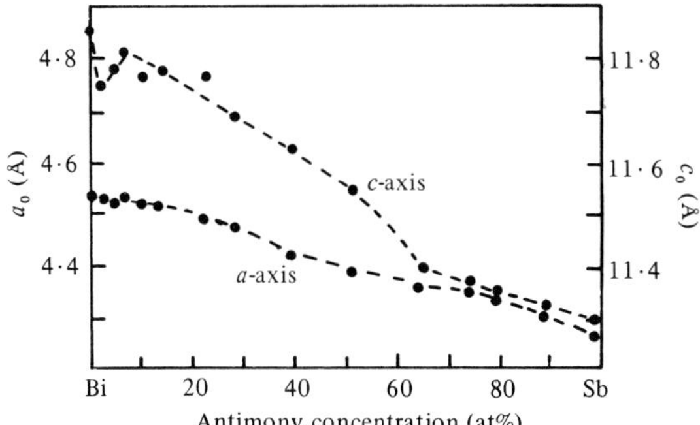


**Figure 6.13.** Lattice parameters $a_0$ and $c_0$ for Bi–Sb alloys as a function of antimony concentration (Jain, 1959).

and assuming that at low concentrations of antimony the conductivity is due to electrons plus T-point holes, Jain obtained (assuming parabolic bands) the variation of effective mobility ($= \sigma/Ne$) as a function of temperature. These curves show that the mobility is very little reduced over that for pure bismuth, nor is the temperature dependence very different (figure 6.15). The electron wavelength in a slightly degenerate semiconductor is of the order of 1000 Å which is much greater than the interatomic spacing; hence alloying has little effect. However, the deductions are only approximate: if the dependence of effective mass is taken into account, the alloying is found to have slightly more effect on the mobility.

In more recent work galvanomagnetic and thermomagnetic effects have been measured in single crystals as a function of orientation. Table 6.5 gives the carrier concentrations and components of the mobility tensors for crystals of $Bi_{95}Sb_5$ doped with tellurium measured at 80 K (Thomas and Goldsmid, 1970). The analysis assumed tilted ellipsoidal surfaces centred at the L-points for the electrons and nontilted spheroidal surfaces for the holes (see figure 6.8 for the Brillouin zone of bismuth). Holes could be detected only in undoped samples. These results are similar to results for bismuth (see section 6.2.4) and also to results for $Bi_{97}Sb_3$ and $Bi_{90}Sb_{10}$ obtained by Yazaki and Abe (1968) who found little variation in orientation or shape of the constant-energy surfaces between 80 and 300 K.

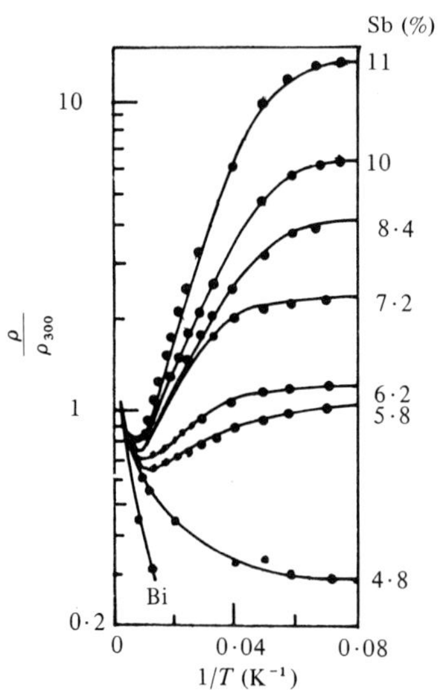


**Figure 6.14.** The ratio of the resistivity of Bi-Sb samples to the resistivity at 300 K for temperatures between 20 and 300 K plotted as a function of $1/T$ (Jain, 1959).

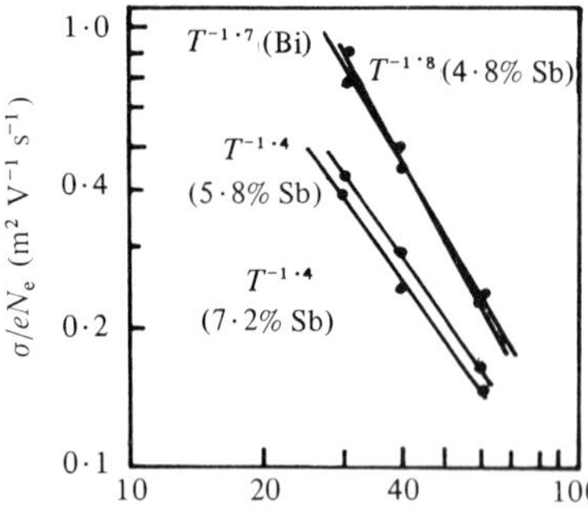


**Figure 6.15.** Variation of effective mobility in the temperature range 30 to 60 K in bismuth samples containing up to 7·2% antimony. (After Jain, 1959.)

From high-field magnetoresistance measurements in pulsed fields up to 45 T Brandt *et al.* (1969) deduced that the gap between the L-point conduction band and the T-point valence band does not vary linearly with composition. This would seem to fit with the nonlinear variation of lattice parameters with composition. The results shown in figure 6.16 were obtained with the current along the bisectrix axis and the magnetic field along the trigonal axis. For alloys containing less than 8·5 at% Sb,

**Table 6.5.** Carrier concentrations and components of the mobility tensors for crystals of $Bi_{95}Sb_5$ doped with tellurium (measured at 80 K). 1 refers to the binary direction, 2 to the bisectrix direction and 3 to the trigonal direction. The relaxation time is assumed isotropic. (Thomas and Goldsmid, 1970.)

| | Tellurium content (at%) | | | | |
|---|---|---|---|---|---|
| | 0 | 0·002 | 0·01 | 0·05 | 0·1 |
| Electron concentration, $n$ ($10^{24}$ $m^{-3}$) | 0·139 | 0·54 | 1·22 | 5·5 | 42 |
| Hole concentration, $p$ ($10^{24}$ $m^{-3}$) | 0·138 | | | | |
| Electron mobility components | | | | | |
| $\mu^n_{11}$ ($m^2$ $V^{-1}$ $s^{-1}$) | 70·5 | 43·5 | 15·5 | 4·3 | 2·7 |
| $\mu^n_{22}$ ($m^2$ $V^{-1}$ $s^{-1}$) | 1·2 | 1·1 | 0·31 | 0·1 | |
| $\mu^n_{33}$ ($m^2$ $V^{-1}$ $s^{-1}$) | 34·1 | 22·8 | 8·03 | 2·3 | insufficient |
| $\mu^n_{23}$ ($m^2$ $V^{-1}$ $s^{-1}$) | −7·7 | −4·5 | −1·9 | | data |
| Hole mobility components | | | | | |
| $\mu^p_{11}$ ($m^2$ $V^{-1}$ $s^{-1}$) | 11·2 | | | | |
| $\mu^p_{33}$ ($m^2$ $V^{-1}$ $s^{-1}$) | 1·2 | | | | |
| $\mu^n_{22}/\mu^n_{11}$ | 0·017 | 0·025 | 0·020 | 0·02 | |
| $\mu^n_{33}/\mu^n_{11}$ | 0·49 | 0·52 | 0·51 | 0·53 | |
| Tilt angle, $\theta$ (°) | −11·0 | −10·3 | −11·4 | assumed −11 | |

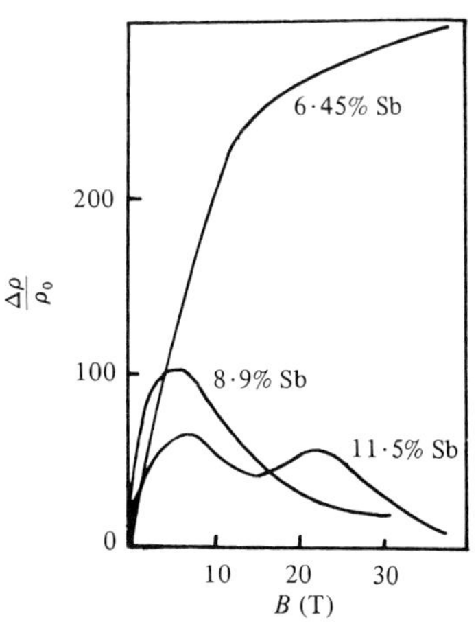


**Figure 6.16.** Transverse magnetoresistivity in Bi–Sb samples in very high magnetic fields at 4·2 K. (After Brandt *et al.*, 1969.)

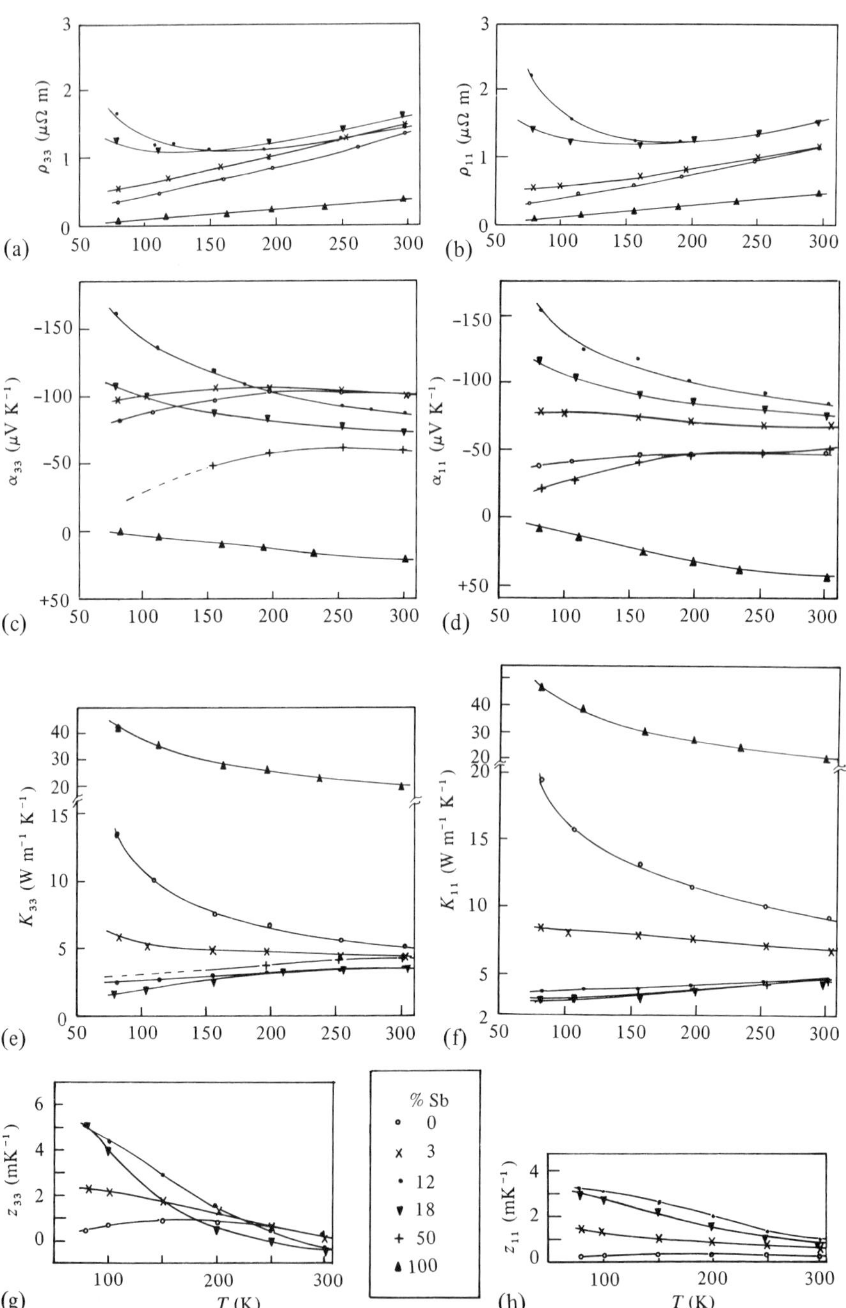


**Figure 6.17.** Resistivity $\rho$, Seebeck coefficient $\alpha$, thermal conductivity $K$, and thermoelectric figure of merit $z$ for undoped Bi–Sb alloys parallel (subscripts 33) and perpendicular (subscripts 11) to the trigonal axis. (After Yim and Amith, 1972.)

the resistivity was found to increase continuously with increase of magnetic field, for concentrations between 8·5% to 9·5% the resistivity curves showed a single maximum, and for concentrations greater than 9·5% (samples containing up to 16% were studied) the curves showed two maxima. The explanation for these curves is based on the fact that movement of the extremum of a band when a magnetic field is applied depends on the difference between $1/m_i^*$ and $1/m_i^\S$ where $m_i^\S$ is the spin effective mass of the carrier and $m_i^*$ the cyclotron effective mass, both masses being positive and applying to band $i$. For $m_i^* < m_i^\S$, the edge of the conduction band is lowered and the edge of the valence band raised (Azbel and Brandt, 1965), which is the case for the Bi–Sb alloys. On this basis the curves have been explained by Brandt *et al.* as follows:

(a) $Bi_{93\cdot55}Sb_{6\cdot45}$—overlap of the conduction and valence bands and ordinary magnetoresistance;

(b) $Bi_{91\cdot1}Sb_{8\cdot9}$—energy gap becomes smaller as the magnetic field increases; at approximately 8 T the alloy undergoes a semiconductor–metal transition as the gap becomes zero;

(c) $Bi_{88\cdot5}Sb_{11\cdot5}$—at $0 < B < 8$ T ordinary magnetoresistance; at 8 T $< B <$ 13 T, transfer of holes from L- to T-points, and hence lower mobility and lower magnetoresistance; at 13 T $< B <$ 23 T, magnetoresistance from holes which have all transferred to T-points; at 23 T $< B <$ 40 T, as the T-point valence band overlaps the L-point conduction band there is a semiconductor–metal transition.

The results would seem to require a band overlap in zero magnetic field to exist up to 8·5% Sb as against the 5% obtained by Jain.

Thermoelectric and thermomagnetic coefficients have recently been measured by Yim and Amith (1972) on highly homogeneous single crystals. The samples were undoped and included a complete range of alloys. Their measurements were made over the temperature range 77 to 300 K and are summarised in figure 6.17.

6.3.2.1 *Electrical resistivity*

Resistivities of the alloys are very similar at room temperature but diverge at lower temperatures. For alloys containing less than 5% Sb, the monotonic decrease characteristic of a semimetal can be seen. For alloys with higher antimony content there is an increase in resistivity with decrease of temperature at the lower end of the temperature range where $k_B T$ is less than the bandgap. Semiconducting behaviour continues up to 35% Sb. Below 8% Sb, $\rho_{33}$ is larger than $\rho_{11}$ ($\rho_{33}$ parallel to the trigonal axis, $\rho_{11}$ perpendicular to the trigonal axis) but this effect is reversed beyond 8% Sb (note that $\rho_{33} > \rho_{11}$ for bismuth and $\rho_{33} < \rho_{11}$ for antimony). There is a change in the electron to hole mobility ratio as the alloying changes and transfer of holes from L- to T-points occurs.

6.3.2.2 *Seebeck coefficient*
For alloys with less than 50% Sb the coefficients are negative with $|\alpha_{33}| > |\alpha_{11}|$ and for alloys with more than 75% Sb the coefficients are positive with $\alpha_{11} > \alpha_{33}$. The anisotropy at any particular composition arises from the difference in electron to hole mobility ratio along the different crystallographic axes. The largest negative (n-type) coefficient is along the trigonal axis where the electron-to-hole mobility ratio is largest and occurs for $Bi_{88}Sb_{12}$; the largest positive (p-type) coefficient is in the trigonal plane (i.e. perpendicular to the trigonal axis).

6.3.2.3 *Thermal conductivity*
For all alloy compositions the thermal conductivity is less along the trigonal axis than in the trigonal plane. Alloying with antimony decreases the thermal conductivity up to a composition of 35% after which the thermal conductivity increases again, the main effect of alloying being to decrease the lattice contribution. Hence the decrease is most pronounced at low temperatures where the lattice contribution is more important.

6.3.2.4 *Thermoelectric figure of merit*
As this is given by $z_{jj} = \alpha_{jj}^2/\rho_{jj}K_{jj}$, curves for the figure of merit can be obtained from the curves already shown. Parameters must be measured along the same crystallographic direction $j$. The values of $z$ are shown as functions of both temperature and antimony content. In the bismuth-rich region (0–35% Sb) the figure of merit as measured along the trigonal axis, $z_{33}$, is larger than for the trigonal plane, and also the coefficient is larger at 80 K than at room temperature. In the antimony-rich region (>75% Sb) the reverse is true, although figure 6.18 does not show this clearly, the values of $z$ being very small. Thus the largest values of $z$ are obtained at low temperatures along the trigonal axis in bismuth-rich alloys. The figures of merit for these alloys are larger than those for pure

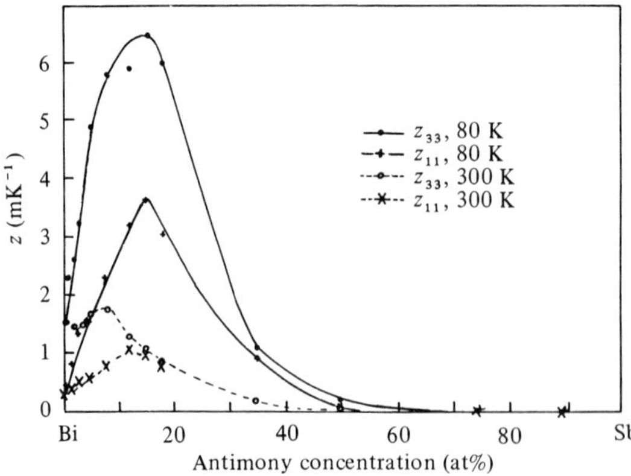


**Figure 6.18.** Thermoelectric figures of merit parallel ($z_{33}$) and perpendicular ($z_{11}$) to the trigonal axis in undoped alloys of Bi–Sb at 80 and 300 K (Yim and Amith, 1972).

bismuth because of the decrease in band overlap and decrease in thermal conductivity which occur on alloying. The largest coefficients are obtained for the $Bi_{85}Sb_{15}$ alloy in the temperature range 80 to 110 K. At higher temperatures the alloy $Bi_{92}Sb_8$ becomes preferable, but the difference in values for the figures of merit for the two alloys is small. The highest figure of merit obtained, $6{\cdot}5 \times 10^{-3}$ $K^{-1}$ at 80 K for the alloy of composition $Bi_{85}Sb_{15}$, compares with the previously reported maximum of $5{\cdot}5 \times 10^{-3}$ $K^{-1}$ at 80 K for the $Bi_{88}Sb_{12}$ alloy (Smith and Wolfe, 1962).

6.3.2.5 *Application of a magnetic field*

The thermoelectric figure of merit can be doubled by application of a magnetic field. The magneto-Seebeck coefficient increases with field and saturates to a limiting value particularly at lower temperatures. Wolfe and Smith (1962) measured a change of Seebeck coefficient from $-130$ $\mu V$ $K^{-1}$ to $-26$ $\mu V$ $K^{-1}$ in a field of $1{\cdot}4$ T for a $Bi_{88}Sb_{12}$ alloy. For a nondegenerate material with one type of carrier, and assuming a parabolic dispersion law, the maximum change of Seebeck coefficient to be expected is 40 $\mu V$. With increasing degeneracy this value is reduced. Hence the large adiabatic coefficients measured arise from an additional contribution from transverse-transverse galvanomagnetic effects, particularly the Hall effect acting on the Nernst effect. One consequence of this is that the measured Seebeck coefficient depends on the shape of the sample (Ertl *et al.*, 1963) just as the measured magnetoresistance of a sample can depend on its shape. The thermal conductivity decreases in a magnetic field owing to decreases in the electronic and ambipolar contributions, and the electrical resistivity increases. Putting these parameters together gives a maximum for the figure of merit. The highest figure of merit obtained by Yim and Amith is $11 \times 10^{-3}$ $K^{-1}$ in $0{\cdot}3$ T at 100 K and in $0{\cdot}13$ T at 80 K.

6.3.2.6 *Umkehr effect*

In Bi–Sb alloys in certain crystallographic orientations the Seebeck coefficient is very dependent on the direction of the magnetic field, forward or reverse; this is the so-called umkehr (meaning 'reversal' in German) effect. For instance, it is possible for a sample in a magnetic field to appear p-type in one direction and n-type in another when the Seebeck coefficient is measured. [From the Kelvin relation, $\alpha(B)T = \Pi(-B)$, it can be seen that the Peltier effect will lead to the converse conclusions.] Smith and Wolfe (1966) have made a theoretical investigation for the case of the temperature gradient along the bisectrix and the magnetic field in the trigonal plane. The umkehr effect vanishes when the magnetic field lies along the trigonal axis. Calculation of the effect has to take into account the tilt of the electron ellipsoids and the presence of holes. In the configuration considered, the effect disappears completely in the absence of tilt of the ellipsoids or becomes very small in the absence of holes.

However, even in the absence of tilt, the effect remains strong for an arbitrarily orientated crystal. The effect is very large in bismuth and Bi–Sb alloys because of the presence of equal numbers of high mobility electrons and holes. These allow large transverse currents which do not reverse direction with reversal of the magnetic field.

6.3.2.7 *Transverse figure of merit*

The Bi–Sb alloys are the most promising materials for energy conversion based on transverse effects in a magnetic field, particularly for use in the liquid-nitrogen temperature region. The product of Nernst effect and magnetic field which has units of V $K^{-1}$ (and hence is equivalent to the Seebeck effect) is very large for these alloys and increases linearly with increasing magnetic fields. However, the effect reduces with increased alloying with antimony, being largest for pure bismuth. Two thermomagnetic figures of merit are usually quoted. These are, for the case of the adiabatic Nernst voltage $N_{31}^{a}$ along the trigonal direction and magnetic field $B_2$ along the bisectrix,

$$z_{31} = \frac{(N_{31}^{a} B_2)^2}{K_{11}^{a}(B_2)\rho_{33}^{i}(B)}$$

and

$$z'_{31} = \frac{z_{31}^{i}}{1 - z_{31}^{i} T},$$

where $z_{31}^{i}$ is the isothermal figure of merit. (In practice $z_{31}$ is often only slightly higher than $z_{31}^{i}$.)

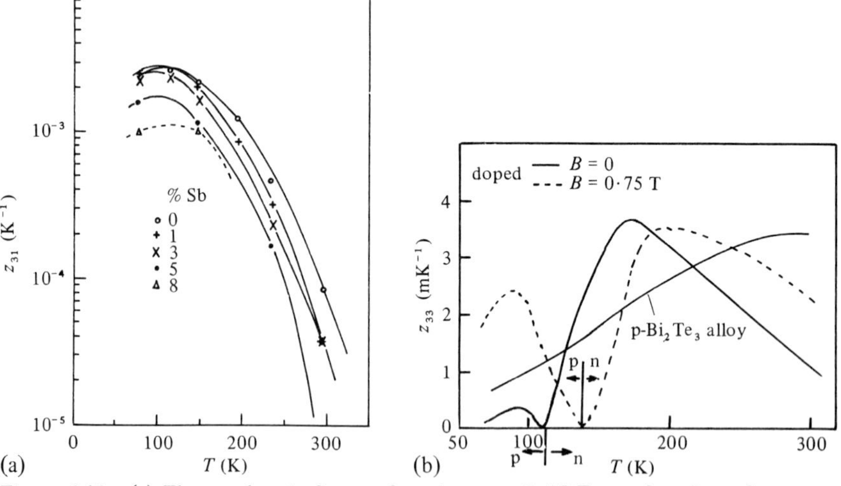


**Figure 6.19.** (a) Thermoelectric figure of merit $z_{31}$ at 0·75 T as a function of temperature for various undoped alloys of Bi–Sb (Yim and Amith, 1972); (b) temperature dependence of the figure of merit $z_{33}$ in zero magnetic field and in 0·75 T ($B_2$) for tin-doped $Bi_{88}Sb_{12}$ and of $z_{33}$ for p-type $Bi_2Te_3$–$Sb_2Te_3$–$Sb_2Se_3$ (Yim *et al.*, 1966).

The figure of merit $z_{31}$ for a $Bi_{99}Sb_1$ alloy (Yim and Amith, 1972) rises rapidly with increasing field but then levels off. However, unlike the thermoelectric figure of merit, it does not fall off beyond the saturation value. The lower the temperature, the lower the saturation value. Figure 6.19a shows the coefficient for different alloys in a field of 0·75 T, at which field strength the coefficient is saturating at the lower temperatures (around 100 K). At this magnetic field strength $z_{31}$ reaches a maximum of $2 \cdot 9 \times 10^{-3}$ $K^{-1}$ at 100 K for an alloy of $Bi_{99}Sb_1$, and this corresponds to a $z'_{31}$ value of $4 \cdot 1 \times 10^{-3}$ $K^{-1}$. The highest value of $z'_{31}$ reported has been for a $Bi_{97}Sb_3$ alloy (Cuff *et al.*, 1963), this being $6 \times 10^{-3}$ $K^{-1}$ in 1 T at 100 K, whereas Harman *et al.* (1964) reported that pure bismuth has a higher figure of merit than the alloys in the temperature range 80 to 300 K. Because the figures of merit are functions of composition and temperature, authors are not necessarily comparing for identical conditions, but, overall, $Bi_{99}Sb_1$ has the higher value of $z'_{31}$ below 130 K and bismuth the higher value in the range 130–300 K.

6.3.2.8 *Doped samples*
Undoped alloys are all n-type but p-type alloys can be produced by doping. Yim and Amith showed that a tin-doped $Bi_{88}Sb_{12}$ sample in a field of 0·75 T gave a p-type figure of merit $z_{33}$ of $2 \cdot 3 \times 10^{-3}$ $K^{-1}$ at 85 K, although in the absence of a magnetic field it was very small ($0 \cdot 3 \times 10^{-3}$ $K^{-1}$). This figure of merit is greater by a factor of 3 than the best p-type figure of merit obtained previously at 80 K on a sample of $(Sb_2Te_3)_{72}(Bi_2Te_3)_{25}(Sb_2Se_3)_3$ (Yim *et al.*, 1966). It is a particular orientation which yields this high magnetothermoelectric figure of merit: current flow along the trigonal axis and magnetic field along the bisectrix. The p-type figure of merit cannot be increased either by further increase of tin content or increase of the magnetic field.

6.3.2.9 *Applications*
The figures of merit both for thermoelectric and thermomagnetic arrangements for Bi–Sb alloys have been discussed already and Bi–Sb alloys and telluride alloys have particular application in Peltier and magneto-Peltier cooling devices, and are most promising in hybrid devices. n-Type $Bi_{88}Sb_{12}$ has been used in a hybrid device of eight stages involving also n- and p-type $Bi_2Te_3$–$Sb_2Te_3$–$Sb_2Se_3$ alloys to obtain 172 K of cooling from room temperature down to 128 K with a pumping capacity of 8 mW (M. S. Crouthamel, 1966, reported by Yim and Amith, 1972). The $Bi_{88}Sb_{12}$ was used in the low temperature stage and was in a magnetic field of 0·3 T.

### 6.3.3 As–Sb alloys

The phase diagram for As–Sb is similar to the type shown in figure 2.2b. Various workers have measured the minimum melting point and the corresponding composition and obtained values for the minimum melting temperature within the range 605 to 612°C and for the composition within the range 19·5 to 25·5 at% As. The most recent determination (Skinner, 1965) gives a temperature of 612°C and a composition of 25·5 at% As. Growth of homogeneous crystals is possible at this point because the liquidus and solidus coincide. At other compositions the same difficulties occur in preparing As–Sb alloys as those which occur in preparing Bi–Sb alloys where there is large separation of the liquidus and solidus over the complete composition range.

The alloys were reported by Saunders *et al.* (1965) to be semimetallic over the complete composition range, although Ohymata (1965) suggested that the alloys are narrow-bandgap semiconductors above 240 K in the composition range 9 to 40 at% As. Further work by Saito and Maezawa (1971) confirms the semimetallic findings. This semimetallic nature of the alloys is to be expected in view of the 0·20 eV band overlap of antimony and 0·37 eV band overlap of arsenic compared with the much smaller overlap of 0·0385 eV for bismuth. The continuous range of solid solutions show the same crystal structure throughout, the so-called A7 rhombohedral structure of arsenic and antimony. The lattice constant $a_0$ and the rhombohedral angle $\alpha$ show an almost linear increase with increase of antimony content (Quensel *et al.*, 1937; Trzebiatowski and Bryjak, 1938).

Akgöz and Saunders (1974) have measured the galvanomagnetic effects in the 25·5 at% As alloy in some detail and interpreted the results in terms of a two-band model as for the elements. They compared the results with measurements made by Jeavons and Saunders (1969) and Öktü and Saunders (1967) and their data are given in table 6.6. It should be noted that some of the data, particularly those for the tilt angle of the hole ellipsoids in antimony, are somewhat different from those in table 6.4 taken from a different series of measurements. The relative magnitudes of the mobilities $\mu_1^n$, $\mu_2^n$ and $\mu_3^n$ for the electrons indicate that the Fermi surface is elongated in the alloy in a similar way to the

**Table 6.6.** Comparison between the model parameters at 77 K of the As–Sb alloy and those of the parent elements (Akgöz and Saunders, 1974).

| Material | Carrier density | Electrons | | | | Holes | | | |
|---|---|---|---|---|---|---|---|---|---|
| | | mobility | | | tilt angle, | mobility | | | tilt angle, |
| | $n = p$ ($10^{25}$ m$^{-3}$) | $\mu_1^n$ | $\mu_2^n$ | $\mu_3^n$ ($10^{-2}$ m$^2$ V$^{-1}$ s$^{-1}$) | $\theta$ (°) | $\mu_1^p$ | $\mu_2^p$ | $\mu_3^p$ ($10^{-2}$ m$^2$ V$^{-1}$ s$^{-1}$) | $\theta$ (°) |
| As[a] | 19·2 | 46·0 | 0·01 | 53·0 | −8 | 87·0 | 8·3 | 66·0 | −51 |
| Sb[b] | 3·8 | 162·0 | 3·8 | 126·0 | −5 | 236·0 | 17·0 | 214·0 | −24 |
| Sb–25·5 at%As | 6·2 | 10·1 | 1·6 | 9·6 | −7 | 11·0 | 0·8 | 22·0 | −34 |

[a] Jeavons and Saunders (1969); [b] Öktü and Saunders (1967).

elongation in the elements. In the case of the holes, the notable feature is the low value of $\mu_2^p$ but the hole surface is warped in all cases and there is a large change in the tilt angles of the hole ellipsoids from one element to the other. All the mobility values are lower in the alloy, almost certainly owing to scattering effects. The mobilities of both the electrons and the holes are independent of temperature below 60 K and vary as $T^{-0\cdot4}$ above 60 K (compared with $T^{-1\cdot7}$ for arsenic, $T^{-1\cdot5}$ for antimony and $T^{-2}$ for Bi–Sb). The mobilities of the carriers are probably dominated by the effect of the disordered lattice in which atomic positions can be occupied by arsenic and antimony somewhat at random. The carrier density is almost independent of temperature, being $6\cdot1 \times 10^{25}$ m$^{-3}$ at liquid helium temperature and $6\cdot4 \times 10^{25}$ m$^{-3}$ at room temperature.

## 6.4 Graphite

### 6.4.1 Introduction

Graphite is a highly anisotropic crystal having an elongated hexagonal cell with $a_0 = 2\cdot456$ Å and $c_0 = 6\cdot696$ Å (Wyckoff, 1963) with space group $C_{6v}^4$ ($P6_3mc$). The four atoms of the cell occupy the positions 0, 0, $u$; 0, 0, $u+\frac{1}{2}$; $\frac{1}{3}, \frac{2}{3}, v$; and $\frac{2}{3}, \frac{1}{3}, v+\frac{1}{2}$. $u$ can be taken as zero and $v$ is almost zero and cannot exceed $0\cdot05$. The structure is shown in figure 6.20. It consists of widely spaced planes of carbon atoms which are hexagonally linked within the planes with the C–C distance equal to $1\cdot42$ Å, this separation being substantially the same as in the benzene ring. Separation of the carbon atoms between planes is considerably greater ($3\cdot40$ Å). Stacking of the layer planes is ABABA ... . The binding between the planes is about one hundred times smaller than the tight binding within the planes; hence the pronounced cleavage shown by graphite. The presence of four valence electrons per atom means that graphite has four filled valence bands but the small overlap between the outermost filled band and the lowest unfilled band gives rise to the semimetallic properties and the free electrons.

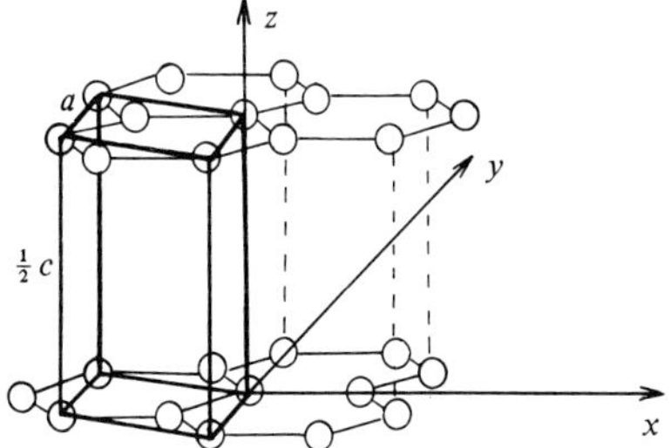


**Figure 6.20.** Crystal structure of graphite.

### 6.4.2 Preparation

Growth from solution has proved a particularly suitable method of obtaining high quality crystals; the crystals are comparable with naturally-occurring (Ticonderago) crystals and better than those of most pyrolytic graphites (see below) which in any case have small crystalline size (1 to $10^3$ μm). This growth can be from iron–carbon solutions or solutions of other systems such as nickel–carbon. Growth of graphite from solution has been reviewed by Austerman (1968). Figure 6.21 shows the iron–carbon phase diagram from 0 to 22 at% C (from Roscoe *et al.*, 1971). The graphite crystals are grown from a hypereutectic solution, a solution in which there is an excess of carbon compared with the eutectic composition. Roscoe *et al.* (1971) estimated a supersaturation of carbon of between 2 and 3% and employed a temperature in the region of 1100 to 1500°C. Growth was in an alumina crucible so that the only carbon coming into contact with the melt was that intended for dissolution, and an atmosphere of $N_2$–10%CO or Ar–10%CO was used in the furnace to prevent any oxidation of the charge. To obtain good crystals, a slow growth rate from a few nucleation centres was found to be necessary. In some runs crystals were removed from the frozen ingots; in other runs cooling was carried out more slowly but the crystals were then removed from the molten iron. Cleavage of crystals obtained by the second method indicated a second phase coexisting with the graphite

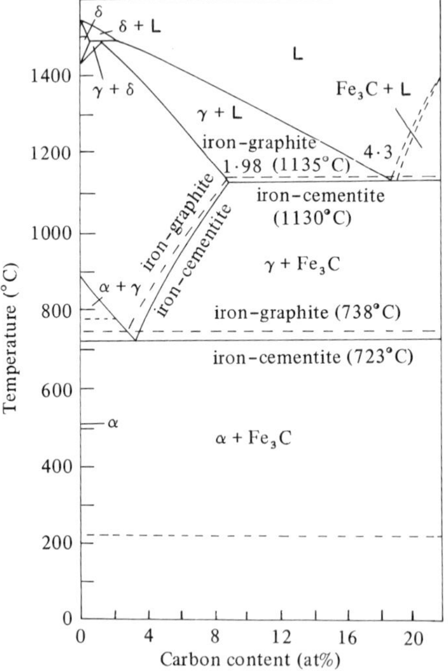


Figure 6.21. Iron–carbon phase diagram.

lattice, this phase being richer in iron than carbon. Although this phase is present in very small amounts, thus making identification difficult, Roscoe *et al.* obtained its x-ray diffraction patterns and identified it as cementite ($Fe_3C$). A small amount of red coloration was found and it was suggested that the iron had acted as a getter for traces of oxygen in the system, and that the oxygen content of the iron probably influences the rate of formation and decomposition of cementite during the growth of crystals, the cementite being retained only within the crystals withdrawn from the melt. It was thought that the carbon crystal platelets nucleated on the prismatic faces of the cementite crystals. Typically, the crystals were thin platelets of size $2 \times 1 \times 0{\cdot}02$ cm$^3$, although columnar crystals were obtained also.

Well-oriented pyrolytic graphite can be prepared by cracking methane or other hydrocarbons onto surfaces which are heated in the region of 2000 to 2500°C. Large pieces can be obtained this way but, although they show marked preferred orientation, they still contain large numbers of structural defects and are generally of low density. In particular, this makes them unsuitable for the measurement of properties in the *c* direction. Higher densities can be obtained to a large extent by the use of stress recrystallisation as carried out by Blackman and Ubbelohde (1962) by pulse heating. Using approximately 5 mm thick deposits, they found that the pulse heating could raise the temperature of the innermost layers to above 3700°C whilst leaving the outer layers sufficiently cool to provide a constraining outer skin. As a consequence, internal stress recrystallisation occurred under *c* axis constraint. This method led to the improved method of Moore *et al.* (1964) of recrystallisation at around 3000°C under a shearing stress obtained by applying a uniaxial stress of about 400 atm to cylindrical specimens. These specimens were embedded in a polycrystalline graphite susceptor which flowed plastically at 3000°C allowing a 12–15% radial expansion. The effect of the resulting axial compression and radial shear was to remove the *c* axis misorientation, and when these samples were annealed at 3200 to 3700°C under a moderate confining pressure ($10^5$ Pa normal to the basal plane) crystallite growth occurred parallel to the *c* axis. As a result blocks of graphite were produced with crystal properties comparable with those of small natural crystals.

### 6.4.3 Band structure

Most theoretical calculations of the electronic band structure have used as a first approximation a single two-dimensional model which neglects any interaction between adjacent layers. The electron states can be separated into $\sigma$ and $\pi$ bands, the former referring to states which are even with respect to reflections within the layer planes and the latter referring to states which are odd. The $\pi$ bands which can be considered as arising from overlap of the $p_z$ orbitals normal to the layer planes are very sensitive to the interlayer interaction which splits them into two closely spaced bands. It is this splitting which produces the $\pi$ valence and conduction band overlap at the Brillouin zone edge that gives rise to the semimetallic nature of graphite and the complex form of the Fermi surface. The overlap is ~0·03 eV and there is only about 1 in $10^4$ free electrons located along the vertical edges of the Brillouin zone. These location positions are indicated in the diagram of the first Brillouin zone for carbon (figure 6.22).

Using the single-layer crystal model and a variational method with linear combination of atomic orbitals (LCAO), Painter and Ellis (1970) obtained an energy band diagram illustrated in figure 6.23. The lattice sum was carried out over forty atoms and the energy converged to within 0·15 eV. (The occupied bands converged to even greater accuracy.) The energy band structure in the multiple-layer crystal is not very dependent on the wavevector $k_z$ along the $z$ axis so that a three-dimensional model

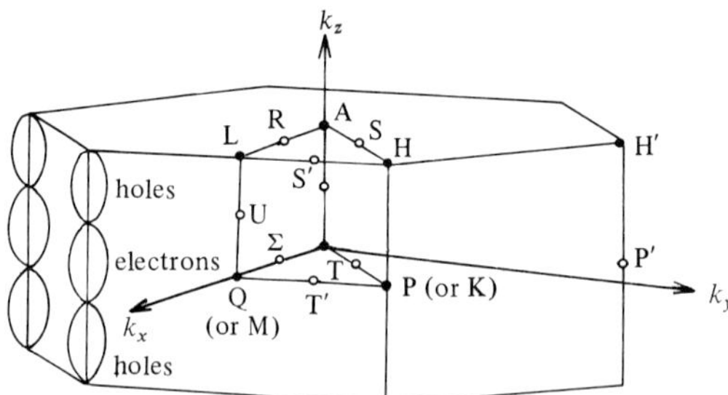


**Figure 6.22.** Brillouin zone for graphite, indicating also the positions of the electrons and holes.

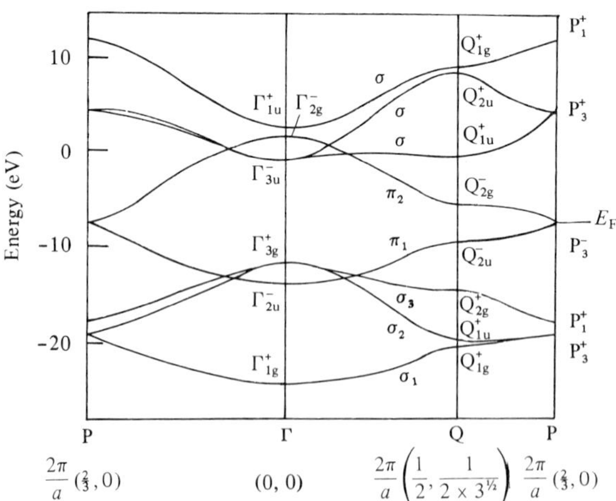


**Figure 6.23.** Energy band structure for graphite in the single-layer model. (After Painter and Ellis, 1970.)

should not give very different results. Haeringen and Junginger (1969), using a pseudopotential method of calculation, obtained qualitatively different results, but a combined orthogonal–plane-wave/tight-binding calculation of Nagayoshi *et al.* (1973) yielded good agreement with Painter and Ellis. Figure 6.24 shows the $\pi$ band structure for points located on the vertical face of the three-dimensional Brillouin zone (note the difference in scale in figures 6.23 and 6.24). The figure is taken from Willis *et al.* (1971) and makes use of the analyses of Doni and Parravicini (1969) and Greenaway *et al.* (1969). The effect of the interaction between adjacent planes was taken into account by considering the overlap between $p_z$ functions and used nearest and next-nearest integrals. The $\sigma$ bands were not considered as they are much less affected by interaction between planes.

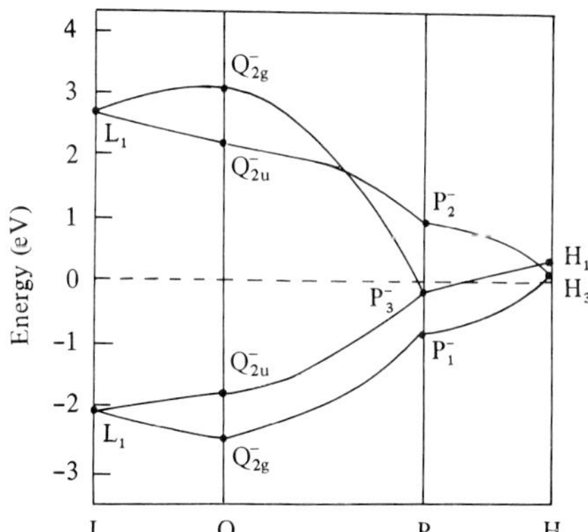


**Figure 6.24.** $\pi$-Band structure of graphite for points located on the vertical face of the three-dimensional Brillouin zone (Willis *et al.*, 1971).

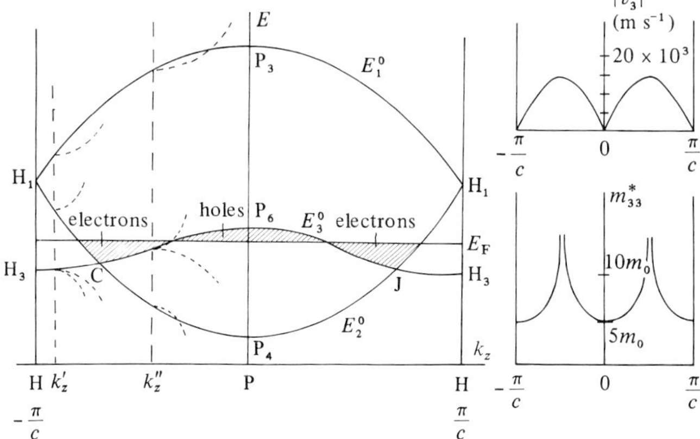


**Figure 6.25.** Band model of Slonczewski and Weiss (1958). Here the conduction and valence bands overlap. $E_F$ is the Fermi energy at 0 K.

Slonczewski and Weiss (1958) proposed for the bands near the vertical zone edge a model which included seven overlap parameters to be determined experimentally. Figure 6.25 shows the variation of electron energy along the vertical zone edge, the solid curves representing the dependence of the band energy on $k_z$ along the $c$ edge of the hexagonal zone, and the dashed lines showing the dependence of energy on $k_x$ and $k_y$ for constant $k_z$ at particular positions along the zone edge. The band overlap is now considered to be ~30 meV. Also shown are the variations of electron velocity and effective mass for variation of position along the zone edge. The electron velocity in the $c$ direction is only $\frac{1}{40}$th that in the basal plane and the effective mass of $\sim 10m_0$ can be compared with the much smaller effective mass of $0{\cdot}05m_0$ in the basal plane. The model is a 'one-electron' model and does not take into account Coulombic interactions between electrons, the importance of which was indicated by McClure (1964). However, McClure indicated that the Slonczewski and Weiss model should be capable of accounting for the electronic specific heat and the de Haas–van Alphen effect but not for cyclotron resonance or steady diamagnetic susceptibility.

#### 6.4.4 Properties

The electrical conductivity of graphite has been reviewed, for instance, by Spain (1971) for both natural and synthetic crystals. Early measurements were of conductivity in the basal plane. Carrier mobilities are almost equal for electrons and holes and can be $\sim 1\ m^2\ V^{-1}\ s^{-1}$ at 300 K and $100\ m^2\ V^{-1}\ s^{-1}$ at 4·2 K; variation of mobility with temperature can be accounted for by assuming electron–phonon scattering at high temperatures and defect scattering at low temperatures. Measurements in the $c$ direction have been more limited and show considerable variation. In particular, there remains uncertainty about the intrinsic anisotropy ratio $(\sigma_{11}/\sigma_{33})$ for the conductivity. Values of the ratio for natural crystals range from $10^2$ (Washburn, 1915) to $10^4$ or $10^5$ (Dutta, 1953; Bhattacharya, 1959) at room temperature. Results on synthetic material tend to show intermediate values. Investigation by Blackman *et al.* (1961) on pyrolytic graphite deposited at 2200°C gave a value of $5{\cdot}5 \times 10^3$, those by Spain *et al.* (1967) on hot-pressed pyrolytic graphite annealed at 3500°C gave a value of $3{\cdot}8 \times 10^3$.

Although the Boltzmann transport equation should be applicable to the case of natural crystals with a conductivity anisotropy ratio of 100, this giving a relaxation time ratio of $(\tau_{11}/\tau_{33}) \sim 1$ and mobility at room temperature $\mu_{33} \sim 10^{-2}\ m^2\ V^{-1}\ s^{-1}$ ($1\ m^2\ V^{-1}\ s^{-1}$ at 4·2 K), this is not so for the synthetic material. This is because the mean free path comes out as 3 Å rather than 1000 Å. To explain the high value of anisotropy of conductivity and a high value of the mean free path in the basal plane, it is necessary to assume some degree of localisation in the $c$-axis. Spain (1971) suggested that to a first approximation the conduction electrons are localised in sheets between the planes of carbon atoms. Within the basal plane the wave-functions are delocalised Bloch-type, whilst in the *c* direction they are localised. The $\sigma$-bonds binding the atoms within the planes have maximum charge cloud density in the plane of the atoms; the $\pi$-bonds delocalise in two dimensions with maximum cloud density between the planes. It may be that a small potential barrier arising from electron correlation separates $\pi$-electrons in neighbouring planes, this requiring thermal excitation across the barrier. The effective number of excited carriers will increase with increasing temperature whereas the mobility will decrease owing to increased scattering. This would lead to the observed weak dependence of $\sigma_{33}$ on temperature. Spain suggested that the measured anisotropy ratio indicates a barrier height at 300 K of 30–50 meV.

Measurements of the longitudinal magnetoresistance in the *c* direction have also been made and compared with similar results for the basal plane (Spain and Woollam, 1971). Results were obtained for magnetic fields up to 8 T. The longitudinal effect, $\rho_{zz}(B_z)$, is much larger than the transverse effect, $\rho_{zz}(B_{x,y})$, but much smaller than the basal magnetoresistance, $\rho_{xx}(B_z)$. At higher temperatures the longitudinal magnetoresistance saturates in the magnetic field and the magnitude of this field is temperature dependent above, but not below, 77 K. As the saturation depends on the value of $\omega\tau$, where $\omega$ is the cyclotron frequency and $\tau$ the effective scattering time, this indicates that $\tau$ is independent of temperature below 77 K. It was found that the majority carrier electrons have a much greater effect, relative to the effect of the majority holes, on the longitudinal magnetoresistance compared with the transverse magneto-resistance. The large value of the longitudinal magnetoresistance could not be predicted theoretically from the Boltzmann transport equation, but as indicated earlier this is not necessarily applicable. Shubnikov–de Haas oscillations were observed for measurements made at $T = 1 \cdot 4$ K.

Thermomagnetic measurements have been made on pyrolytic graphite, with hope that the material might be suitable for Ettingshausen refrigeration (the production of a transverse temperature gradient in the presence of orthogonal current and magnetic field) because of the high intrinsic carrier concentration and high mobilities at low temperatures. However, the figure of merit is limited by high values of the lattice thermal conductivity which is approximately $10^3$ W m$^{-1}$ K$^{-1}$ at 80 K (Mills *et al.*, 1965). At 60 K the high-field figure of merit was measured as $5 \times 10^{-5}$. Increase of electrical conductivity and carrier mobility with improved material perfection also increases the lattice thermal conductivity.

**References**

Abeles, B., Meiboom, S., 1956, *Phys. Rev.*, **101**, 544.
Abrikosov, A. A., Fal'kovskii, L. A., 1963, *Sov. Phys. JETP*, **16**, 769.
Akgöz, Y. C., Saunders, G. A., 1974, *J. Phys. C.*, **7**, 1655.
Aubrey, J. E., 1971, *J. Phys. F.*, **1**, 493.

Austerman, S. B., 1968, *Chemistry and Physics of Carbon,* **4**, Ed. P. L. Walker (Marcel Dekker, New York), p.137.
Azbel, M. Ya., Brandt, N. B., 1965, *Sov. Phys. JETP,* **21**, 804.
Bansal, S. K., Duggal, V. P., 1973, *Phys. Status Solidi,* **B56**, 717.
Bednarski, S., 1970, *J. Cryst. Growth,* **6**, 193.
Bhargava, R. N., 1967, *Phys. Rev.,* **156**, 785.
Bhattacharya, R., 1959, *J. Ind. Phys.,* **33**, 407.
Blackman, L. C. F., Saunders, G. A., Ubbelohde, A. R., 1961, *Proc. R. Soc. London, Ser. A.,* **264**, 19.
Blackman, L. C. F., Ubbelohde, A. R., 1962, *Proc. R. Soc. London, Ser. A.,* **266**, 20.
Booth, B. L., Ewald, A. W., 1968, *Phys. Rev.,* **168**, 805.
Brandt, N. B., Minana, N. Ya., Chen-Kang, Chu, 1967, *Sov. Phys. JETP,* **24**, 73.
Brandt, N. B., Svistova, E. A., Valeev, R. G., 1969, *Sov. Phys. JETP,* **28**, 245.
Bresler, M. S., Red'ko, N. A., 1972, *Sov. Phys. JETP,* **35**, 973.
Brown, R. D., 1970, *Phys. Rev.,* **B2**, 928.
Brown, R. D., Hartman, R. L., Koenig, S. H., 1968, *Phys. Rev.,* **172**, 598.
Brown, R. D., Mavroides, J. G., Lax, B., 1963, *Phys. Rev.,* **129**, 2055.
Cohen, M. H., 1961, *Phys. Rev.,* **121**, 387.
Cooper, G. S., Lawson, A. W., 1971, *Phys. Rev.,* **B4**, 3261.
Cuff, K. F., Horst, R. B., Weaver, J. L., Hawkins, S. R., Kooi, C. F., Enslow, G. M., 1963, *Appl. Phys. Lett.,* **2**, 145.
Datars, W. R., Vanderkooy, J., 1964, *IBM J. Res. Dev.,* **8**, 247.
Datars, W. R., Vanderkooy, J., 1966, *Proceedings of the International Conference on Physics of Semiconductors, Kyoto, J. Phys. Soc. Jpn.,* **21**, supplement 657.
Doni, E., Parravicini, G. P., 1969, *Nuovo Cimento,* **64B**, 117.
Dresselhaus, M. S., 1971, *Proceedings of the International Conference on Semimetals and Narrow Gap Semiconductors, Dallas (1970), J. Phys. Chem. Solids,* Supplement 3, **32**.
Dutta, A. K., 1953, *Phys. Rev.,* **90**, 187.
Ertl, M. E., Pfister, G. R., Goldsmid, H. J., 1963, *Br. J. Appl. Phys.,* **14**, 161.
Ertl, M. E., Rahman, I. U., 1971, *Phys. Status Solidi,* **45**, 85.
Esaki, L., 1966, *Proceedings of the International Conference on Physics of Semiconductors, Kyoto, J. Phys. Soc. Jpn.,* **21**, supplement 589.
Ewald, A. W., Tufte, O. N., 1958, *J. Appl. Phys.,* **29**, 1007.
Falicov, L. M., Golin, S., 1965, *Phys. Rev.,* **137**, A871.
Falicov, L. M., Lin, P. J., 1966, *Phys. Rev.,* **141**, 562.
Friedman, A. N., 1967, *Phys. Rev.,* **159**, 553.
Fukase, T., 1969, *J. Phys. Soc. Jpn.,* **26**, 964.
Goldsmid, H. J., 1970, *Phys. Status Solidi,* **A1**, 7.
Golin, S., 1968, *Phys. Rev.,* **166**, 643.
Greenaway, D. L., Harbeke, G., Bassani, F., Tosatti, E., 1969, *Phys. Rev.,* **178**, 1340.
Groves, S. H., Paul, W., 1963, *Phys. Rev. Lett.,* **11**, 194.
Haeringen, W. van, Junginger, H. G., 1969, *Solid State Commun.,* **7**, 1723.
Harman, T. C., Honig, J. M., Fischler, S., Paladino, A. E., 1964, *Solid-State Electron.,* **7**, 705.
Hartman, R., 1969, *Phys. Rev.,* **181**, 1070.
Herman, F., 1955, *J. Electron.,* **1**, 103.
Huntley, D. A., Shah, J. S., 1970, *J. Cryst. Growth,* **6**, 216.
Ih, C. S., Langenberg, D. N., 1970, *Phys. Rev.,* **B1**, 1425.
Ih, C. S., Langenberg, D. N., 1971, *Proceedings of the International Conference on Semimetals and Narrow Gap Semiconductors, Dallas (1970), J. Phys. Chem. Solids,* supplement 74, **32**.
Isaacson, R. T., Williams, G. A., 1969, *Phys. Rev.,* **185**, 682.
Ishizawa, Y., Tanuma, S., 1965, *J. Phys. Soc. Jpn.,* **20**, 1278.

Jacobson, D. M., Ertl, M. E., 1972, *J. Phys. D.*, **5**, 1358.
Jain, A. L., 1959, *Phys. Rev.*, **114**, 1518.
Jain, A. L., Koenig, S. H., 1962, *Phys. Rev.*, **127**, 442.
Jain, A. L., Suri, S. K., Tanaka, K., 1968, *Phys. Lett.*, **28A**, 435.
Jeavons, A. P., Saunders, G. A., 1968, *Br. J. Appl. Phys. (J. Phys. D.), Ser. 2,* **1**, 869.
Jeavons, A. P., Saunders, G. A., 1969, *Proc. R. Soc. London, Ser. A,* **310**, 415.
Ketterson, J. B., Eckstein, Y., 1965, *Phys. Rev.*, **140**, A1355.
Koenig, S. H., Lopez, A. A., Smith, D. B., Yarnell, J. L., 1968, *Phys. Rev. Lett.*, **20**, 48.
Korenblit, J. Ya., 1967, *Sov. Phys. - Semicond.*, **2**, 1192.
Kowalski, J., Zawadski, W., 1973, *Solid State Commun.*, **13**, 1433.
Lavine, C. F., Ewald, A. W., 1971, *J. Phys. Chem. Solids,* **32**, 1121.
Lin, P. J., Falicov, L. M., 1966, *Phys. Rev.*, **142**, 441.
Lopez, A. A., 1968, *Phys. Rev.*, **175**, 823.
McClure, J. W., 1964, *IBM J. Res. Dev.*, **8**, 255.
Michenaud, J-P., Issi, J-P., 1972, *J. Phys. C.*, **5**, 3061.
Mills, J. J., Morant, R. A., Wright, D. A., 1965, *Br. J. Appl. Phys.*, **16**, 479.
Moore, A. W., Ubbelohde, A. R., Young, D. A., 1964, *Proc. R. Soc. London, Ser. A.*, **280**. 153.
Nagayoshi, H., Tsukado, M., Nakao, K., Uemura, Y., 1973, *J. Phys. Soc. Jpn.*, **35**, 396.
Noothoven van Goor, J. M., 1971, *Philips Res. Rep.* supplement 4.
Ohymata, M., 1965, *J. Phys. Soc. Jpn.*, **20**, 1538.
Okada, J., 1957, *J. Phys. Soc. Jpn.*, **12**, 1327.
Öktü, Ö., Saunders, G. A., 1967, *Proc. Phys. Soc.*, **91**, 156.
Painter, G. S., Ellis, D. E., 1970, *Phys. Rev.*, **B1**, 4747.
Pikus, G. E., Bir, G. L., 1960, *Sov. Phys. Solid State,* **1**, 1502.
Priestley, M. G., Windmiller, L. R., Ketterson, J. B., Eckstein, Y., 1967, *Phys. Rev.*, **154**, 671.
Quensel, P., Ahlborg, K., Westgren, A., 1937, *Geol. Foeren. Stockholm Foerh.*, **59**, 135.
Rau, H., Rabenau, A., 1968, *J. Cryst. Growth,* **3/4**, 417.
Roscoe, C., Nagle, D., Austerman, B., 1971, *J. Mater. Sci.*, **6**, 998.
Saito, Y., Maezawa, K., 1971, *Proceedings of the Twelfth International Conference on Low Temperature Physics, Kyoto (1970)* (Academic Press of Japan, Tokyo), p.329.
Saunders, G. A., Cooper, G., Miziumski, C., Lawson, A. W., 1965, *J. Phys. Chem. Solids,* **26**, 533.
Saunders, G. A., Lawson, A. W., 1965, *J. Appl. Phys.*, **36**, 1787.
Shetty, M. M. Taylor, J. B., Despault, J. G., 1969, *J. Appl. Phys.*, **40**, 898.
Skinner, B. J., 1965, *Econ. Geol.*, **60**, 228.
Slonczewski, J. C., Weiss, P. R., 1958, *Phys. Rev.*, **109**, 272.
Smith, A. W., 1911, *Phys. Rev.*, **32**, 178.
Smith, G. E. Baraff, G. A., Rowell, J. M., 1964, *Phys. Rev.*, **135**, A1118.
Smith, G. E., Wolfe, R., 1962, *J. Appl. Phys.*, **33**, 841.
Smith, G. E., Wolfe, R., 1966, *J. Phys. Soc. Jpn.*, **21**, supplement 651.
Spain, I. L., 1971, *Proceedings of the International Conference on Semimetals and Narrow Gap Semiconductors, Dallas (1970), J. Phys. Chem. Solids,* supplement 177, **32**.
Spain, I. L., Ubbelohde, A. R., Young, D. A., 1967, *Philos. Trans. R. Soc., Ser. A.*, **262**, 345.
Spain, I. L., Woollam, J. A., 1971, *Solid State Commun.*, **9**, 1581.
Sümengen, Z., Saunders, G. A., 1972, *J. Phys. C.*, **5**, 424.
Sümengen, Z., Türetken, N., Saunders, G. A., 1974, *J. Phys. C.*, **7**, 2204.
Tanaka, K., Suri, S. K., Jain, A. L., 1968, *Phys. Rev.*, **170**, 664.
Tanenbaum, M., Goss, A. J., Pfann, W. G., 1954, *Trans. AIME,* **200**, 762.
Thomas, C. B., Goldsmid, H. J., 1970, *J. Phys. C.*, **3**, 696.

Trzebiatowski, W., Bryjak, E., 1938, *Z. Anorg. Allg. Chem.,* **238**, 255.
Uher, C., Goldsmid, H. J., 1974, *Phys. Status Solidi,* **B63**, 163.
Vanderkooy, J. Datars, W. R., 1968, *Can. J. Phys.,* **46**, 1935.
Van der Planken, J., 1970, *J. Cryst. Growth,* **6**, 352.
Vechi, M. P., Dresselhaus, M. S., 1974, *Phys. Rev.,* **B9**, 3257.
Washburn, G. E., 1915, *Ann. Phys.,* **48**, 436.
Willis, R. F., Feuerbacher, B., Fitton, B., 1971, *Phys. Rev.,* **B4**, 2441.
Wilson, A. H., 1953, *The Theory of Metals,* 2nd edition (Cambridge University Press, Cambridge).
Windmiller, L. R., 1966, *Phys. Rev.,* **149**, A472.
Wolfe, R., Smith, G. E., 1962, *Appl. Phys. Lett.,* **1**, 5.
Wyckoff, R. W. G., 1963, *Crystal Structures,* 2nd edition (Interscience, New York), 1.
Yim, W. M., Amith, A., 1972, *Solid-State Electron.,* **15**, 1141.
Yim, W. M., Fitzke, E. V., Rosi, F. D., 1966, *J. Mater. Sci.,* **1**, 52.
Yazaki, T., Abe, Y., 1968, *J. Phys. Soc. Jpn.,* **24**, 290, and **25**, 633.

7

# Chalcogenide binaries and their mixed systems

Compounds of selenium and tellurium, and to a lesser extent sulphur, and mixed systems of the compounds are a very important group of semiconductors exhibiting a complete range of possible narrow-bandgap values. Because of this extensive range of materials which is available, a complete chapter will be devoted to their discussion. First to be considered are the relevant chalcogenide compounds involving mercury.

## 7.1 Mercury selenide, HgSe, and mercury telluride, HgTe

The band structure of these two compounds around the Γ point is now considered to be similar to that of α-Sn so that they are both zero-bandgap materials. Originally mercury telluride was thought to be a narrow-bandgap semiconductor with an energy gap of between 0·01 and 0·02 eV. Both compounds are of zinc blende structure with similar lattice parameters (6·08 Å for HgSe and 6·44 Å for HgTe).

Although mercury sulphide in its usual room temperature form is of hexagonal structure (a distorted NaCl structure) and constitutes a large-bandgap semiconducting material, there is a zinc blende modification, β-HgS, which occurs in stable form above 600 K. Zallen and Slade (1970) have made optical measurements on natural polycrystalline material (metacinnabar) which remains in the metastable zinc blende form because of the presence of approximately 1 at% Fe. These measurements indicated the samples to be n-type with a carrier concentration of $n \approx 5 \cdot 3 \times 10^{24}$ $m^{-3}$ and that the band structure is of similar form to that of α-Sn and to that of the other mercury chalcogenides of zinc blende structure.

### 7.1.1 Preparation

Figure 7.1 shows the phase diagrams for the two compounds HgSe and HgTe. That for HgSe was determined by Strauss and Farrell (1962). The mercury-rich side of the system was not investigated because of the risk of explosions on this side of the diagram. The selenium-rich side was found to be similar to the selenium-rich side of the Pb–Se system. The melting point for the congruently melting HgSe is 799°C. In the diagram there is a monotectic at 680°C attributed to the formation of a region of two immiscible liquid phases, and a eutectic at 220°C (melting point of selenium) occurring between mercury selenide and selenium. This eutectic at 220°C shows in the heating curves but not in the cooling curves. This behaviour is similar to that which occurs in the case of selenium, and arises because selenium transforms to a crystalline form when heated above 100°C but when it is slowly cooled the liquid changes direct to the amorphous form.

Strauss and Farrell made no attempt to determine the solubility of mercury or selenium in mercury selenide. However, study of the electrical properties of mercury selenide indicated that the compound is slightly rich in mercury. The concentration of excess mercury is about $10^{23}$ atoms $m^{-3}$ for samples equilibrated with selenium-rich liquid and $10^{24}$ atoms $m^{-3}$ for samples equilibrated with mercury-rich liquid. Blue and Kruse (1962) also found that the properties of mercury selenide were controlled by the lattice defects arising from excess mercury.

The liquidus diagram for mercury telluride is shown in figure 7.1b as obtained by Strauss and Brebrick (and taken from Harman, 1967), who found an upper limit of 50·5 at% Te for the composition of tellurium-saturated mercury telluride at temperatures up to 578°C. They suggested that electrical measurements on annealed mercury telluride indicate a homogeneity of within 0·01 at%, considerably less than the upper limit of 0·6 at% obtained by partial-pressure measurements. However, shown inset on the diagram is an expanded part of the diagram according to Delves and Lewis (1963), who found the composition of mercury telluride at its congruent melting point to be 51·25 to 52·0 at% Te and to Delves (1965) who found a two-liquid region similar to that for mercury selenide. Electrical properties are very dependent on growth conditions and, as will be discussed later, annealing in mercury has a particularly marked effect.

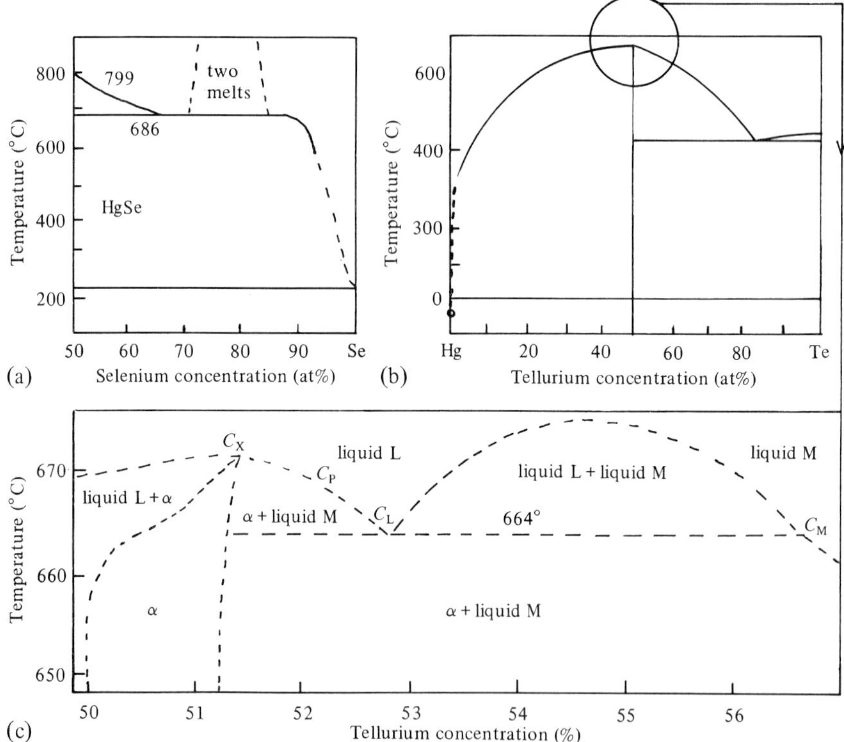


**Figure 7.1.** Phase diagrams for the Hg-Se and Hg-Te systems: (a) Hg-Se; (b) Hg-Te; (c) Hg-Te expanded.

Brebrick and Strauss (1965) have measured the partial vapour pressures of mercury and tellurium in equilibrium with mercury-saturated and tellurium-saturated solid HgTe over a temperature range of 400 to 700 K. This was achieved by measuring as a function of wavelength the optical density of the vapour. Figure 7.2 shows schematically the results of the study. The solid lines are three-phase lines, the vapour being in equilibrium with solid and liquid, whereas the dotted lines indicate the variation of partial pressure with temperature when the vapour is in equilibrium with molten material only. A similar plot was obtained for HgSe.

Both mercury selenide and mercury telluride have been prepared by direct synthesis from the elements. The usual procedure is for the reaction to take place in silica tubes sealed under vacuum at a pressure of about 1 $\mu$Torr. Blue and Kruse discuss the preparation of mercury selenide. At the melting point of HgSe (799°C), free mercury has a vapour pressure of 100 atm. It is therefore important to avoid explosions resulting from unreacted mercury. Hence an excess of selenium was used and reaction of mercury with selenium carried out at 600°C for a period of 24 h before the material was heated on up to 800°C for a few hours. Without removing the resultant mercury selenide from the tube, a silica rod was added and the capsule lowered through a Bridgman–Stockbarger furnace from an upper zone temperature of 850°C to a lower zone temperature of 750°C. The resultant material remained polycrystalline

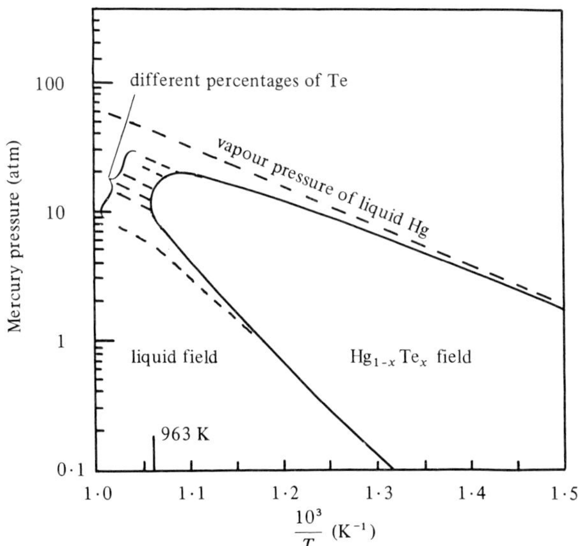


**Figure 7.2.** Partial pressures of mercury in the gaseous state in equilibrium with Hg–Te liquid and solid phases, plotted against reciprocal absolute temperature (Brebrick and Strauss, 1965).

with crystallites of a few millimetres size. Preparation of single-crystal material by the Bridgman method has been used as a fairly general technique and an improved version of the method for both mercury selenide and mercury telluride is discussed below. Growth from the vapour is also a well-tried technique [see Hamilton (1958) for instance].

In the case of the preparation of mercury telluride, the components can be heated to its melting point rather more safely as it has a lower melting point of 670°C. Harman *et al.* (1958) discuss preparation in some detail. They used a two-zone furnace, one zone to heat the components for reaction and the other zone for controlling the vapour pressure of mercury at ~15 atm. After heating for at least 4 h for reaction to take place, slow cooling was allowed with the mercury pressure maintained equal to or greater than the dissociation pressure of HgTe. If this were not done, if instead the mercury telluride were quenched to room temperature, then there would be a tendency for liquid mercury to separate out.

To produce single crystals, a zone melting technique was used incorporating three zones (see section 2.6.5 for description of the technique). The temperature profile used is shown in figure 2.22. $T_1$ is set so as to melt the mercury telluride, $T_2$ must not be so high as to prevent its crystallisation nor so low as to encourage alloying, while $T_3$ controls the mercury pressure at 15 atm. Initial experiments showed that if the mercury pressure was maintained at a lower pressure value of 5–10 atm, only polycrystalline mercury telluride was produced. The zone was passed at a velocity of $1 \cdot 2$ cm $h^{-1}$.

Harman also used a Bridgman technique for growing single crystals of both mercury telluride and mercury selenide. An excess of mercury was added to stoichiometric quantities of the elements to maintain mercury pressure, and the following method used in order to prevent explosions (Harman, 1967). (i) The temperature of the centre of furnace A (figure 7.3) was raised to 550°C. (ii) The temperature of the centre of furnace B was raised to 580°C. (iii) The centre of furnace A was raised to 50 K below the melting point. (iv) The centre of furnace B was raised to 50 K above the melting point. Once this had been done, the temperature equilibrated between zones A and B at a value about 80 K below the melting point and the temperature dip between furnaces helped keep the mercury vapour pressure sufficiently low to prevent bursting of the capsule.

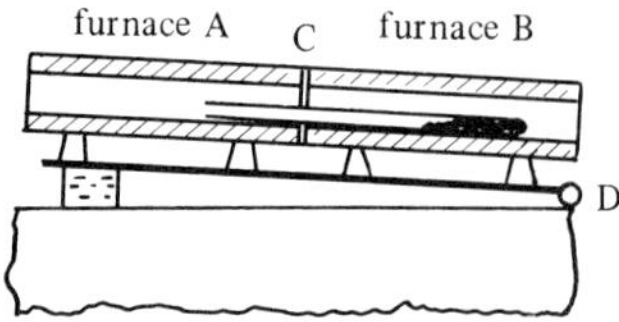


**Figure 7.3.** Furnace for the growth of mercury selenide and mercury telluride. (After Harman, 1967.)

Finally, (v) the furnace was rotated 90° about D to a vertical position, and (vi) the capsule was lowered at 1 mm $h^{-1}$ into a tank of flowing water. The method has also been used for $Hg_{1-x}Cd_xTe$ and $Hg_{1-x}Cd_xSe$.

As already mentioned, Delves (1965) found a two-liquid zone which occurred between $Hg_{1\cdot 0}Te_{1\cdot 12}$ and $Hg_{1\cdot 0}Te_{1\cdot 25}$ and he used this feature of the phase diagram to investigate an alternative method of crystal growth. At composition $C_X$ (figure 7.1c), the solidus and liquidus touch so that solid mercury telluride can be grown at this composition without any segregation. However, the solid will be nonstoichiometric if the phase diagram is as suggested by Delves and Lewis. Growth from composition $C_P$ will result in segregation of tellurium in the liquid and, unless the growth rate is very slow or the temperature gradient steep, constitutional supercooling will occur. However, if growth from $C_L$ is carried out, the composition of the solid and two liquids is fixed. So liquid L turns into solid $\alpha$ and second liquid M (i.e. a monotectic reaction), and the latter carrying the excess tellurium is rejected from the freezing surface. No segregation occurs and, provided liquid M is formed quickly and removed (by gravity for example) from the interface, no constitutional supercooling takes place. (The stoichiometry of the resulting solid may then be different to that formed at $C_X$.) Samples grown by a directional freeze from a $Hg_{1\cdot 0}Te_{1\cdot 1}$ composition melt showed a 'veinous' substructure thought to be due to low-angle grain boundaries. Crystals grown from melts of composition $Hg_{1\cdot 0}Te_{1\cdot 125}$ (growth speed 0·3 to 1·5 cm $h^{-1}$) did not show this substructure to start with, but did display an abrupt change at a particular distance (about $\frac{3}{4}$ of the length) to a cellular structure for samples grown at the low speed or a columnar structure for samples grown at the high speed. This coincided with the end of the two-liquid region for growth from these particular melts.

### 7.1.2 Band structure

The band structures of mercury selenide and mercury telluride have been studied extensively both experimentally and theoretically. Since 1967 it has been accepted that the band structures for both chalcogenides about the Γ-point are similar to that for $\alpha$-Sn. A range of results from magnetooptical, Shubnikov–de Haas, de Haas–van Alphen, and reflectivity measurements all support the semimetal form of band structure rather than the semiconductor form in which HgTe was thought to have a very narrow energy gap of 0·01 to 0·02 eV.

There are two main differences between the band structures of mercury selenide and mercury telluride and the band structure of $\alpha$-Sn. The lack of inversion symmetry in the compounds allows the splitting of the heavy hole valence band, although this splitting is very small. Also there are no additional conduction band minima as there are in $\alpha$-Sn at the L-point. The forms of the energy bands in the vicinity of the Fermi surface are shown in figures 7.4a and 7.4b. These are taken from Overhof (1971a), who calculated band structures for mercury selenide and mercury telluride using the relativistic Green's function method. In his calculation, Overhof particularly attempted to reproduce by calculation the very small valence band overlap of less than 1 meV reported by Vérié and Decamps (1965) and this he achieved. The band overlap is caused by the inversion symmetry splitting. Overhof also gives more extensive band energy diagrams for both chalcogenides. The bands arise as follows:

lower valence bands: Hg d-electrons and Se or Te s-electrons;

higher valence bands: Hg s-electrons and Se or Te p-electrons (spin–orbit split);

lower conduction bands: p-type (this band touches the highest valence band at the Γ-point);

upper conduction bands: mainly Hg p-states.

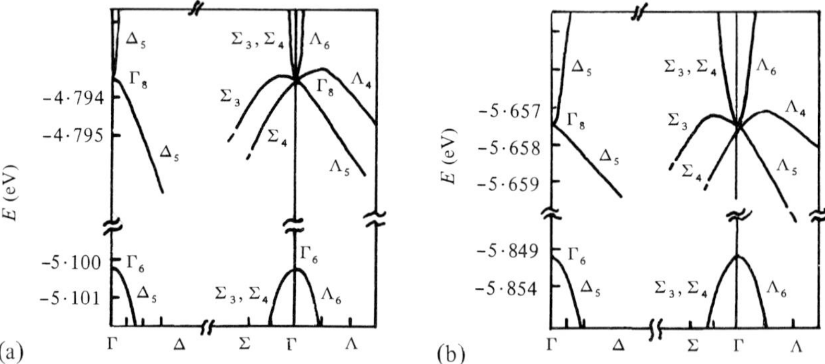


**Figure 7.4.** Band structure near the Fermi surfaces for mercury telluride and mercury selenide. (After Overhof, 1971a.)

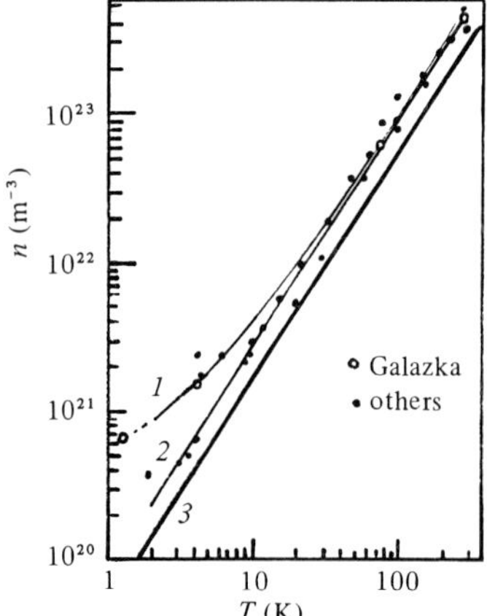


**Figure 7.5.** Temperature dependence of the intrinsic concentration in mercury telluride showing theoretical curves and experimental points. (After Galazka, 1970.) *1*, 1 meV overlap. *2*, zero overlap. *3*, for comparison, straight line $n_i \sim T^{3/2}$ dependence.

s- and p-type bands can be distinguished at the Γ-point but are mixed on the axes.

Galazka (1970) calculated the temperature dependence of intrinsic conduction assuming zero bandgap, and also small overlap, and compared the calculated values with the experimental values for mercury telluride (figure 7.5). Parameters other than the band overlap $\Delta E$ were fixed. Kane's model was assumed for the conduction band, a spherical heavy-hole band was assumed and the heavy-hole effective mass taken to be $0{\cdot}55m_0$. Best agreement was obtained for an overlap of 1 meV.

### 7.1.3 Properties

Although the electrical properties depend closely on the conditions of growth, both compounds can exhibit extremely high electron mobilities, particularly at low temperatures. This may indicate an unusually large static or low-frequency dielectric constant as is found in lead telluride, thus leading to reduced impurity scattering.

#### 7.1.3.1 *Mercury selenide*

The electrical properties of mercury selenide have been less widely measured than those of mercury telluride but galvanomagnetic and thermomagnetic measurements have been made by Harman (1964) on mercury selenide for magnetic fields up to 11 T. For an analysis of the results, the Kane model was assumed with $E_g \ll \Delta$ (i.e. an energy gap small compared with spin-orbit splitting, see page 73). With these assumptions it was shown that the dominant scattering mechanism is polar scattering due to optical phonons (whereas an earlier analysis using a parabolic band model had indicated scattering due to acoustical phonons). Table 7.1 gives calculated parameters for samples of different carrier concentrations.

The carrier concentration $n$ was calculated from $n = 1/R_H e$. Conductivity mobility $\mu$ and Hall mobility $\mu_H = R_H \sigma$ agreed within 5%. Assuming the presence of polar scattering is a correct deduction, the decrease in mobility with increasing carrier concentration is a consequence of the change of effective mass rather than a change in relaxation time. The Hall coefficient was found to be independent of magnetic field (to within 2–4%) for magnetic fields up to 11 T. For the highest magnetic field used, a magnetoresistance $(\Delta\rho/\rho_0)$ of $1{\cdot}6$ was obtained and this was attributed to a combined effect of electrons and holes. Experimental

**Table 7.1.** Transport parameters for mercury selenide (Harman, 1964).

| $n$ ($10^{24}$ m$^{-3}$) | $x_g$ | $E_F^*$ | $\mu$ (m$^2$ V$^{-1}$ s$^{-1}$) | $r$ | $K_L$ (W m$^{-1}$ K$^{-1}$) |
|---|---|---|---|---|---|
| 0·4 | 4·0 | 4·0 | 1·8 | 1·0 | 0·011 |
| 1·5 | 4·0 | 8·0 | 0·9 | 1·0 | 0·011 |
| 4·3 | 4·0 | 12·0 | 0·52 | 1·0 | 0·011 |

results for a range of room temperature measurements are summarised in figure 7.6. Some of the results included were taken from Whitsett (1961). Because of the large carrier concentration and high mobility, the Righi–Leduc effect is particularly large and is shown in figure 7.6d.

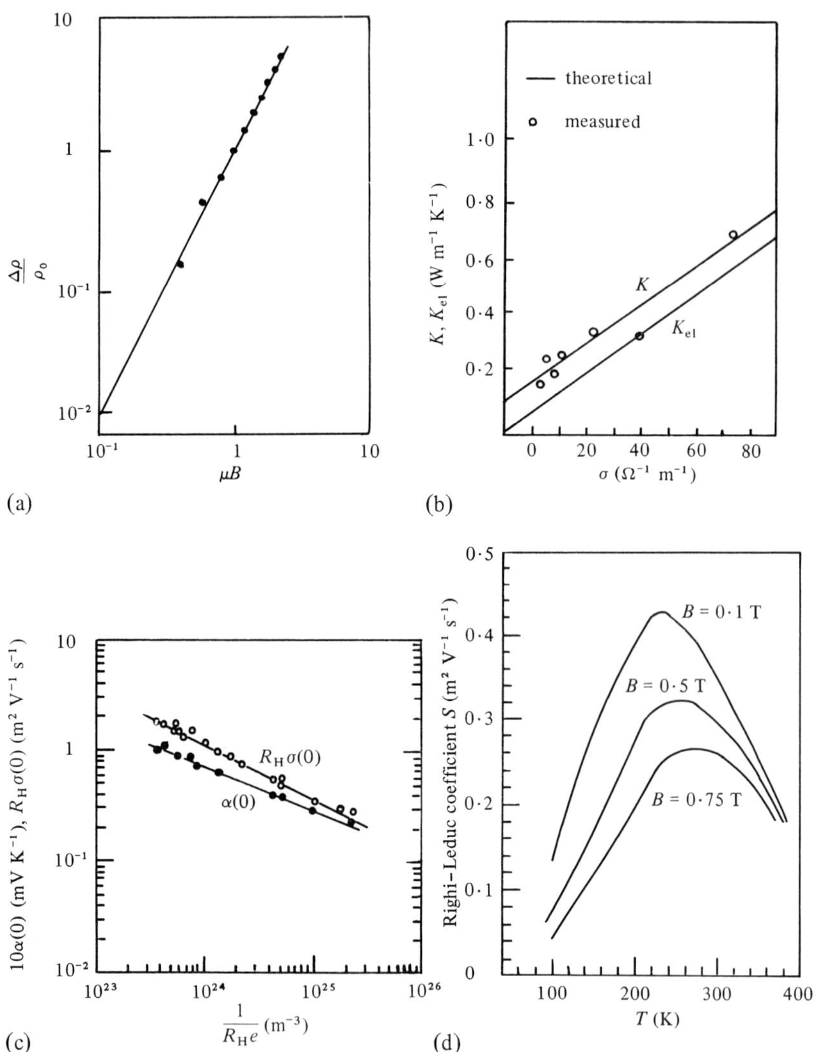


**Figure 7.6.** Various transport effects in mercury selenide: (a) electrical resistance of single-crystal material versus magnetic field strength (Corbino disc geometry) (Harman, 1964); (b) total thermal conductivity, $K$, and electronic part $K_{el}$ of the thermal conductivity versus electrical conductivity $\sigma$ (Harman, 1964); (c) electron Hall mobility $R_H\sigma(0)$ and Seebeck coefficient $\alpha(0)$ versus carrier density (Harman, 1964); (d) Righi–Leduc coefficient as a function of magnetic field (Whitsett, 1961).

7.1.3.2 *Mercury telluride*

The electrical resistivity and Hall coefficient of mercury telluride have been measured by numerous workers. Harman *et al.* (1958) investigated the variation of electrical properties on annealing the compound under varying mercury vapour conditions. Giriat (1964) made measurements on undoped samples and samples doped with lead, and found the electrical properties to be independent of doping up to concentrations of $6 \times 10^{26}$ atoms $m^{-3}$, but in any case the samples had excess tellurium of up to 1% owing to the fact that in the method of preparation used some free mercury remained in the tube. However, electrical properties did vary with annealing under a vapour pressure of mercury, because the samples absorbed mercury to approach stoichiometry, and also microprecipitates within the samples dissolved. Annealing at 250°C for about 100 h maximised the Hall coefficient and electrical conductivity. The Hall mobility increased with decrease of temperature to give a value of 7 $m^2$ $V^{-1}$ $s^{-1}$ at 77 K. The relation $\mu \propto T^{-n}$ was found to be satisfied with $n = 2 \cdot 1$ for 330–400 K, and $n = 0 \cdot 65$ near 77 K. From this, phonon scattering was deduced at the higher temperatures and impurity scattering at 77 K. However, Dziuba and Zakrzewski (1964) have suggested polar scattering at the higher temperatures. They obtained an electron effective mass of $0 \cdot 02 m_0$ and a Seebeck coefficient of 135 $\mu$V $K^{-1}$.

Harman (1967) investigated mercury telluride for a wide range of acceptor and donor concentrations at different temperatures and varying magnetic fields. The analysis of the results was based on a two-band model with a nonparabolic conduction band and an assumed parabolic valence band. Deviations from the model occurred at both 4·2 K and 78 K and at low values of magnetic field (i.e. low $\mu B$), indicating the need for a more detailed model. For the as-grown n-type sample, the Hall mobility at 78 K was found to be 7 $m^2$ $V^{-1}$ $s^{-1}$, whereas a highly doped p-type sample had a mobility of 0·03 $m^2$ $V^{-1}$ $s^{-1}$. Values of mobility for samples with intermediate dopings ranged between these limits. Certain samples showed sufficiently large magnetoresistance to require the simultaneous presence of electrons and holes, consistent with the semimetal model. Similarly, the presence of electrons and holes explains the change of sign of Hall coefficient with magnetic field for certain samples. The systematic variation of galvanomagnetic properties with doping ruled out the possibility that certain of the features which failed to fit the two-band model arose from inhomogeneities.

In a further paper, Harman *et al.* (1967) ascertained that Hall effect and magnetoresistance data at 4·2 K could be fitted to a three-carrier model with two sets of electrons and one set of holes. Table 7.2 shows the calculated data for three samples (signs as from the computer calculation). Figures 7.7a and 7.7b show the experimental and theoretical curves for Hall effect and magnetoresistance for the samples. Sample 50 LB exhibited a negative Hall coefficient and the minimum in Hall coefficient (maximum on the graph) is not possible for a two-carrier system. The magnetoresistance does not agree well with theory although the trends agree with the calculation. The results agree with the off-axis maxima shown in the band structure (figure 7.4a). The value of electron mobility of 65 $m^2\ V^{-1}\ s^{-1}$ at 4·2 K is notable.

**Table 7.2.** Carrier concentrations and mobilities for mercury telluride (Harman *et al.*, 1967).

| Sample number | Carrier concentrations ($10^{21}\ m^{-3}$) | | | Carrier mobilities ($m^2\ V^{-1}\ s^{-1}$) | | |
|---|---|---|---|---|---|---|
| | $n_1^e$ | $n_2^e$ | $n^h$ | $\mu_1^e$ | $\mu_2^e$ | $\mu^h$ |
| 50 LB | −2·1 | −7·0 | 0·62 | −65 | −0·95 | 7 |
| 300-2A | −1·2 | −3·3 | 64 | −10 | −1·3 | 0·042 |
| 4A | −1·2 | −10 | 2500 | −1·2 | −0·17 | 0·016 |

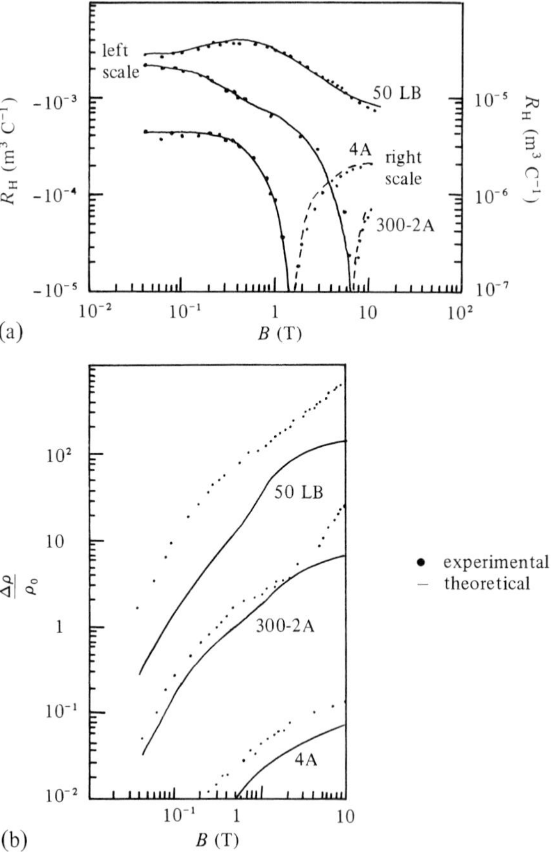


**Figure 7.7.** (a) Hall coefficient and (b) magnetoresistance for mercury telluride. (After Harman *et al.*, 1967.)

More recent results obtained by Zakrzewski and Dziuba (1972) were explained also by assuming low-mobility electrons (assigned to an impurity band) in addition to the light-electron and heavy-hole carriers. Samples used had been prepared by the Bridgman technique and annealed in mercury at 520 K. The mobility of the light electrons was found to be 6 $m^2 V^{-1} s^{-1}$, and that of the heavy electrons to be 0·5 $m^2 V^{-1} s^{-1}$ at 4·2 K. Hence, the mobilities were much lower than those for samples used by Harman *et al.*

### 7.1.4 $HgTe-In_2Te_3$

Alloying of $In_2Te_3$ with $Hg_3Te_3$ shows that single phase material exists up to 20 mol% $In_2Te_3$ (Ray and Spencer, 1967). Addition of $In_2Te_3$ causes a linear decrease in lattice parameter, an increase in optical energy gap to 0·35 eV at 20 mol% $In_2Te_3$ and a decrease in thermal conductivity (Spencer, 1964). The limit of the semimetallic region was found to depend on the heat treatment given to the samples. The room temperature Hall mobility showed a steady decrease from the value for pure mercury telluride to 0·2 $m^2 V^{-1} s^{-1}$ at 15 mol% $In_2Te_3$. Seebeck coefficients varied between −40 and −100 $\mu V K^{-1}$. Pitt *et al.* (1972) found that addition of $In_2Te_3$ caused the $\Gamma_6$ and $\Gamma_8$ bands to close up such that there is zero gap for 4–7 mol% $In_2Te_3$. They carried out high pressure measurements and found that at 14 kbar a very rapid increase in resistivity occurs because of a transition from the zinc-blende structure to a cinnabar (HgS) structure. At compositions greater than 20 mol% $In_2Te_3$, the alloys pass through a number of ordered phases with gradually increasing energy gaps.

## 7.2 Lead salts: lead selenide, PbSe, lead telluride, PbTe (and lead sulphide, PbS)

These IV–VI compounds are highly polar compounds crystallising in the NaCl structure with $a_0$ = 6·124 Å (PbSe), 6·462 Å (PbTe) and 5·936 Å (PbS) (from Dalven, 1969). As the structure consists of two interpenetrating face-centred cubic lattices, the Brillouin zone has the same shape as that for diamond or zinc blende (p.44). In general for a group of compound semiconductors of type MX, as the atomic number of the atom X increases, the energy gap $E_g$ decreases. However, a monotonic decrease has not been observed for the series PbS, PbSe, and PbTe, and it has been reported by Dalven (1971) that it is the energy gap for lead telluride that is anomalous. Values of the energy gaps, $E_g$, at 4 K for the series are: 0·286 ± 0·003, 0·190 ± 0·002, and 0·165 ± 0·05 eV, respectively (Mitchell *et al.*, 1964). Thus the bandgaps for all these compounds are not particularly narrow, and this especially applies to that of lead sulphide. However, as the materials are best considered together as a family of compounds, and as they are of importance alloyed with other compounds to form ternary compounds of variable bandgap, they will be considered in some detail. Energy calculations have also been made on a further member of the group, lead polonide, and Rabii and Lasseter (1974) have shown that it is a semimetal with a band overlap of 0·2 eV rather than a direct-gap semiconductor as predicted previously.

### 7.2.1 Preparation

Figure 7.8 shows the relevant parts of the binary phase diagrams for the three compounds as taken from Hansen and Anderko (1958) and supplement. Usually the compounds are prepared by reaction between stoichiometric proportions of the elements. Reaction is exothermic and

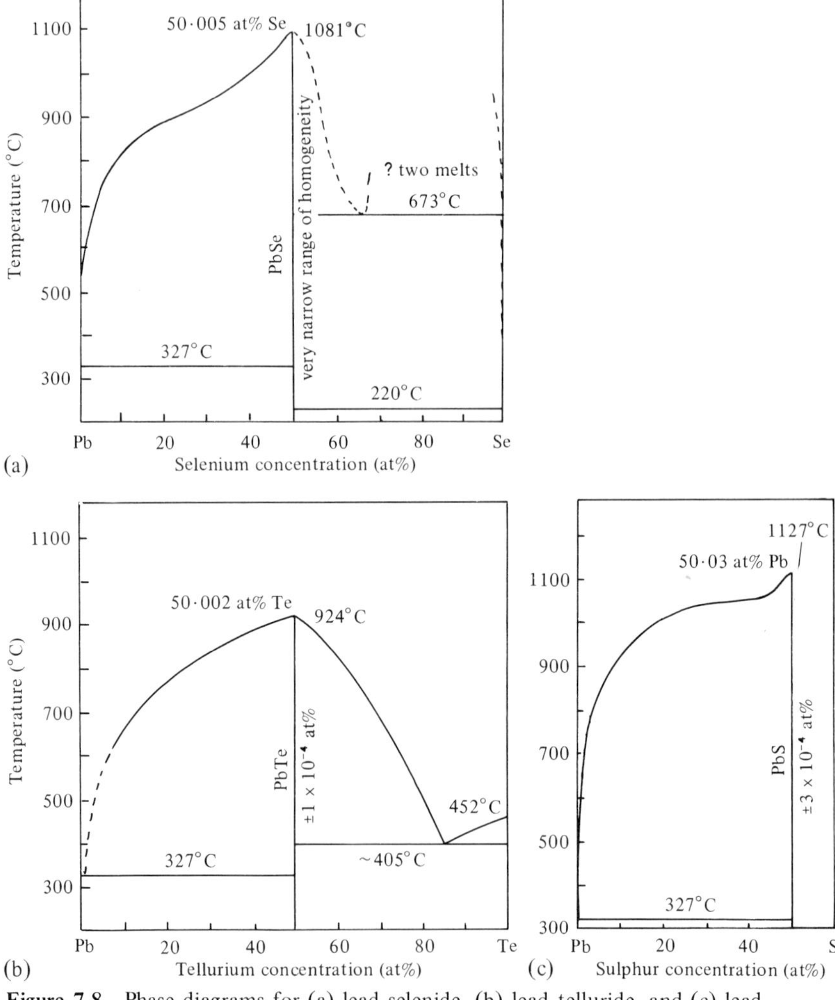


**Figure 7.8.** Phase diagrams for (a) lead selenide, (b) lead telluride, and (c) lead sulphide. (After Hansen and Anderko, 1958.)

hence heating of the mixture is carried out carefully to prevent excessive vaporisation of the anion.

Single crystals of all the compounds can be grown by the Bridgman–Stockbarger method. This was first carried out by Lawson (1951, 1952) who used a two-zone furnace in which the hotter zone was held at 50 K above the melting point and the lower zone at 50 K below. A downwards movement of a few millimetres per hour was used. The first solid to freeze has a composition corresponding to the invariant melting point of the compound. This produces n-type material in the case of lead sulphide and p-type for the other two. However, as the melt becomes richer in one of the components, the sign of the carrier can change. If large deviations from stoichiometry occur in the melt, then segregation of a second phase occurs.

Both lead selenide and lead telluride have been grown by the Czochralski method, but, because of their high vapour pressures, the liquid encapsulation method using boric oxide is required. Single crystals have also been grown from the vapour, particularly lead selenide (see page 26). Control of selenium pressure enables n- and p-type crystals to be grown as required.

### 7.2.2 Band structure

Considerable attention has been given to the band structure of the three compounds but uncertainty remains regarding the band ordering at the L point.

Much experimental data is available for the lead salts. Bandgaps, effective masses and $g$ factors (Landé splitting factors) have been measured from magnetooptical studies, Shubnikov–de Haas experiments, Raman spin-flip scattering experiments, photoemission, cyclotron resonance, and tunnelling. Calculation of the band structures has also included a wide range of methods. Lin and Kleinman (1966) developed a pseudopotential scheme for obtaining the energy bands in the lead salts, because of the difficulty in obtaining crystal potentials. The pseudopotential scheme used included relativistic effects and spin–orbit interactions. It required five parameters which were chosen to fit the ordering around the energy gap at the $(\frac{1}{2}, \frac{1}{2}, \frac{1}{2})$ point in the Brillouin-zone face, and also to fit the optical energy gaps measured experimentally and considered to be at the centre of the zone. The energy bands so obtained fit well with the experimental optical and photoelectric data and also with the pressure dependence of the forbidden bandgap. Also, the schemes give good agreement with experimental values for effective masses and $g$ values. But, in particular, Lin and Kleinman show a different band-ordering at the L point from that obtained by previous workers. Even with this ordering, the choice of parameters is not unique, and Bernick and Kleinman (1970) repeated the calculation concentrating on obtaining agreement with energy gaps rather than optical gaps. Good agreement for the effective masses and $g$ values was again obtained. Tung and Cohen (1969, 1970) also used empirical pseudopotential techniques and figure 7.9 shows the band structure which they obtained. Similar techniques have been used for lead selenide and lead sulphide (Kohn *et al.*, 1973) but as the general features are similar they are not shown here. Tung and Cohen's energy diagram shows the 'standard' band ordering of $L_6^-$, $L_6^-$, $L_{45}^-$. Tsang and Cohen (1971b) attempted to confirm this conduction band ordering in lead telluride to enable them to decide whether the band ordering agrees with Lin and Kleinman, $L_6^-$ ($L_3'$), $L_{45}^-$ ($L_3'$), $L_6^-$ ($L_2'$), or with the earlier results using APW, OPW, and EPM methods, $L_6^-$ ($L_2'$), $L_6^-$ ($L_3'$), $L_{45}^-$ ($L_3'$). The calculations proved inconclusive, although the Lin–Kleinman ordering was considered less attractive. Martinez *et al.* (1975) have used an improved pseudopotential technique and claim that their band-structure model

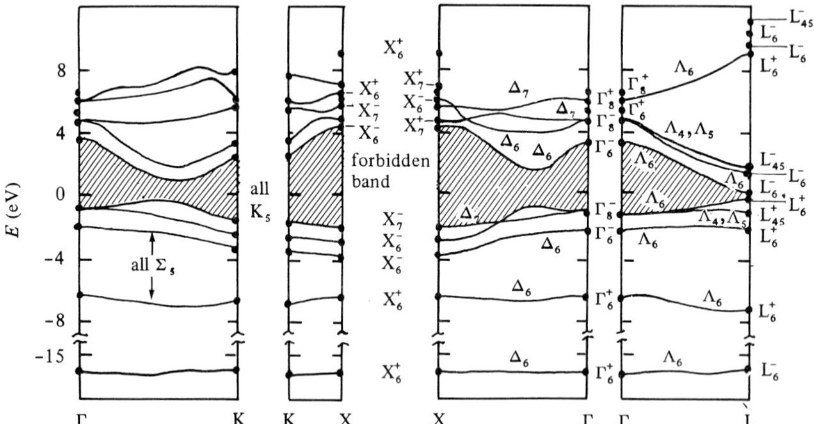


**Figure 7.9.** Band structure of lead telluride calculated by the empirical pseudopotential method (Tung and Cohen, 1969, 1970).

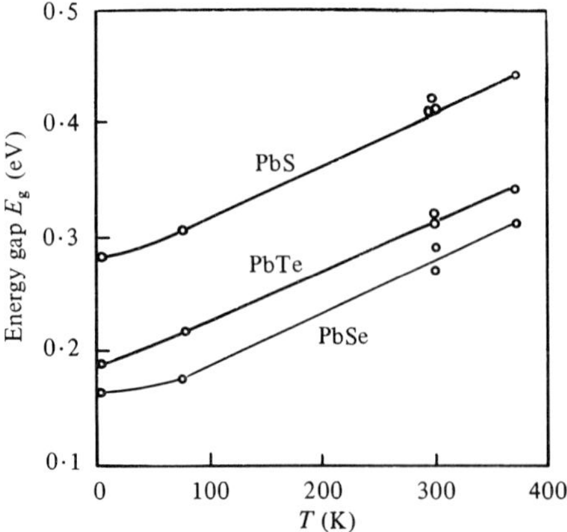


**Figure 7.10.** Energy gap $E_g$ as a function of temperature for lead sulphide, lead selenide, and lead telluride (Dalven, 1973).

explains for the first time experimental results for effective masses, optical properties, Knight shift and photoemission. Their band ordering is the 'standard' $L_6^-$, $L_6^-$, $L_{45}^-$, although their band structure shows some variation with respect to that in figure 7.9. In figure 7.10 is illustrated the variation of energy gap with temperature for the three compounds. The variation of the bandgap is linear with temperature above approximately 75 K but is nonlinear at lower temperatures (Dalven, 1973). The temperature coefficient $(\partial E/\partial T)_p \approx 4{\cdot}7 \times 10^{-4}$ eV $K^{-1}$ for all three salts.

For each of the compounds and for both electrons and holes, the energy surfaces at the L points are approximately prolate spheroids with the major axes in the $\langle 111 \rangle$ directions. Therefore, within the first Brillouin zone there are eight half-spheroids. However, the spheroidal approximation is only good for low carrier concentrations. At higher concentrations, with $k$ values well off those at the band edges, the cross-section takes on a cylindrical shape and the $E-k$ dependence is no longer parabolic.

**7.2.3 Properties**

The electrical properties have been studied in detail. Putley (1955a) measured the Hall coefficient, electrical conductivity, and magnetoresistance of all three materials and the thermoelectric power of lead telluride and lead selenide (1955b) for a wide range of samples. The information from these and other sources is reviewed by Scanlon (1959). Because of the quality of material then available, bandgap and mobility values have been revised in the light of later measurements, but early measurements showed the compounds to possess remarkably high values of mobility in view of the large effective masses and high carrier concentrations which were usually present. Lead telluride will be considered in greatest detail with some comment on lead selenide and lead sulphide afterwards.

Detailed discussion of the scattering mechanisms involved in all the lead chalcogenides is contained in a review article by Ravich *et al.* (1971). Earlier assumptions of acoustical phonon scattering together with impurity scattering at low temperatures were unable to explain the temperature dependence of various coefficients such as Hall and Nernst–Ettingshausen coefficients and of the mobility. Hence it is necessary to take into account collisions between carriers and polar scattering by optical phonons. Also, the presence of a nonparabolic energy band leads to an energy dependence of the matrix element of carrier interaction with acoustical phonons.

7.2.3.1 *Lead telluride*

Interpretation of measurements has been very dependent on the band structure model used and on the assumed scattering mechanisms. Crocker and Rogers (1967) carried out measurements of Hall coefficient, electrical resistivity and Seebeck coefficient for undoped and sodium-doped p-type single crystals and from their measurements determined parameters for a simple two-valence-band model assuming scattering by lattice acoustical phonons. A simple parabolic band model did not account for a low thermoelectric power in high-carrier-density material nor did it give a realistic value of density-of-states mass for the principal valence band. The results were reinterpreted by Rogers (1967) with the aid of a nonparabolic principal valence band and, again, scattering by acoustical phonons. Figure 7.11 shows the Hall mobility as a function of temperature and figure 7.12 (which will be referred to again later) the Seebeck coefficient as a function of carrier concentration. The conductivity mobility has a complicated energy dependence so that it was not found possible to give a quantitative explanation of the temperature dependence of the Hall mobility. However, both n- and p-type materials show an approximate $T^{-5/2}$ dependence of the Hall mobility. An effective density-of-states mass of $0{\cdot}45m_0$ was obtained for the high carrier concentration p-type samples and this value was found to remain constant below room temperature.

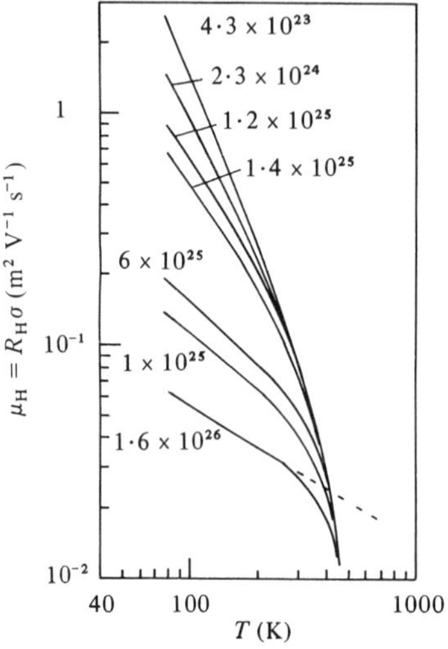


**Figure 7.11.** Hall mobility $\mu_H$ as a function of temperature for lead telluride with different carrier concentrations ($m^{-3}$) (Crocker and Rogers, 1967).

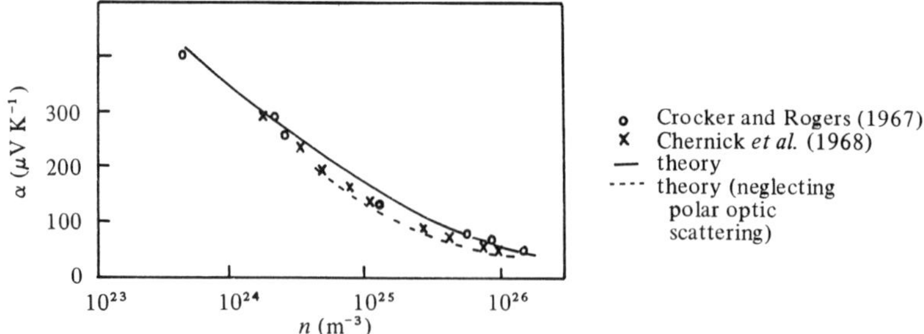


**Figure 7.12.** Seebeck coefficient as a function of carrier concentration. (After Harris and Ridley, 1972a.)

The room temperature Hall mobility remains constant as a function of carrier concentration up to a carrier concentration of $10^{25}$ (Allgaier and Houston, 1962). Other points noted by Rogers were: at low carrier concentrations, an electron to light-mass hole mobility ratio of 2·0 at room temperature, a light-mass to heavy-mass hole mobility ratio at 430 K of 6 ± 1 : 1 and at room temperature a density-of-states mass ratio for heavy to light holes of 6 ± 1 : 1.

Harris and Ridley (1972a) re-examined the room temperature transport properties of p-type lead telluride incorporating various scattering mechanisms as discussed by Ravich *et al.* (1971). At low carrier densities, acoustical phonon and optical phonon scattering were of equal importance whereas at high carrier densities acoustical phonon and carrier-carrier scattering dominated. The contribution from polar optical phonons is reduced by screening. In the lead chalcogenides generally, the number of free carriers present gives rise to considerable screening.

Although the bandgap between conduction and light-hole valence band is only 0·19 eV at 0 K, it increases considerably at higher temperatures, whereas the heavy-hole valence band remains approximately 0·36 eV below the conduction band over the complete temperature range as shown in figure 7.13 (Tauber *et al.*, 1966). Hence the energy separation $\Delta E$ between valence bands should be approximately 0·05 eV at room temperature. Optical measurements, however, have yielded a value of 0·08 eV (Riedl, 1962). The bandgap at room temperature is 0·31 eV and hence sufficiently narrow for the bands to be nonparabolic. The Kane model rather than the Cohen model is thought to be applicable and was used by Harris and Ridley in their analysis. Parameters used by them are shown in table 7.3. The values of density-of-states effective masses of $0{\cdot}20m_0$ and $1{\cdot}0m_0$ for light and heavy holes are not in agreement with earlier work which had been based on a parabolic band and different scattering mechanisms. Because subsidiary conduction minima lie well

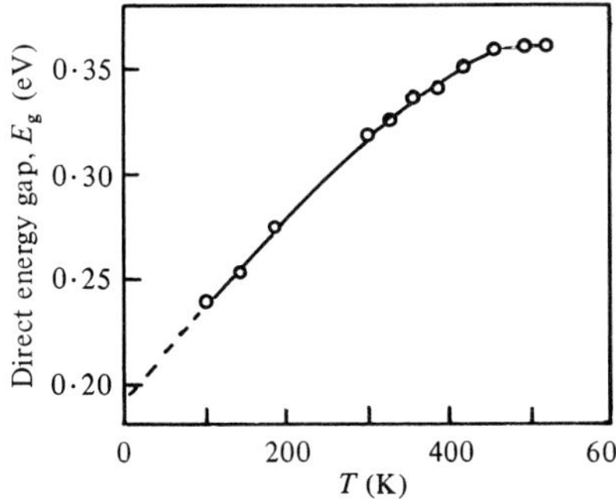


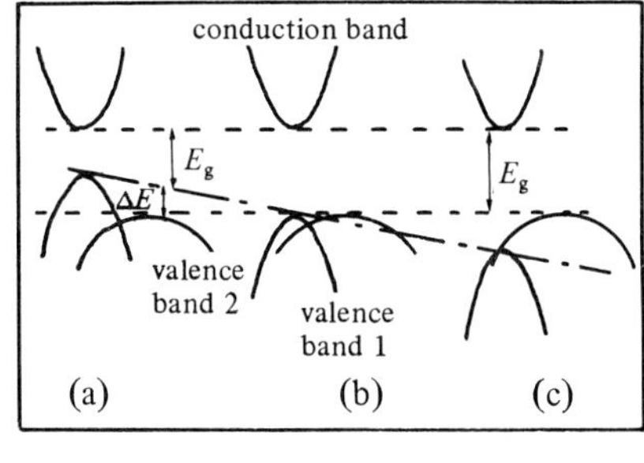


**Figure 7.13.** Energy gap of lead telluride as a function of temperature (Tauber *et al.*, 1966) and band structure proposed by Harris and Ridley (1972a): (a) 0 K, $E_g$ = 0·19 eV, $\Delta E$ = 0·17 eV; (b) ~450 K, $E$ = 0; (c) >450 K, $E_g$ = 0·36 eV.

above the main conduction band minimum (as shown from results on n-type material), the heavy-hole valence band can be assumed parabolic. Room-temperature Hall mobility for these p-type samples was found to be 0·085 $m^2 V^{-1} s^{-1}$ for carrier densities below $10^{25}$ $m^{-3}$ but decreased to 0·04 $m^2 V^{-1} s^{-1}$ for densities of $10^{26}$ $m^{-3}$. The mobility values are weighted averages for the two valence holes, the population in the heavy-hole valence band becoming more dominant at the higher carrier concentrations. There is a decrease in mobility of both carriers with increasing concentration as the carrier distribution becomes more degenerate and also as hole–hole scattering becomes more important.

Using the assumptions regarding scattering and band-structure outlined above, Harris and Ridley obtained good agreement between calculated and measured transport coefficients such as Hall coefficient, Seebeck coefficient (see figure 7.12) and transverse Nernst–Ettingshausen coefficient. Harris and Ridley (1972b) have also investigated high-field transport in p-type lead telluride.

7.2.3.2 *Lead selenide and lead sulphide*

Lead selenide and lead sulphide show very similar behaviour to lead telluride. The intrinsic carrier concentration $n_i$ is $3 \times 10^{22}$ $m^{-3}$ for lead selenide and $2 \times 10^{21}$ $m^{-3}$ for lead sulphide at room temperature compared with $1 \cdot 5 \times 10^{22}$ $m^{-3}$ for lead telluride (Putley, 1965). The temperature dependence of resistivity is similar in all cases and the Hall coefficients are constant over the temperature range 300 to 4·2 K. Electrons and holes have similar mobilities in all three materials at high temperatures but the mobilities are slightly lower in lead selenide and lead sulphide at low temperatures than in lead telluride (Allgaier and Scanlon, 1958).

**Table 7.3.** Room temperature parameters of p-lead telluride.

| Parameter | Value | Reference |
|---|---|---|
| Density-of-states effective mass of valence band 1 | $0 \cdot 20 m_0$ | Harris and Ridley (1972a) |
| Effective interaction gap of valence band 1 | 0·27 eV | Zhitinskaya *et al.* (1966) |
| Density-of-states effective mass of valence band 2 | $1 \cdot 0 m_0$ | Harris and Ridley (1972a) |
| Energy separation between valence bands | 0·08 eV | Riedl (1962) |
| Mass anisotropy of valence band 1 | 13 | Burke *et al.* (1970) |
| Mass anisotropy of valence band 2 | 1 | Harris and Ridley (1972a) (assumption) |
| Mobility ratio $\mu_2/\mu_1$ | $1 \cdot 113 - 0 \cdot 05 \lg p$ | Crocker and Rogers (1967) |

Table 7.4 compares the band-edge parameters for the three compounds at 0 K, the surfaces of the bands for lead selenide and lead sulphide being more spherical than in the case of lead telluride (Dalven, 1969). The mass anisotropy $K$ gives a measure of the anisotropy of the energy surfaces. For ellipsoidal constant-energy surfaces, the conductivity effective mass $m_c$ is given by

$$m_c = \frac{3m_\ell m_t}{2m_\ell + m_t} .$$

**Table 7.4.** The band edge parameters of lead sulphide, lead selenide, and lead telluride at 4 K (Dalven, 1969).

| Compound | Band | $\frac{m_t}{m_0}$ | $\frac{m_\ell}{m_0}$ | $K = \frac{m_\ell}{m_t}$ | $\frac{m_c}{m_0}$ |
|---|---|---|---|---|---|
| PbS | conduction | 0·080 | 0·105 | 1·3 | 0·087 |
| | valence | 0·075 | 0·105 | 1·4 | 0·083 |
| PbSe | conduction | 0·040 | 0·07 | 1·75 | 0·047 |
| | valence | 0·034 | 0·068 | 2·0 | 0·041 |
| PbTe | conduction | 0·024 | 0·24 | 10 | 0·034 |
| | valence | 0·022 | 0·31 | 14 | 0·032 |

### 7.2.3.3 *PbTe–PbSe alloys*

Use of dilute alloys can reduce the thermal conductivity without significantly affecting the electrical properties, which can lead to improved performance as in thermoelectric applications. This effect has been investigated in p-type PbTe–PbSe alloys by Kudman (1972). A series of samples was prepared by the Bridgman technique for the composition range 0 to 25 mol% PbSe and containing 0·5 to 1·0 mol% Na as the p-type dopant. Alloying increased the electrical resistivity, as well as the thermal resistivity but an increased figure of merit, $z = \alpha^2/(\rho K)$ (see p.93)

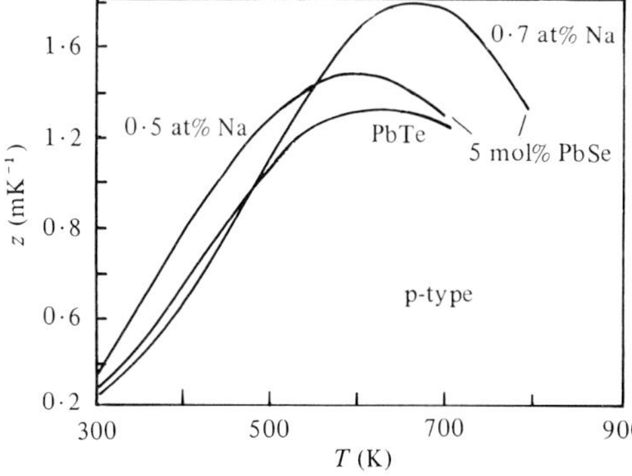


**Figure 7.14.** Figure of merit for PbTe–PbSe alloys as a function of temperature (Kudman, 1972).

over that for lead telluride was nevertheless obtained at higher temperatures (see figure 7.14). The high-temperature figures of merit for alloys in the composition range 5 to 15 mol% PbSe were found to be very similar because increase in electrical resistivity balanced the decrease in thermal conductivity. The improvement in the figure of merit occurs for alloys containing at least 0·7 mol% Na and hence arises not from the lowering of the thermal conductivity but from higher values of $\alpha^2/\rho$ at high temperatures. In the alloys it is possible to have a higher solubility of p-type dopant and hence a higher value of $\alpha^2/\rho$ before intrinsic conduction sets in.

## 7.3 Lead–tin chalcogenides: $Pb_{1-x}Sn_xTe$ and $Pb_{1-x}Sn_xSe$

### 7.3.1 Introduction

As these two systems of solid solutions are very similar in their properties, they will be considered together. Lead telluride and lead selenide have been discussed already as narrow-bandgap semiconductors. Tin telluride will be discussed later. It has a bandgap of approximately 0·3 eV at 4 K but this decreases to 0·18 eV at room temperature. Tin selenide with a bandgap of 0·88 eV lies outside the range of this book. In each of the mixed systems there is a cross-over between the valence and conduction bands as the composition changes, and hence it is possible to have a gradation of bandgap widths with change of either composition or temperature.

In the phase diagrams for each of the binaries the compounds exist in a very narrow phase at 50 at%; although it is finite, this width is particularly important when electrical properties of the compound are considered as it leads to the existence of n- and p-type material. As grown, the lead–tin chalcogenides tend to have high carrier concentrations ($\sim 10^{25}$ $m^{-3}$), but annealing has reduced concentrations to as little as $10^{21}$ $m^{-3}$.

### 7.3.2 Crystal growth

Vapour growth techniques and also the Bridgman–Stockbarger method have been used for growing single crystals of these chalcogenides.

Butler and Harman (1968) used a specially designed fused silica ampoule for carrying out growth from the vapour followed up by an annealing process. The ampoule was tapered with three indentations inserted a few centimetres up from the tip. The source material, a course powder obtained from a quenched ingot of $(\text{metal})_{0\cdot 51}(\text{chalcogenide})_{0\cdot 49}$ or $(\text{metal})_{0\cdot 49}(\text{chalcogenide})_{0\cdot 51}$ prepared from the elements was placed in the inner tube within the ampoule resting on the indentations (figure 7.15) and the ampoule was evacuated to $10^{-7}$ Torr. Growth was allowed to take place for $\sim$20 h, resulting in crystal faces of $mm^2$ size and the tube was then removed to allow it to cool in the air. The bulk carrier concentrations for these crystals corresponded to points on the metal-saturated solidus line as indicated in figure 7.16. (Typical curves for carrier concentrations as a function of annealing temperature will be shown in figure 7.19.) Concentration of the crystals varied up to $x = 0 \cdot 27$ for the telluride and $x = 0 \cdot 13$ for the selenide. Butler and Harman were interested in producing n–p type layers and these they achieved for $Pb_{1-x}Sn_xTe$ by isothermally annealing the crystals within the tapered ampoule in a horizontal furnace. Annealing thus took place in the presence of the remaining charge. For the achieved values of $x$, annealing at a lower temperature resulted in surface layers changing from p-type to n-type. Conversion temperatures for $x$ of 0·17, 0·20, 0·27 were 580, 525, and 425°C, respectively. However, for $Pb_{1-x}Sn_xSe$, the capsule had to be opened and selenium-saturated powder substituted for the metal-saturated charge in order to obtain n–p layers by annealing.

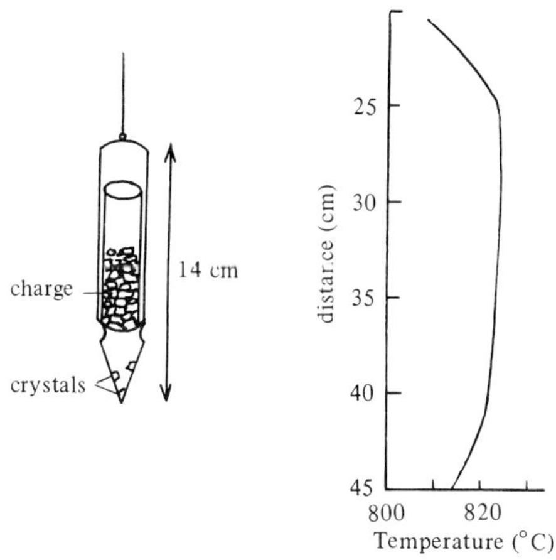


**Figure 7.15.** Schematic diagram of the ampoule used for vapour growth and diffusion annealing of $Pb_{1-x}Sn_xTe$ and the furnace temperature profile (Butler and Harman, 1968).

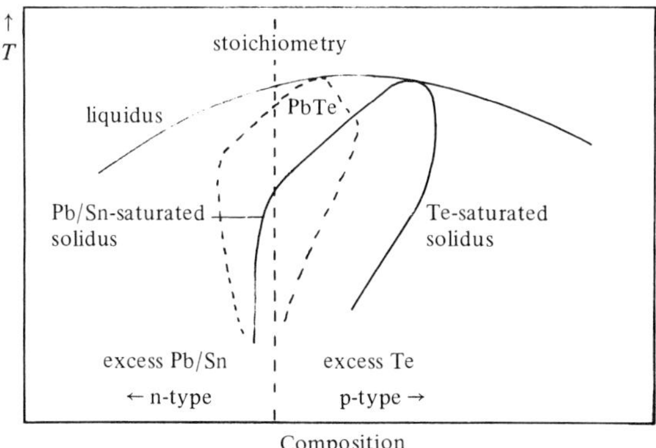


**Figure 7.16.** The equilibrium phase diagram of $Pb_{1-x}Sn_xTe$ near stoichiometry compared with that for PbTe.

Calawa *et al.* (1968) used the Bridgman method to prepare homogeneous samples with a constant Pb : Sn ratio for both the telluride and the selenide systems. The value of $x$ down the sample showed some variation. They found that the ratio of the tin content in the part of the ingot frozen first to that in the starting liquid alloy was 0·5 in the case of the selenide system and between 0·65 and 0·9 for the telluride system. Starting material consisted of stoichiometric proportions of the elements inserted in a carbon-coated and evacuated silica ampoule which was narrowed and tapered in order to encourage single crystal growth. For the actual crystal growth the furnace and temperature profile were as shown in figure 2.17b when the Bridgman method was discussed. For the selenide compounds the ampoule was enclosed in a protecting outer tube because of cracking of the ampoule on cooling the samples. Before growth was carried out the furnace was held horizontal and the charge heated to approximately 50 K above the liquidus temperature for the alloy in question (see figures 7.17a and 7.17b for the relevant phase diagrams) and then the furnace rotated to the vertical position to start the directional freeze. The samples were p-type with high carrier densities, probably due to micro-precipitates of electrically neutral selenium or tellurium. This agreed with observation of etched crystals which showed that inclusions existed in crystals grown from metal-rich melts. The presence of metal inclusions and also low-angle grain boundaries is considered to be due to constitutional cooling (see p. 19).

For samples grown by the Bridgman technique it has been shown to be necessary to reduce carrier concentrations by annealing. For the tellurium system, isothermal annealing has proved suitable. Crystalline wafers of $Pb_{1-x}Sn_xTe$ were placed either on a shelf or in an inner tube within a

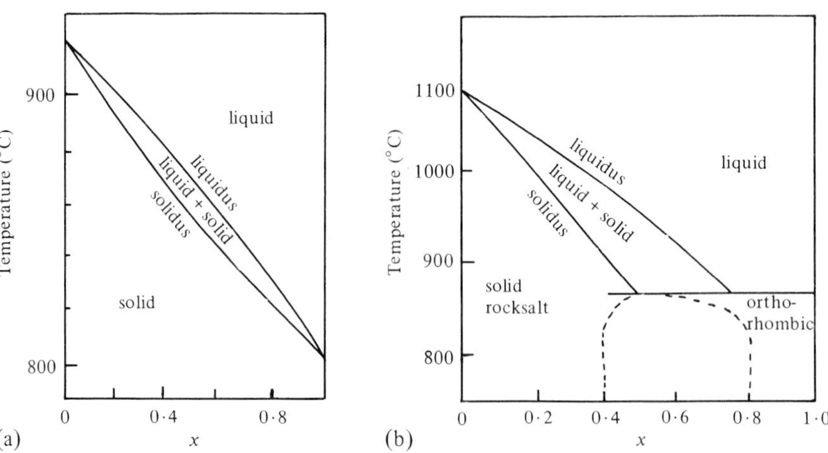


**Figure 7.17.** $(T, x)$ phase diagrams for (a) $Pb_{1-x}Sn_xTe$ and (b) $Pb_{1-x}Sn_xSe$ (Wagner and Willardson, 1968, and Strauss, 1968, respectively).

larger ampoule (figure 7.18a). For runs where the annealing temperature was below 600°C, the ampoule was filled with argon at 100–300 Torr after evacuation. This was to reduce thermal etching of the samples. After etching (for a range of total times) the crystals were quenched in water. In the case of the selenium system, p-type and n-type material could be prepared; it was important to quench the material rapidly in a brine solution at 0°C to prevent precipitation. However, rather lower carrier concentrations could be obtained for the selenium system by annealing in a two-zone furnace (figure 7.18b). Carrier concentrations down to $5 \times 10^{23}$ $m^{-3}$ for p-type and $1{\cdot}5 \times 10^{24}$ $m^{-3}$ for n-type material were achieved by this method. The variation of carrier concentration as a function of the temperature at which annealing took place is summarised for the two systems in figure 7.19.

Large crystals of $Pb_{1-x}Sn_xTe$ (with $0 \leqslant x \leqslant 0{\cdot}26$) were grown by Parker *et al.* (1974) in sealed tubes by self-transport on to single crystal seeds. Temperature differences greater than 5 K between source and seed tended to produce polycrystalline growth, whereas temperature differences less than 2 K resulted in very slow growth rates. Seed temperatures between 800 and 850°C were used. Crystals of 9 and 26 mm diameter weighing up to 125 g were grown. The crystals were found to be free of voids and inclusions and with dislocation densities as low as $10^3$ $cm^{-2}$. Similarly, crystals of $Pb_{1-x}Sn_xSe$ with $0 \leqslant x \leqslant 0{\cdot}07$ have been obtained by Kasai *et al.* (1974). Typical growth temperatures were around 800°C although good crystals were not obtained above 830°C and sublimation was allowed down a temperature gradient of approximately 1 K $cm^{-1}$, the Piper and Polich technique having been found to be unsuitable. Dislocation densities were as low as $5 \times 10^4$ $cm^{-2}$ and crystals of $cm^3$ size were obtained.

Laugier *et al.* (1974) have calculated ternary phase diagrams for $Pb_{1-x}Sn_x(Te, Se)$ by computing the conditions for equilibrium between the pseudobinary solid solutions Pb(Te, Se)–Sn(Te, Se) and the ternary liquid solutions $Pb_{1-x}Sn_x(Te, Se)$. A similar computation has been carried out by Harris *et al.* (1975), but there are major discrepancies between their

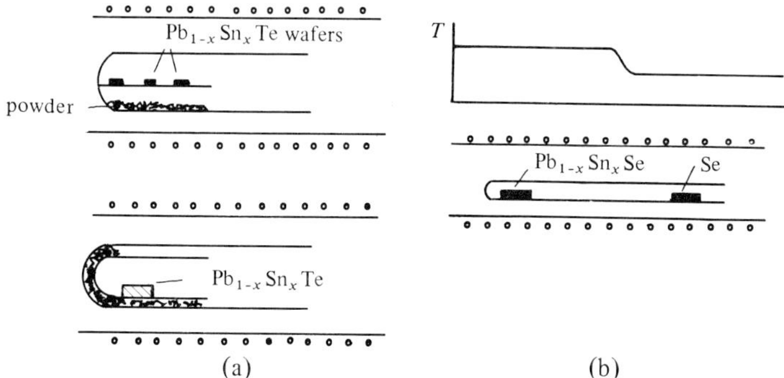


**Figure 7.18.** (a) Isothermal annealing and (b) two-zone annealing.

results and those of Laugier *et al.* at low temperatures. Figure 7.20a gives the diagram obtained by Laugier *et al.* for $Pb_{1-x}Sn_xSe$; figure 7.20b shows the diagram for $Pb_{1-x}Sn_xTe$ with the low temperature corrections of Harris *et al.* Solid lines are the computed liquidus isotherms. Also given are the calculated iso-(solid composition) lines which are the loci of liquid compositions in equilibrium with a given composition measured in mole fractions of Sn(Te, Se) (i.e. $x$). The hatched areas represent monotectic regions which were not considered. The analysis used assumed

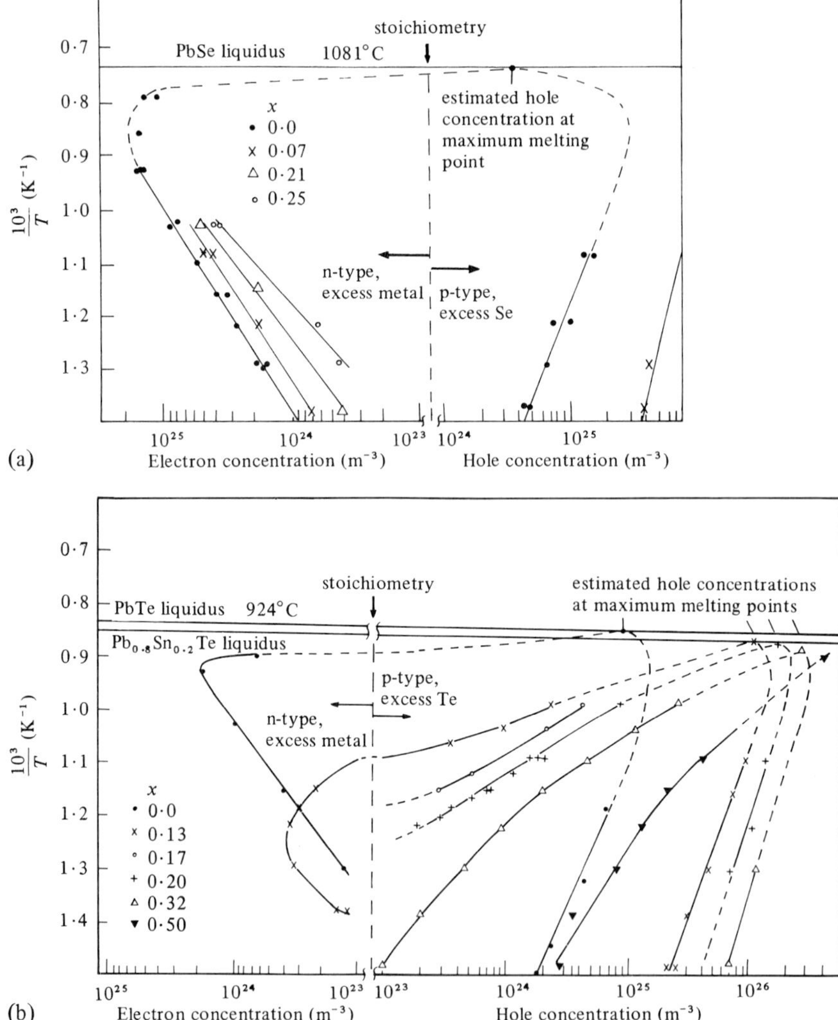


**Figure 7.19.** Carrier concentration at 77 K as a function of isothermal annealing temperature for (a) $Pb_{1-x}Sn_xSe$ and (b) $Pb_{1-x}Sn_xTe$. (After Harman, 1974.)

a rocksalt structure for tin selenide (instead of the orthorhombic structure which is a distortion of the rocksalt structure), experimentally obtained thermodynamic values such as entropy, and pseudo-binary data for Pb(Te, Se)–Sn(Te, Se).

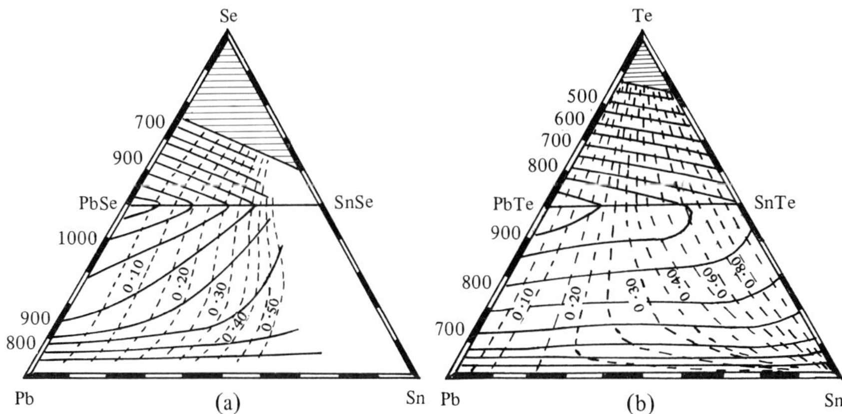


**Figure 7.20.** Calculated ternary diagrams of (a) $Pb_{1-x}Sn_xSe$ and (b) $Pb_{1-x}Sn_xTe$. Solid lines are isotherms, dashed lines iso-(solid) compositions lines labelled in mole fraction of SnSe or SnTe. The hatched area (monotectic region) was not taken into account (Laugier *et al.*, 1974; Harris *et al.*, 1975).

### 7.3.3 Band structure

The four constituent binary compounds all have the NaCl structure except that in the case of SnSe the structure is distorted into an orthorhombic lattice. Alloying gives rise to a linear change of the lattice parameter with composition. Both chalcogenide systems show inversion of the conduction and valence bands as the composition changes, and this gives rise to a narrower bandgap for the alloys than exists for the pure compounds.

Both lead telluride and lead selenide have $L_6^-$ conduction bands and $L_6^+$ valence bands, whereas for tin telluride and tin selenide the bands are $L_6^+$ and $L_6^-$ respectively. As alloying occurs, these bands close up, cross over (figure 7.21) and separate out again. It had been thought that for lead telluride and tin telluride the minimum bandgap is at the L point. Whereas this is true for lead telluride, Tsang and Cohen (1971a) have shown that this is not so for tin telluride. Although the valence band energy decreases along the $\Omega$ direction from L (see figure 3.2 for nomenclature), the energy increases perpendicular to the $\Omega$ axis, and in fact the valence and conduction band extrema lie in the hexagonal face of the Brillouin zone displaced from L by $\frac{1}{25}$th of the [111] reciprocal lattice vector. As alloying takes place, the bands do not cross at this point because of interaction between them, but cross at the L point. The band structure of the selenide system is less precisely known although crossover of the bands occurs in the same way. The directions of change of the band structure under the effects of temperature, pressure and magnetic field are shown in figure 7.21. The selenide system is exactly analogous [see Martinez (1973) for a discussion of band inversion in $Pb_{1-x}Sn_xSe$, particularly under hydrostatic pressure, or Strauss (1967)].

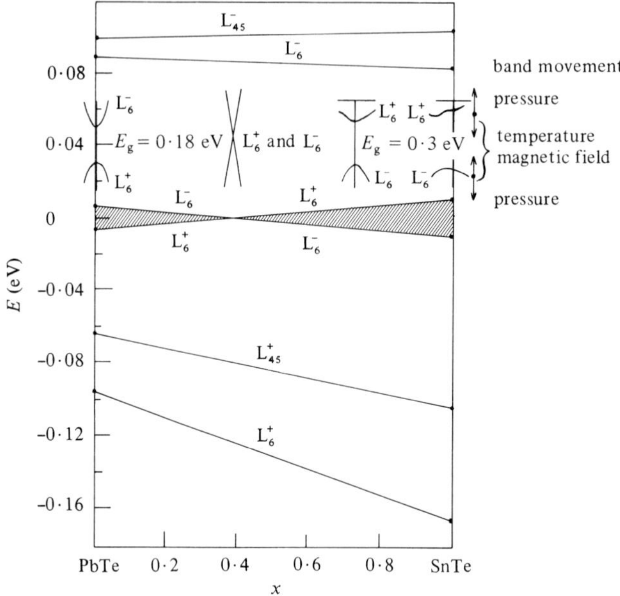


**Figure 7.21.** Approximate positions of the energy bands near the L-point for $Pb_{1-x}Sn_xTe$. (After Harman, 1971.)

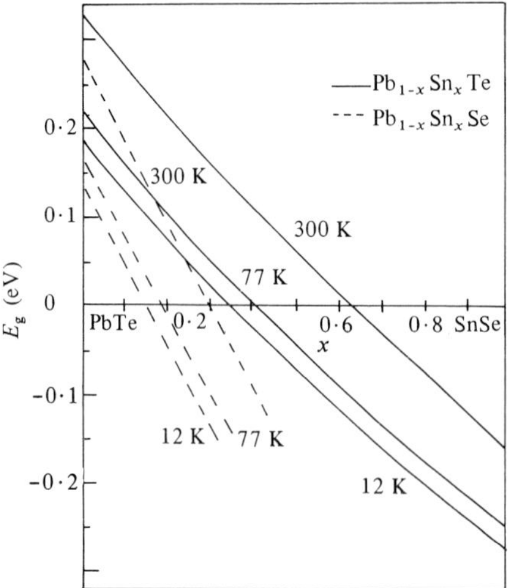


**Figure 7.22.** Energy gaps of $Pb_{1-x}Sn_xTe$ and $Pb_{1-x}Sn_xSe$ as a function of the mole fraction of SnSe, $x$.

Positive temperature dependence of the energy gap, such as is found in the cases of lead telluride and lead selenide, is very unusual for common semiconductors. Variation of the energy gaps for the two systems with composition at different temperatures is shown in figure 7.22 (Dimmock *et al.*, 1966; Strauss, 1967). In the case of lead telluride, the Fermi surface consists of ellipsoidal surfaces centred at the L point with the major axis along the [111] axis. However, for tin telluride, Cohen and Tsang have shown that, because of the band structure indicated, the pocket of holes centred about L consists of multiply-connected surfaces.

### 7.3.4 Properties

Certain properties, such as variation of carrier concentration with isothermal annealing, have already been discussed for both mixed systems. In general, the properties of the telluride ternary have been more investigated than those of the selenide.

Electrical conductivity and Hall data for $Pb_{1-x}Sn_xTe$ have been measured on a wide range of samples of differing compositions and carrier concentrations. Wagner *et al.* (1971) have investigated particularly these electrical properties for hole concentrations of $\sim 1 \times 10^{25}$ $m^{-3}$ when the transport properties are dominated by ionised impurity scattering. Hall mobilities are shown as a function of composition at 77 K in figure 7.23. Figure 7.23a shows the hole mobilities for samples grown from stoichiometric melts, whereas figure 7.23b shows the variation for samples having the same carrier concentration. Detailed explanation of the results is difficult because of the number of factors involved. However, from 2 to 5 mol% SnTe there is a sharp decrease in mobility for a given carrier concentration. Efimova and Kolomoets (1965) have shown that there is

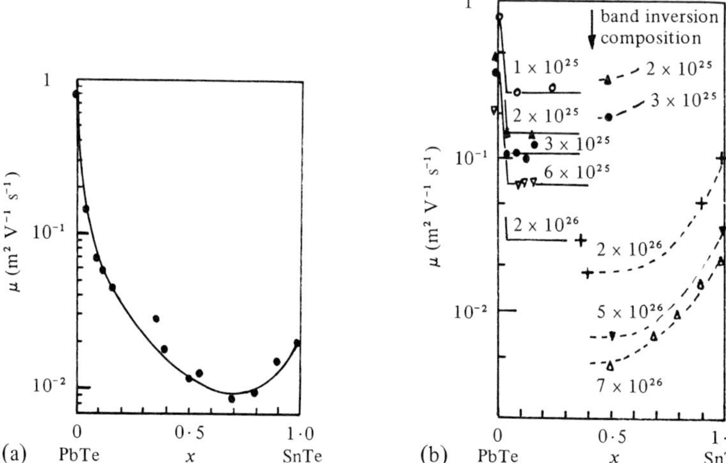


**Figure 7.23.** (a) Hole mobilities in $Pb_{1-x}Sn_xTe$ single crystals grown from stoichiometric melts versus $x$ at 77 K and (b) isocarrier-concentration hole mobilities versus $x$ at 77 K (Wagner *et al.*, 1971).

no significant change of effective mass in this region and so it is deduced that the change of mobility arises from a sharp change in the static dielectric constant (by a factor of ~2). From 5 to 37 mol% SnTe, the mobility is independent of composition so that, if there is any change of effective mass, it is balanced by a change of dielectric constant. (A decrease in effective mass can be expected because the bands are converging; therefore this implies that the dielectric constant is decreasing also.) From 37 to 40 mol% SnTe there is a discontinuity in each of the mobility curves, just before band inversion occurs, and the point on the graph in figure 7.23a at $x = 0{\cdot}37$ does not constitute an experimental error. Beyond 40 mol% SnTe, the results are explicable in terms of a gradual change in static dielectric constant and/or effective mass. On the basis of their data close to 40 mol% SnTe, Wagner *et al.* suggested band forms of the type shown in figure 7.24, so that as the bands invert the valence band takes on a form with a steeper slope to the edges on the $E$–$k$ curve. This would produce a smaller effective mass and the mobility would vary more with carrier concentration because of a more rapid change of Fermi level with composition.

Electrical properties also show considerable variation with temperature. Figures 7.25 and 7.26 show the variation of Seebeck and Hall coefficients with temperature for a number of compositions. Undoped samples of lead telluride can be either p- or n-type, as homogeneous samples can be obtained over a composition range which includes 50 at%. However, for the ternary compositions the samples are p-type even for a composition of $x = 0{\cdot}1$, the Seebeck coefficient being positive at low temperatures. This sample with $x = 0{\cdot}1$ shows a negative coefficient at 100 K as intrinsic electrons are contributing to the coefficient and have a higher mobility than the holes. With increasing value of $x$ the samples become more p-type. The results plotted in figure 7.25 are for polycrystalline material, no attempt having been made to control the stoichiometry. Although the Hall coefficient is negative for samples of $x = 0{\cdot}1$ and $x = 0{\cdot}15$ at room temperature, it becomes positive at lower temperatures, these temperatures depending on the value of $x$. Dionne and Woolley (1974) used effective-mass values as obtained from susceptibility measurements to compare the application of two-band Kane and Cohen models and a six-band model of Dimmock (1971), and found the two-band model to be more suitable.

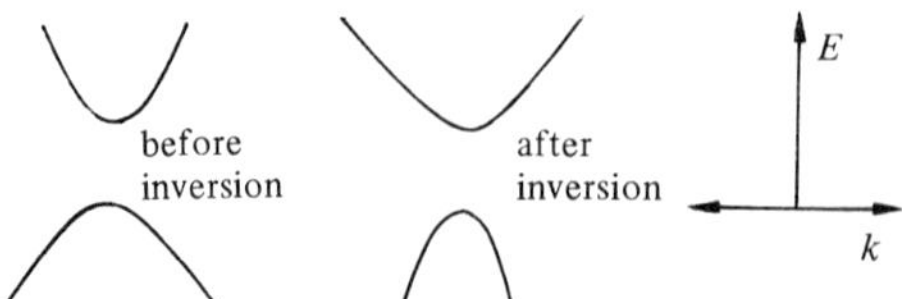


**Figure 7.24.** Hypothetical energy band representation for $Pb_{1-x}Sn_xTe$ as proposed by Wagner *et al.* (1971).

Hall coefficient data indicated that the energy surfaces in p-type $Pb_{1-x}Sn_xTe$ are highly distorted.

Just as breaks in mobility (and hence resistivity) were noted at the inversion temperature for $Pb_{1-x}Sn_xTe$, so also are these observed for $Pb_{1-x}Sn_xSe$. In particular, Hoff and Dixon (1972) investigated the composition range $0{\cdot}18 \leqslant x \leqslant 0{\cdot}29$ on polycrystalline samples composed of crystallites of 1–5 mm size. As-grown samples were annealed in the presence of selenium-rich alloys of the same Pb : Sn ratio in order to control stoichiometry and, hence, defect concentration. Figure 7.27 shows the Hall coefficient and resistivity data for a sample of composition $Pb_{0\cdot75}Sn_{0\cdot25}Se$ over a temperature range either side of the inversion temperature. No significant change is noticeable at the inversion temperature, but the resistivity values were measured with considerably greater accuracy than shown in the figures; a function $\Delta\rho$ could therefore be plotted such that

$$\Delta\rho = \rho_{exp} - \rho_{linear}$$

where $\rho_{exp}$ represents the experimental values and $\rho_{linear}$ is a linear function approximating to the line in figure 7.27b,

$$\rho_{linear} = 1{\cdot}456 \times 10^{-2}T + 1{\cdot}514\ \mu\Omega\ m\ .$$

Graph 7.27c now shows the linear dependence of $\Delta\rho$ on $T$ from 25 to 180 K. The deviation in the low temperature region is due to residual

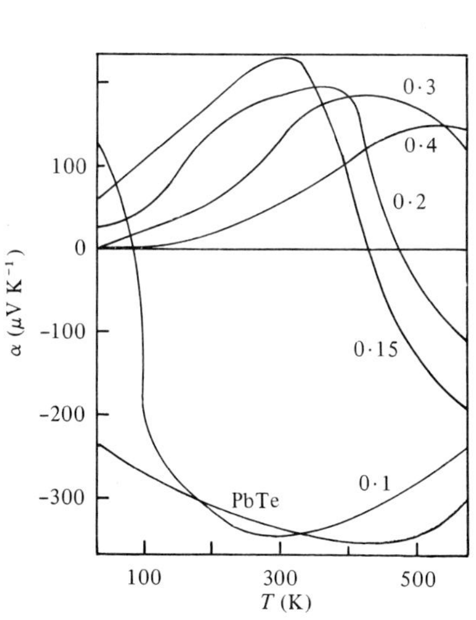


Figure 7.25. Temperature dependence of the Seebeck coefficient for $Pb_{1-x}Sn_xTe$ alloys (Machonis and Cardoff, 1964). Values of $x$ are given on the curves.

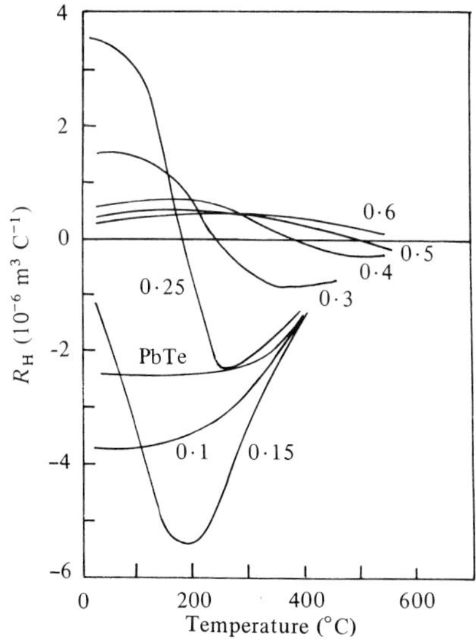


Figure 7.26. Temperature dependence of the Hall coefficient for $Pb_{1-x}Sn_xTe$ alloys (Borde *et al.*, 1966). Values of $x$ are given on the curves.

resistivity whereas that at the higher temperature $T_B$ is close to the transition temperature. Results of measurements on a further sample of similar composition and different carrier concentration shows that the temperature $T_B$ is independent of carrier concentration. However, as can be expected, $T_B$ is a function of composition as is shown in figure 7.28, the results being in close agreement with the dependence (shown by a solid line) expected from the band inversion model.

Shubnikov–de Haas measurements have been made on both types of alloy. Melngailis *et al.* (1971) and Melngailis *et al.* (1972) obtained the following data:

$Pb_{1-x}Sn_xTe$
Composition range, $x$: 0·16 to 0·32,
Carrier concentration, $n$: $3 \times 10^{22}$ to $1{\cdot}8 \times 10^{24}$ m$^{-3}$, mainly p-type,
Fermi surface: prolate ellipsoids orientated along the ⟨111⟩ axes at the L points.
Ellipsoid anisotropy, $K$: values between 10·5 and 11·2, independent of $n$ and $x$.
Transverse effective mass, $m^*_{111}$: e.g. $0{\cdot}0225m_0$ for $x = 0{\cdot}216$ and $n = 8{\cdot}5 \times 10^{23}$, but ranges from $0{\cdot}012m_0$ to $0{\cdot}028m_0$ for various $x$ and $n$.

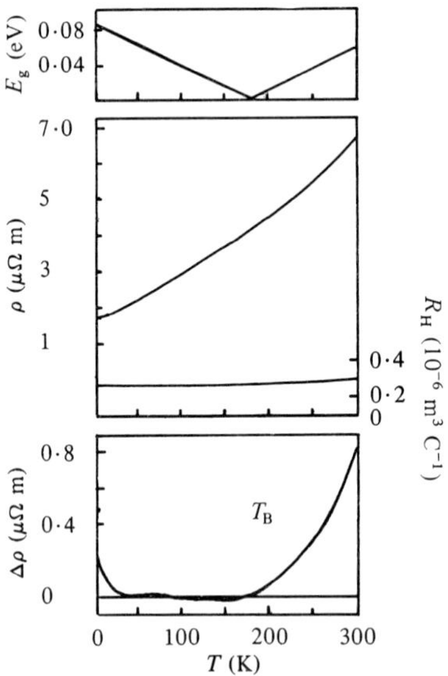


Figure 7.27. Temperature dependence in $Pb_{0{\cdot}75}Sn_{0{\cdot}25}Se$ of (a) energy gap, $E_g$, (b) resistivity, $\rho$, in a sample of free hole concentration of $4{\cdot}0 \times 10^{25}$ m$^{-3}$, and (c) deviation from linear dependence, $\Delta\rho$ (Hoff and Dixon, 1972).

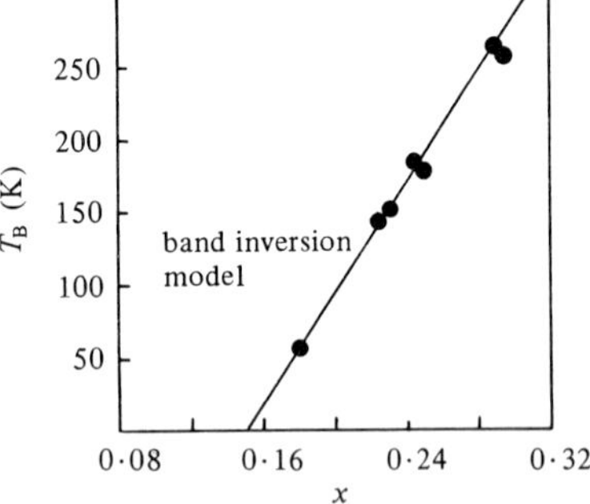


Figure 7.28. Temperature of the resistivity break $T_B$ in $Pb_{1-x}Sn_xSe$ as a function of $x$. The solid line represents the temperature $T_0$ of band crossing as predicted on the basis of the band-inversion model (Hoff and Dixon, 1972).

$Pb_{1-x}Sn_xSe$
Composition values, $x$: 0·08, 0·17, 0·20.
Carrier concentration, $n$: $2 \times 10^{23}$ to $3 \times 10^{24}$ $m^{-3}$.
Fermi surface: pockets of holes or electrons at the L points on the Brillouin zone faces; for $x = 0{\cdot}17$ and 0·20 (p- and n-type) the Fermi surface is nearly spherical;
$x = 0{\cdot}08$ and $n = 1{\cdot}4 \times 10^{24}$ (p-type), anisotropy with $K = 1{\cdot}71 \pm 0{\cdot}1$.
Transverse effective mass, $m^*_{111}$: e.g. $0{\cdot}038m_0$ for $x = 0{\cdot}20$ at $n = 3{\cdot}6 \times 10^{23}$, but ranges from $0{\cdot}030m_0$ to $0{\cdot}068m_0$ for various $x$ and $n$.

### 7.3.5 Applications

#### 7.3.5.1 *Photovoltaic detectors*

These ternaries have been of particular interest as materials for photovoltaic detectors (see p.83) and these detectors are in practical use. Sensitive detectors to operate particularly at 77 K and in the wavelength range 8 μm to 14 μm have been fabricated from both alloy systems by diffusion techniques. The usual procedure is to change the stoichiometry close to the surface so that a 10 μm thick n-type layer is produced on a p-type substrate. The active area of the diode is defined by a $SiO_2$ mask and diffusion can be carried out at around 400°C over a period of 3 h in an evacuated tube. So that the diodes do not show deterioration with time, appropriate contacts must be chosen. Gold contacts are suitable and a copper heat sink is located on the gold contact on the p-side. Indium-plated leads are then used on both sides. Before attaching the contacts it is sometimes advantageous to etch electrolytically the surfaces to reduce

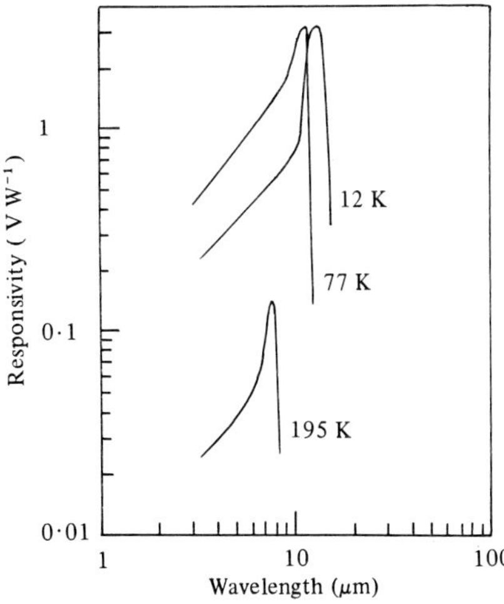


**Figure 7.29.** Responsivity spectra of a $Pb_{0·936}Sn_{0·064}Se$ diode at 12, 77, and 195 K (Melngailis and Harman, 1968).

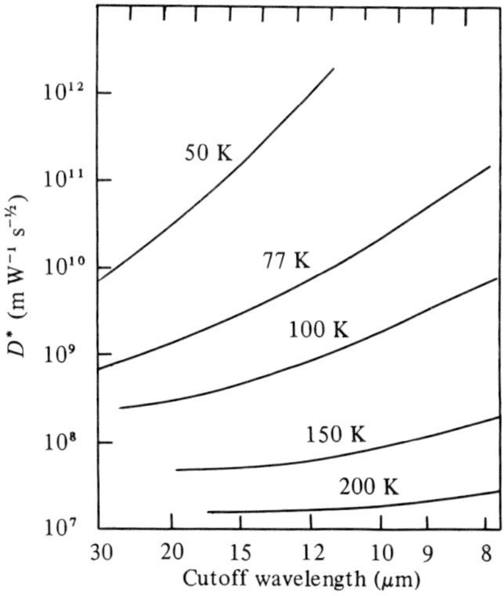


**Figure 7.30.** Maximum detectivity, $D^*$, of $Pb_{1-x}Sn_xTe$ photovoltaic detectors as a function of cutoff wavelength (energy gap) for various temperatures (Melngailis and Harman, 1968).

surface recombination of carriers. It has been shown (Melngailis and Harman, 1968) that the application of an antireflection coating can increase the responsivity by a factor of 2; for instance, coating a diode of composition $x = 0\cdot19$ can increase the peak responsivity from 40 V W$^{-1}$ to nearly 80 V W$^{-1}$, the cutoff of the response being at 12 $\mu$m at 77 K. With an increased tin content, higher wavelength cutoffs can be achieved (up to 20 $\mu$m at 77 K and 30 $\mu$m at 12 K). The $Pb_{1-x}Sn_xSe$ diodes show less responsivity because of low values of zero-bias resistance (typically 1 to 10 Ω) and this resistivity, and hence the responsivity, is lower at lower temperature (figure 7.29).

Although the cutoff wavelength can be varied by changing composition, this also affects the detectivity. Figure 7.30 shows the maximum detectivity $D^*$ which is obtainable for a particular cutoff frequency for the case of $Pb_{1-x}Sn_xTe$. The response time for the diodes varies from approximately 10 ns to 100 ns so that the photovoltaic detectors combine good sensitivity with high speed.

7.3.5.2 *Photoconductive detectors*

Samples of $Pb_{1-x}Sn_xTe$ which have been Bridgman-grown and then annealed to give carrier concentrations of $2 \times 10^{21}$ to $8 \times 10^{21}$ m$^{-3}$ have been used for photoconductive detectors (Melngailis and Harman, 1968). For samples of thickness 10 to 50 $\mu$m (obtained by etching), the detectivities range from $10^6$ to $10^7$ mW$^{-1}$ s$^{-1/2}$ at $\lambda = 15$ $\mu$m and a temperature of 77 K. Lifetimes of approximately $10^{-8}$ s have been observed so that speeds are similar to those for the photovoltaic detectors. Figure 7.31 shows the responsivity curve for typical $Pb_{1-x}Sn_xTe$ detectors.

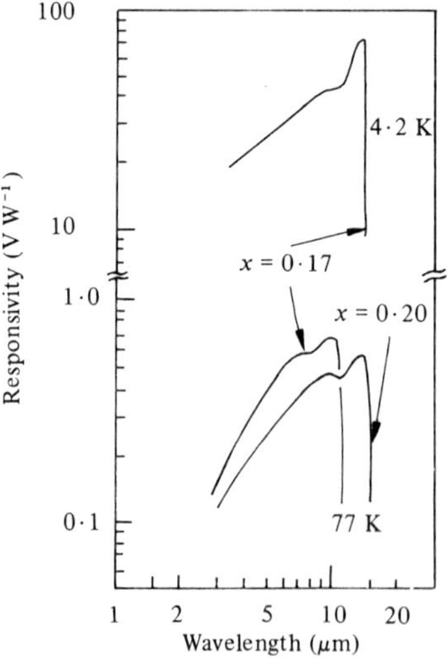


**Figure 7.31.** Responsivity spectra of $Pb_{1-x}Sn_xTe$ photoconductors. (After Melngailis and Harman, 1968.)

The composition of $x = 0{\cdot}17$ is of particular interest because at 77 K the responsivity peaks close to $10{\cdot}6\ \mu m$, the wavelength for the $CO_2$ laser. A composition of $x = 0{\cdot}20$ gives a peak of $14\ \mu m$ corresponding to the upper limit of the atmospheric window.

7.3.5.3 *Diode lasers*

$Pb_{1-x}Sn_xSe$ has been used in diode lasers to give output power as high as 80 mW (Butler and Harman, 1967). A particular composition studied has been $x = 0{\cdot}055$, the material having a carrier concentration of $\sim 10^{23}\ m^{-3}$ and a Hall mobility greater than $2\ m^2\ V^{-1}\ s^{-1}$ at 77 K. p-Type and n-type materials were found to give similar results. Junctions of 10 to 20 $\mu m$ were achieved by diffusion into {100} surfaces. Ohmic contacts were made with silver on the p-side and indium on to the n-side. Figure 7.32 shows spectra at 12 K, one for a current which is less than the threshold current of $9{\cdot}5 \times 10^6\ A\ m^{-2}$ and one for a current which is greater than the threshold current. Above the threshold current, the spectrum is resolution-limited with a line width less than $10^{-5}$ eV. Operation was found to be single mode for currents up to well above the threshold value.

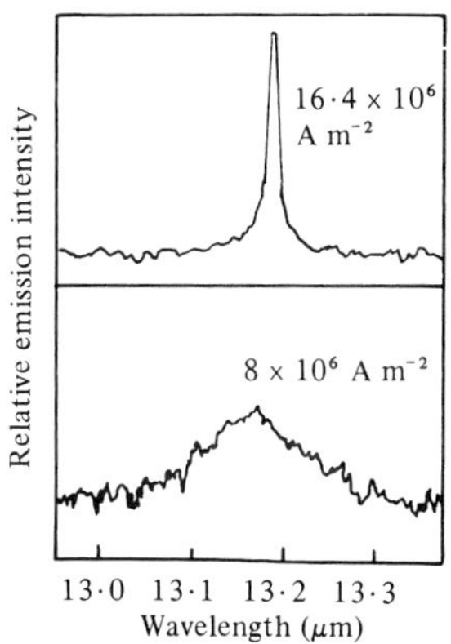


**Figure 7.32.** Emission spectra of a $Pb_{0{\cdot}945}Sn_{0{\cdot}055}Se$ diode laser at 12 K (Butler and Harman, 1967).

## 7.4 The mercury–cadmium chalcogenides: $Hg_{1-x}Cd_xTe$ and $Hg_{1-x}Cd_xSe$

### 7.4.1 Introduction

Just as a gradation of bandgaps is possible with the lead–tin chalcogenides, so this gradation can be achieved with these two chalcogenide systems. Mercury telluride and mercury selenide have been discussed as semimetals. Cadmium telluride with an energy gap of 1·605 eV and cadmium selenide with an energy gap of 1·84 eV fall well outside the range of narrow-bandgap materials being considered, but, as with the lead–tin chalcogenide systems, mixed systems give rise to a crossover between the valence and conduction bands.

The telluride system has been studied in considerably more detail than the selenide system. Mercury and cadmium tellurides both have zinc blende structures and, although there is deviation from Vegard's law for the mixed system, this deviation from linearity between the lattice constant $a_0$ and composition is very small. The phase diagram for the system has been studied by a number of workers but not with complete agreement. The results are summarised in figure 7.33 where the shaded areas represent the regions of uncertainty for the liquidus and solidus lines. The large separation between the liquidus and the solidus is of considerable disadvantage when growing crystals of particular composition. This puts the system at a disadvantage when compared with the lead–tin chalcogenides which have a smaller separation of the liquidus and solidus. (However, mercury–cadmium chalcogenides possess a much smaller dielectric constant leading to smaller capacitive effects at n–p junctions and smaller $RC$ constants when used in n–p devices.)

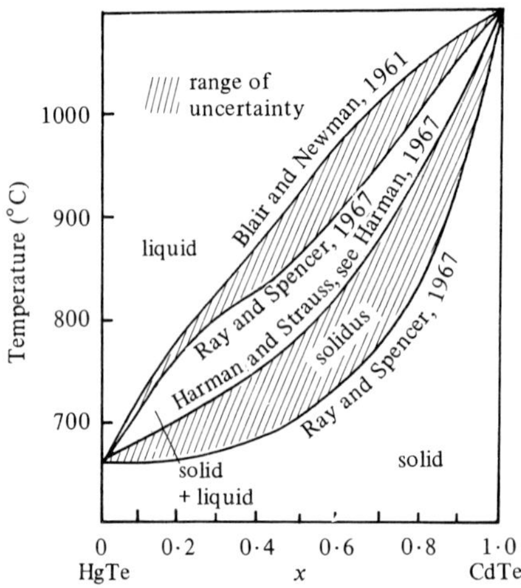


**Figure 7.33.** $(T, x)$ phase diagram for $Hg_{1-x}Cd_xTe$.

### 7.4.2 Crystal growth

$Hg_{1-x}Cd_xTe$ is usually grown in crystalline form from a molten charge of the alloy. The alloy itself can be obtained either by heating the individual compounds in the required proportions or by heating the elements in a carbon-coated thick-walled quartz tube to above the melting point of the alloy. Heating is carried out very slowly in order to prevent any explosion due to unreacted mercury. In fact, an alternative is to insert a side tube into the reacting capsule to establish a set mercury pressure by having excess mercury at a lower temperature in the side-arm. The reaction capsule is best rocked so as to produce a homogeneous ingot. Once this ingot of starting material has been obtained, single-crystal growth can be achieved by the Bridgman method. The growth rate must be carefully selected: a very fast rate leads to constitutional supercooling and dendritic growth, a slow rate can lead to macroscopic segregation.

Often the material is required in single-crystal layers for use as detectors. In this case, epitaxial growth is satisfactory. The usual procedure is to use a cadmium telluride substrate and transport either mercury telluride or an alloy of the mixed system. Even with zero temperature gradient, transport occurs because of a difference in chemical potential between mercury telluride and cadmium telluride. As can be expected, composition profiles show a rapid fall-off of cadmium content with the thickness of the layer (Cohen-Solal *et al.*, 1965). Increasing the vapour pressure of mercury reduces the rate of growth but does not change the composition. For a given set of conditions, including specific temperature, source material, source-to-substrate spacing, and excess mercury pressure, the composition of the layer asymptotically approaches a particular value whatever time is allowed for deposition. Layers have constancy of composition of within 0·5 mol% over an area of 1 $cm^2$ but can be very nonuniform with depth because of the diffusion profile.

High-temperature annealing can remove the dendritic structure which tends to occur in samples prepared by the Bridgman technique, whereas lower-temperature annealing can be used to adjust the stoichiometry. This is discussed in a review article by Long and Schmit (1970). By annealing in a mercury atmosphere it is possible to diffuse mercury into p-type material to provide an n-type surface; n–p junctions for use as photovoltaic infrared detectors can be produced in this way.

### 7.4.3 Band structure

Most of the available information on band structure concerns the telluride system. Figure 7.34 based on that of Overhof (1971b) shows the form of the energy bands near the bandgap as calculated either side of the zero gap position. The calculation used the KKR method and although this does not give variation of the bandgap with composition, the diagrams correspond to approximately the $x = 0{\cdot}2$ region. The bands represented

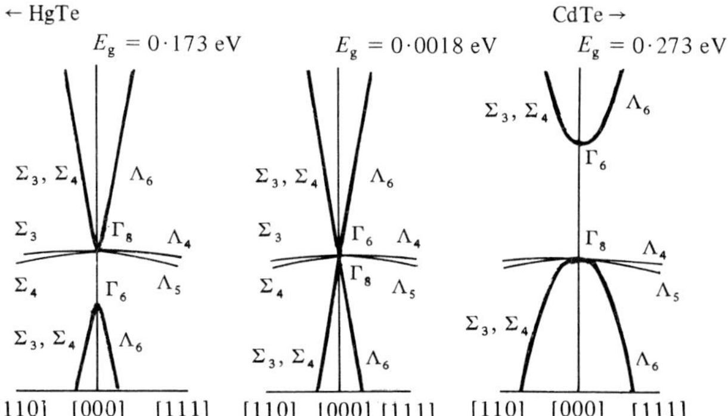


**Figure 7.34.** Energy bands near the bandgap for $Hg_{1-x}Cd_xTe$ as the composition changes. (After Overhof, 1971b.)

by thick lines should be symmetrical in shape by applying the Kane model. However, the $\Gamma_6$ band has s-type wave-functions and the $\Gamma_8$ has p-type. Variation of energy with composition as found experimentally (Long, 1968a) is shown in figure 7.35. Good experimental results were available for the range $0 < x < 0{\cdot}4$ only, and the region beyond this has been obtained by extrapolation to the value of the bandgap for cadmium telluride. This has been done for 0 K and 300 K. There is likely to be some nonlinearity—a very slight drop in the middle of the curve. An accidental degeneracy occurs when $E(\Gamma_6) = E(\Gamma_8)$, which corresponds to the semimetal-to-semiconductor transition.

As can be seen, the change of bandgap with temperature is significant and can be either positive or negative according to composition, the reversal occurring at a composition of approximately $x = 0{\cdot}5$. The positive change of gap with temperature is most unusual, although it has already been noted that it occurs for the entire composition ranges of the lead–tin chalcogenides. The change is approximately linear with temperature right down to 0 K. The variation of $\partial E/\partial T$ is thought to be a lattice potential effect (Long and Schmit, 1970), probably because there is a gradual change of crystal lattice potential and hence of lattice vibrational properties with $x$.

From figure 7.35 and with the use of $\boldsymbol{k \cdot p}$ theory it is possible to calculate the effective mass corresponding to the conduction band edge. Figure 7.36 shows this variation (Long, 1968b) for an assumed value for the momentum matrix $P$ of $9 \times 10^{-10}$ eV m.

The energy gap for $Hg_{1-x}Cd_xSe$ has been investigated as a function of composition in the range $x < 0{\cdot}2$ by Stankiewicz *et al.* (1974). It can be seen that the variation, shown in figure 7.37, is very similar to that for the telluride system. At 4·2 K the energy gap goes through zero at $x = 0{\cdot}11$.

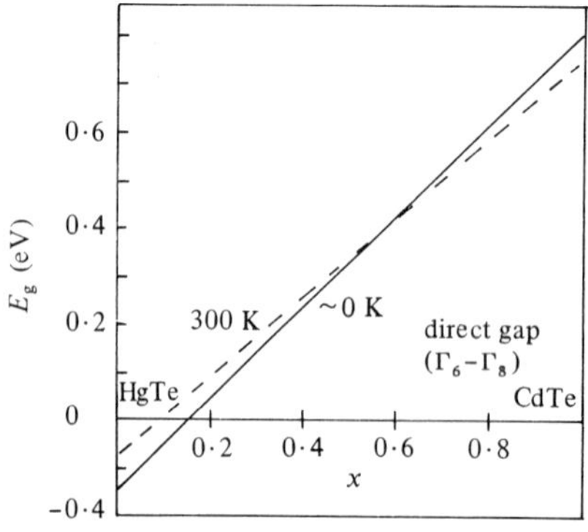


**Figure 7.35.** Energy gap versus composition in $Hg_{1-x}Cd_xTe$ deduced from optical experiments. (After Long, 1968a.)

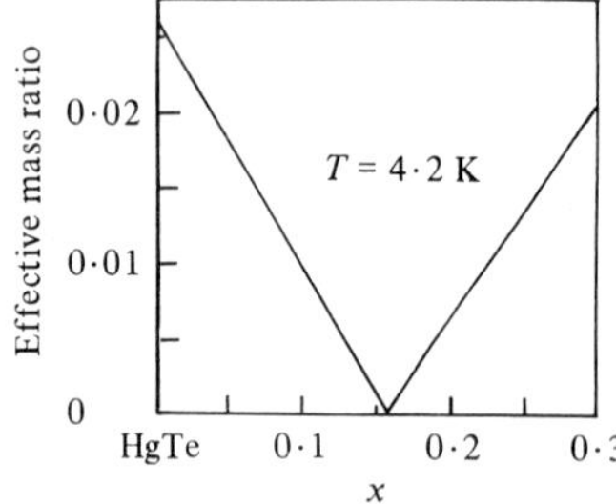


**Figure 7.36.** Effective-mass ratio at the conduction-band edge versus alloy composition in $Hg_{1-x}Cd_xTe$ at 4·2 K in the range $0 \geqslant x \geqslant 0{\cdot}3$ (Long, 1968b).

The effective mass shows good dependence on $n^{2/3}$ as should be expected from a Kane-type band-form. This is shown in table 7.5 where the effective mass values listed have been taken from Stankiewicz *et al.*

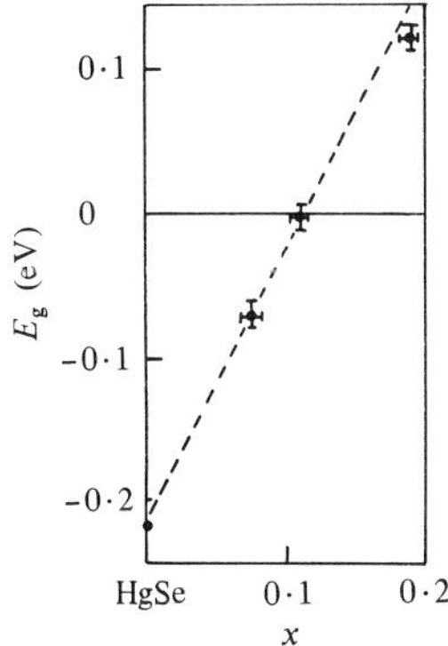


**Figure 7.37.** Bandgap at 4·2 K for $Hg_{1-x}Cd_xSe$ as a function of composition (Stankiewicz *et al.*, 1974).

**Table 7.5.** Cyclotron effective mass data for $Hg_{1-x}Cd_xSe$ from Shubnikov-de Haas oscillations. (Selected data from Stankiewicz *et al.*, 1974.)

| $x = 0{\cdot}05$ | | $x = 0{\cdot}075$ | | $x = 0{\cdot}11$ | | $x = 0{\cdot}19$ | |
|---|---|---|---|---|---|---|---|
| $n$ ($10^{23}$ m$^{-3}$) | $\frac{m^*}{m_0}$ | $n$ ($10^{23}$ m$^{-3}$) | $\frac{m^*}{m_0}$ | $n$ ($10^{23}$ m$^{-3}$) | $\frac{m^*}{m_0}$ | $n$ ($10^{23}$ m$^{-3}$) | $\frac{m^*}{m_0}$ |
| | | 0·83 | 0·162 | | | 0·596 | 0·171 |
| 1·52 | 0·216 | 1·39 | 0·178 | 1·55 | 0·169 | 1·51 | 0·206 |
| 2·33 | 0·224 | 2·67 | 0·233 | 2·89 | 0·233 | 2·69 | 0·242 |
| | | 4·58 | 0·277 | 5·18 | 0·256 | 6·05 | 0·293 |
| | | 13·50 | 0·369 | | | | |

### 7.4.4 Properties

The transport properties of $Hg_{1-x}Cd_xTe$ have been measured in some detail and in certain cases show complex behaviour. The behaviour of semiconducting alloys with electron concentrations greater than $10^{21}$ m$^{-3}$ is the more straightforward and will be considered first.

Long (1968b) particularly investigated this type of alloy in the composition region $0 < x < 0{\cdot}3$ and obtained conventional n-type material curves for the temperature dependence of the Hall coefficient and resistivity, corresponding to a semiconductor in which the donor activation energy is vanishingly small. Hence no freeze-out of extrinsic carriers occurs at low temperatures and the carrier concentration is independent of temperature. The Hall mobility was shown to increase with decreasing temperature but to become almost independent of temperature in the liquid helium temperature range (figure 7.38) where statistical degeneracy sets in and the mobility is limited by ionised-impurity scattering. Long combined figures 7.35 and 7.36 to calculate the expected variation of mobility as a function of composition under the assumption of ionised impurity scattering and obtained good experimental agreement (figure 7.39). A linear variation of the static dielectric constant with composition was also assumed.

Scott (1972) obtained Hall and resistivity data on samples of a wider range of compositions (up to $x = 0{\cdot}6$) with carrier concentrations generally less than $2 \times 10^{21}\ \mathrm{m^{-3}}$. The calculations of Long were extended to remove the restriction of extreme degeneracy and good agreement was obtained with the assumption of scattering from singly ionised donors. Figure 7.40 shows the variation of Hall mobility with temperature for a range of compositions. As all the samples had carrier concentrations less than $2 \times 10^{21}\ \mathrm{m^{-3}}$, the high-temperature mobility (i.e. that close to room temperature) is the intrinsic mobility for pure crystals. At the lower values of $x$, where the energy gap is approximately zero, electron–hole scattering is probably important and is accounting for the decrease in slope of the curves. Long and Schmit (1970) calculated the intrinsic carrier concentrations for $\mathrm{Hg_{1-x}Cd_xTe}$ for various temperatures and compositions and their results, shown in figure 7.41, are of special use for determining the suitability of the alloys as infrared detectors. The dependences on alloy composition shown are good, although agreement between absolute values as calculated and as measured is less good. At other compositions it is optical-mode scattering that is important. Room temperature mobilities plotted against composition showed similar peaks

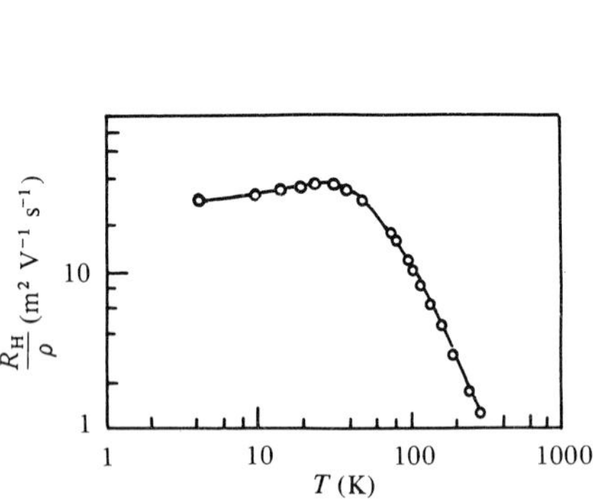


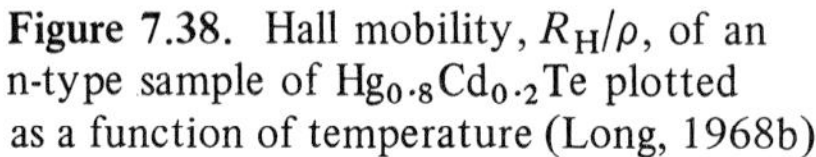
**Figure 7.38.** Hall mobility, $R_H/\rho$, of an n-type sample of $\mathrm{Hg_{0\cdot8}Cd_{0\cdot2}Te}$ plotted as a function of temperature (Long, 1968b).

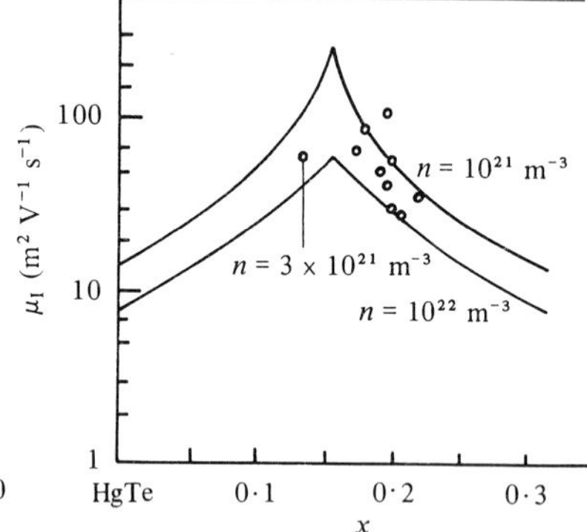


**Figure 7.39.** Ionised-impurity scattering mobility $\mu_I$ versus alloy composition in $\mathrm{Hg_{1-x}Cd_xTe}$ at 4·2 K. The curves are theoretical and the points represent experimental data. The data points are from samples with $n$ within a factor of 2 of $10^{21}\ \mathrm{m^{-3}}$ except for the point $3 \times 10^{21}\ \mathrm{m^{-3}}$ as marked. (After Long, 1968b.)

(at $x = 0{\cdot}01$) to those shown in figure 7.39. Elliott (1971) determined the intrinsic carrier concentration of p-type samples of composition close to $x = 0{\cdot}25$. With the accepted band parameters the temperature dependence of the carrier concentration was found to differ slightly from the theoretical value, and it was deduced that the discrepancy arose from tails of states on the band edges; this effect, if present, would be important at the lower carrier concentrations.

Dornhaus *et al.* (1974) compared galvanomagnetic effects in n-type $Hg_{0\cdot8}Cd_{0\cdot2}Te$ in the temperature range 1·8 to 77 K for low-mobility and high-mobility samples and observed distinct differences in behaviour. The respective mobilities were 6·4 and 21·6 $m^2\ V^{-1}\ s^{-1}$ at 77 K. The low-mobility samples showed an oscillatory dependence of Hall coefficient on temperature up to 50 K, whereas the high-mobility samples showed a constant Hall coefficient. Variation of magnetoresistance with magnetic field was also different. The high-mobility samples showed $\rho(B) \propto B$ in a transverse field and a decrease in a longitudinal field; the low-mobility samples showed an increase in magnetoresistance for both transverse and longitudinal fields. Although a negative longitudinal magnetoresistance can be explained by screened-ionised-impurity scattering, the dominant scattering below 77 K was thought to be polar-optical scattering, and the variation needs further investigation.

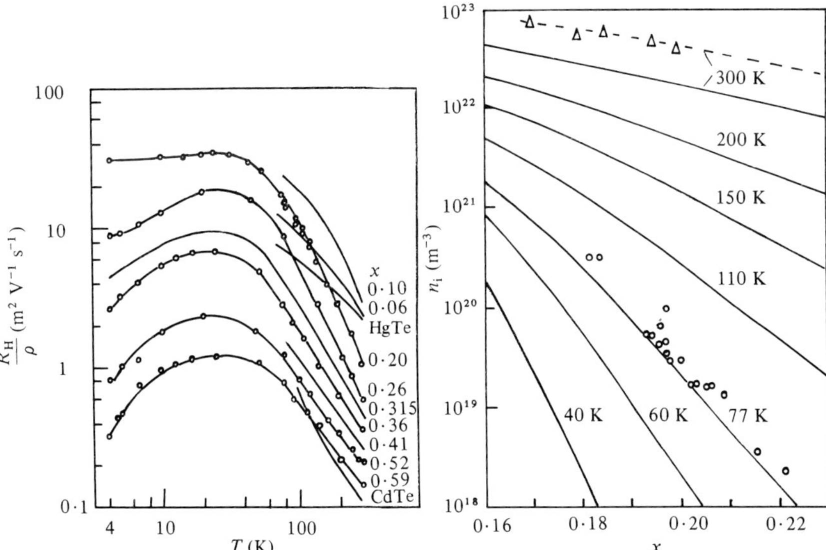


**Figure 7.40.** Temperature dependence of Hall mobility, $R_H/\rho$, for $Hg_{1-x}Cd_xTe$ samples (Scott, 1972).

**Figure 7.41.** Intrinsic carrier concentration, $n_i$, versus composition for $Hg_{1-x}Cd_xTe$ for different temperatures (Long and Schmit, 1970). The solid curves are theoretical and the points represent experimental values determined from the Hall effect.

Measurements on $Hg_{1-x}Cd_xSe$ have been limited largely to optical measurements and measurements of the Shubnikov–de Haas effect. Data from the latter type of measurement have been given already (table 7.5), and results generally indicate considerable analogy with the $Hg_{1-x}Cd_xTe$ system.

#### 7.4.5 Applications

The suitability of $Pb_{1-x}Sn_xTe$ and $Pb_{1-x}Sn_xSe$ as detectors in the photovoltaic and photoconductive modes has already been discussed, and the mercury cadmium chalcogenides are similarly suitable, particularly the telluride system. Preparation of the p–n junctions usually involves the diffusion of mercury into p-type crystals to produce an n-type layer (the mercury atoms compensate the acceptors and contribute an excess of donors). The composition range normally used for $Hg_{1-x}Cd_xTe$ detectors is $0{\cdot}15 \geqslant x \geqslant 0{\cdot}40$. The samples can be cut easily but, because the material is soft, care must be taken to avoid strain and the introduction of dislocations. Surface damage can be removed by etching ($HCl:HNO_3:H_2O$ in the ratio 1:2:3 for 30 s at room temperature). The detectors act in the intrinsic mode; therefore they have a high absorption and need be only a few micrometers thick. It is difficult to make good electrical contacts; soldering with indium, evaporation of an indium pad to which contacts are bonded, and electroplating gold contacts on to the samples are all methods which have been used. Response curves for both the photoconductive and photovoltaic modes are very similar to those already shown for $Pb_{1-x}Sn_xTe$ (or $Pb_{1-x}Sn_xSe$). The mercury–cadmium systems have the advantage over the lead–tin systems in that the static dielectric constant is much smaller (approximately 16 compared with 400), hence giving a much smaller capacitance in the depletion layer and so a smaller *RC* constant. Mercury cadmium telluride detectors have been reviewed by Long and Schmit (1970).

### 7.5 $Mn_xHg_{1-x}Te$

Solid solutions of HgTe–MnTe can be prepared in the range 0 to 80 mol% MnTe (Delves and Lewis, 1963) and a semimetal–semiconductor transition occurs near $x = 0{\cdot}13$ (Kharakhorin *et al.*, 1971).

#### 7.5.1 Preparation

Delves and Lewis formed alloys of composition $0 \leqslant x \leqslant 0{\cdot}8$ by sintering powders of mercury telluride and manganese telluride, the binaries having been produced from the purified elements and then powdered by milling the ingots of the compounds to a particle size of less than 75 $\mu$m diameter. The powders were mixed in the correct proportions and then compressed at a pressure of $5 \times 10^8$ Pa. Sintering was carried out in evacuated glass capsules at various temperatures, although treatment at 450°C for three days was usual. Free space within the capsules was filled with glass discs in order to minimise loss of volatile components.

x-Ray examination, differential thermal analysis and vapour-phase examination was carried out on the sintered samples in order to study the phases present. The results are summarised in figure 7.42. Single-phase alloys can be prepared up to 35 mol% MnTe, and hence either side of the semimetal–semiconductor transition, but above this value the α-phase is metastable and there is decomposition into $MnTe_2$ and a mercury-rich α-phase. No trace of free mercury was found.

Single crystals of the alloys have been grown by Delves (1966) and Kharakhorin *et al.* The latter started from sintered samples (composition $0 \leqslant x \leqslant 0{\cdot}3$) and obtained their crystals by the Bridgman–Stockbarger method with subsequent annealing under mercury vapour pressures,

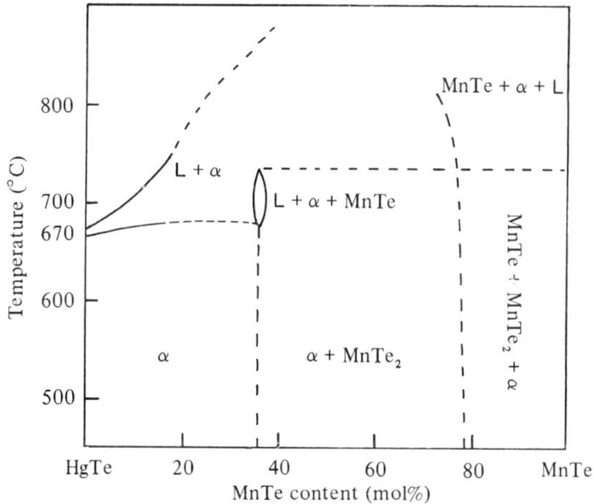


**Figure 7.42.** HgTe–MnTe section of the Mn–Hg–Te ternary phase diagram (Delves and Lewis, 1963).

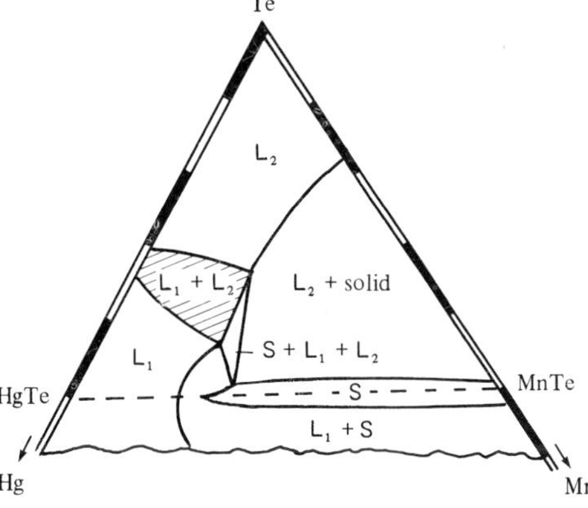


**Figure 7.43.** Schematic ternary isotherm for the Mn–Hg–Te system. (From Delves, 1965.)

whereas the former obtained crystals of composition $0 \leqslant x \leqslant 0{\cdot}15$ grown from a two-liquid melt to avoid constitutional supercooling (Delves, 1965). This method has already been discussed for pure mercury telluride (section 7.1.1) and requires knowledge of the ternary phase diagram. Figure 7.43 shows a typical isothermal cross section (schematic) indicating a two-liquid region on the tellurium-rich side of the diagram.

### 7.5.2 Properties

The electrical properties have been investigated by Delves both for sintered samples (1963) and single crystals (1966). Results for the two types of material were in close agreement except that, as is to be expected, mobilities in the single crystals were higher. Replacement of mercury in mercury telluride by manganese is similar to replacement of mercury in mercury telluride by cadmium, and change of band structure for the $Mn_xHg_{1-x}Te$ system is similar to that for $Hg_{1-x}Cd_xTe$ (q.v.); i.e. there is a point where the $\Gamma_7^-$ and $\Gamma_8^+$ bands change over and the solid solution develops a positive bandgap (see figure 7.34). Unless specially doped, all the samples possessed an acceptor concentration of 1 to $3 \times 10^{22}$ centres $m^{-3}$. This may have arisen from antistructure disorder with mercury atoms occupying sites on the tellurium sublattice. Samples containing varying amounts of manganese were investigated.

*Type I: 12–15% MnTe, p-type.* At 77 K the hole concentration was found to be $2 \times 10^{22}$ $m^{-3}$ with a mobility of $0{\cdot}05$ $m^2$ $V^{-1}$ $s^{-1}$, whereas at $4{\cdot}2$ K the holes were found to freeze on to acceptor levels. Mercury-doped samples showed an anomalous dependence on magnetic field for magnetoresistance (longitudinal and transverse) and Hall coefficient, and this was attributed to minority carriers, electrons with a mobility of $0{\cdot}5$ $m^2$ $V^{-1}$ $s^{-1}$. As the conduction-band electrons have a mobility greater than 3 $m^2$ $V^{-1}$ $s^{-1}$, it was suggested that these electrons come from a negative mass region of the valence band. The excess amount of mercury involved was $5 \times 10^{26}$ atoms $m^{-3}$ and Delves suggested that the strain involved in this number being incorporated in the lattice emphasised the negative mass region (probably owing to low-lying d bands being available for mixing into the valence band).

*Type II: 10–12% MnTe, intrinsic.* These samples were found to be intrinsic at room temperature but p-type at 77 K with less than $10^{21}$ $m^{-3}$ intrinsic electrons at that temperature, so that although the electrons contributed to the resistivity and Hall coefficient, they did not dominate it. As with the 12–15% MnTe composition, the holes were found to freeze on to acceptor centres at $4{\cdot}2$ K. One particular feature was a decrease in Hall coefficient and resistivity values with applied magnetic field, showing that the holes unfreeze and pass into the valence band.

*Type III: 8–10% MnTe, semimetallic.* In this composition range the energy gap was found to be either zero or negative. At 77 K, Hall coefficient and resistivity are dominated by the intrinsic electrons of mobility approximately 20 $m^2$ $V^{-1}$ $s^{-1}$, although in large magnetic fields the Hall coefficient became p-type. At 4·2 K, the acceptor centres were filled with holes and the material was semimetallic with equal numbers of electrons and holes. Because the electron mobility was higher than the hole mobility, the material behaved like an n-type semiconductor, as long as the magnetic field was small. At this temperature, the carrier concentration (electrons or holes) ranged from $10^{21}$ to $2 \times 10^{19}$ $m^{-3}$ depending on the band overlap, with mobilities as high as 50 $m^2$ $V^{-1}$ $s^{-1}$. The high mobilities arose from a small effective mass rather than a long relaxation time, and it was suggested that higher values should be obtainable with better material.

*Type IV: 0–8% MnTe, semimetallic, n-type.* Although semimetallic, these samples were heavily n-doped and results were largely similar to those for mercury telluride.

Kharakhorin *et al.* (1971) investigated the optical and photoelectric properties of the system for compositions up to $x = 0{\cdot}3$ and, as already mentioned, found the semiconductor–semimetal transition to be approximately at $x = 0{\cdot}13$. Increasing the manganese content increased the energy gap from 0·13 eV for $x = 0{\cdot}2$ to 0·19 eV for $x = 0{\cdot}3$, although results may have been affected by a Burstein shift due to a high carrier concentration together with a small effective mass. Optical transition measurements were not possible for $x < 0{\cdot}2$.

Stankiewicz *et al.* (1975) have measured the Hall coefficient and electrical conductivity of samples having $x = 0{\cdot}012$, 0·05, 0·07 and 0·09 corresponding to type IV (semimetallic) as listed above. They confirm the semimetallic properties and show band overlaps at 77 K of 235, 134, 65, and 18 meV respectively, this representing a linear change with composition. A two-carrier model was used and Kane-type conduction bands assumed. Variation of band structure with hydrostatic pressure was also measured and maximum values of mobility shown to occur close to and consistent with the expected semimetal–semiconductor transitions.

### 7.6 Bismuth telluride, $Bi_2Te_3$, and bismuth selenide, $Bi_2Se_3$

Bismuth telluride and bismuth selenide are narrow-bandgap semiconductors which resemble each other in many ways (although there are important differences in band structure and defect properties), and they and their alloys are of particular interest for possible application in thermoelectric devices.

### 7.6.1 Structure and preparation

The compounds have a rhombohedral unit cell consisting of five atoms per cell (two bismuth, three chalcogen) and have the point group $R\bar{3}m$. They have a layered structure and are very anisotropic. Because the bonding between tellurium (selenium) atoms in different layers is much less than the bonding within the layers (the spacing between the tellurium atoms is abnormally large), crystals show pronounced cleavage planes. The sequence of layers can be represented by $Te^{[1]}$ Bi $Te^{[2]}$ Bi $Te^{[1]}$. The superscripts refer to the two types of bonding. The $Te^{[1]}$ atoms are bound by van der Waals forces to the $Te^{[1]}$ atoms of the adjacent tellurium layer and by mixed covalent–ionic bonds to the Bi atoms, whereas the Bi–$Te^{[2]}$ bond is less ionic and weaker (Nakajima, 1963). The structure which applies to both chalcogenides is shown in figure 7.44 (from Caywood and Miller, 1970). Dotted lines indicate the primitive trigonal unit cell whilst the solid lines show a hexagonal unit cell which contains 18 bismuth atoms and 27 chalcogen atoms. The symmetry elements corresponding to the point group are an inversion centre, a threefold axis, three reflection planes parallel to the threefold axis and three twofold axes situated halfway between the reflection planes. The first Brillouin zone is shown in figure 7.45. Two coordinate systems have been used in the literature to describe the zone; that shown in figure 7.45, or one with the $x$ and $y$

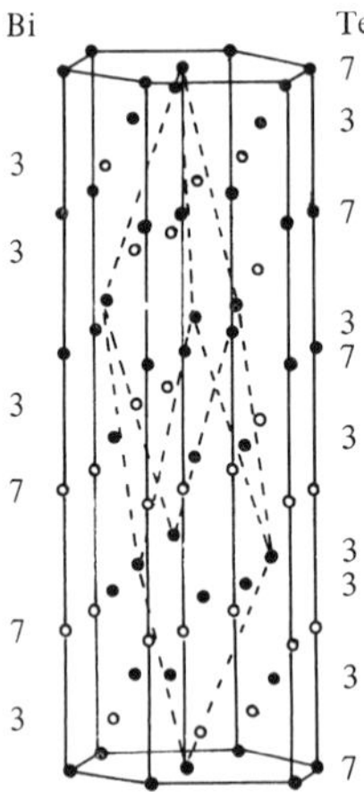


**Figure 7.44.** Crystal structure of bismuth telluride and bismuth selenide. The solid circles represent tellurium or selenium atoms and the open circles represent bismuth. The hexagonal unit cell is indicated by solid lines and the primitive trigonal unit cell by dashed lines. The numbers of atoms in each layer are indicated alongside the hexagonal cell.

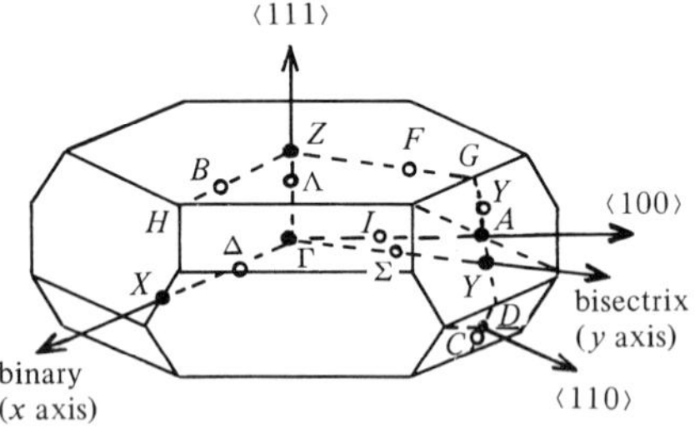


**Figure 7.45.** Brillouin zone for bismuth telluride and bismuth selenide with lengths in the $z$ direction exaggerated.

axes interchanged. The coordinate system shown is the one used also for bismuth. Crystallographic data for the unit cells and Brillouin zones for bismuth telluride and bismuth selenide are shown in table 7.6. Data are mainly available for bismuth telluride and those for bismuth selenide are similar. The Brillouin zone for each is similar to that for bismuth but is more distorted because the unit cell has a large dimension in the trigonal direction.

The phase diagrams for bismuth telluride and bismuth selenide are shown in figure 7.46. It has been found that bismuth telluride has a melting point maximum close to the stoichiometric composition (Brown and Lewis, 1962) but there is considerable solubility between bismuth and bismuth telluride. Single-phase solid solutions ($\beta$) occur for tellurium concentrations ranging from 30 at% to 60 at%, although preparation of these alloys is difficult. They are probably metallic. Only a slight solubility of bismuth and selenium in bismuth selenide is thought to occur.

The compounds are prepared in polycrystalline form by heating stoichiometric quantities of the elements (of 99·999% purity) to above the melting temperature for the particular compound. Woollam *et al.* (1973) in preparing bismuth selenide in this way heated the charge for 12 h at 850°C and used a furnace vibrating at 8 Hz. Single crystals are usually grown by a vertical Bridgman technique at a rate of approximately 1 cm $h^{-1}$ although a horizontal method has been used successfully (Ashworth *et al.*, 1971). It is easier to grow good single crystals of bismuth telluride than of bismuth selenide. Samples of bismuth telluride grown from stoichiometric melts are p-type with a hole concentration of about $2 \times 10^{25}$ $m^{-3}$, but addition of excess tellurium or dopant such as iodine can produce n-type samples. Bismuth selenide samples tend to be n-type but doping with copper reduces the carrier concentration. Woollam *et al.* (1972) have grown p-type samples directly by the Bridgman technique.

**Table 7.6.** Crystallographic data for bismuth telluride and bismuth selenide.

| Symbol | Magnitude | | | Definition |
|---|---|---|---|---|
| | $Bi_2Te_3$ [a] | $Bi_2Se_3$ [b] | units of $b$ | |
| $a_0$ | 10·418 Å | 9·841 Å | | rhombohedral vector at 0 K |
| $\alpha$ | 24°12′40″ | 24°16′ | | rhombohedral angle at 0 K |
| $b_0$ | 1·6731 $Å^{-1}$ | | 1·000 | reciprocal lattice vector |
| $\beta$ | 61°30′37″ | | | rhombohedral angle for reciprocal lattice |
| $V$ | 169·11 Å | | | unit cell volume |
| ΓA | 0·8366 $Å^{-1}$ | | 0·5000 | $\frac{1}{2}$(100) |
| ΓD | 0·8556 $Å^{-1}$ | | 0·5114 | $\frac{1}{2}$(110) |
| ΓZ | 0·3108 $Å^{-1}$ | | 0·1858 | $\frac{1}{2}$(111) |
| $\theta_1$ | 7°6′50″ | | | angle between ΓA and ΓY |
| $\theta_2$ | 14°0′50″ | | | angle between ΓD and ΓY |

[a] Mallinson *et al.* (1968); [b] Wyckoff (1964).

Cutting of samples is best achieved with a wire saw and a 600-mesh silicon carbide slurry. Least cracking occurs if an acid saw is used: 1 : 1 $HCl:HNO_3$ for bismuth telluride and 1 : 3 $HCl:HNO_3$ solution for bismuth selenide (Caywood and Miller, 1970). The crystals can be etched in a dilute solution of methanol. By these techniques it has proved possible for various workers to obtain crack-free samples of 10 mm × 2 mm × 1 mm or slightly larger. Samples grown perpendicular to the cleavage plane tend to be mechanically weak.

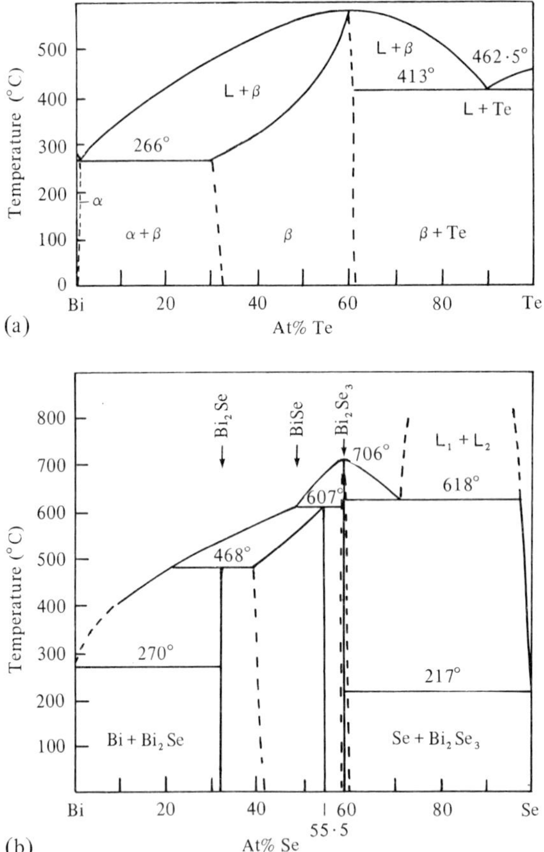


**Figure 7.46.** Phase diagrams for (a) Bi–Te (from Brown and Lewis, 1962) and (b) Bi–Se (from Elliott, 1965) systems.

### 7.6.2 Band structure

The most general electron constant-energy surfaces consistent with the crystal symmetry constitute a 12-ellipsoidal-valley model (Drabble and Wolfe, 1956), but early on it was shown that magnetoresistance results for bismuth telluride (Drabble *et al.*, 1958) fitted a 6-ellipsoidal model, with the ellipsoids centred on the reflection planes and having one axis perpendicular to the reflection plane. The axes of the representative ellipsoid are of arbitrary length with none of the axes parallel to the trigonal axis. In calculations of the band structure the energy gap arises on introduction of the spin–orbit coupling which is particularly important in bismuth telluride (bismuth and tellurium having high atomic numbers). Various experimental determinations of the bandgap have been made. Li *et al.* (1961) measured a value of 0·17 eV whereas Greenaway and Harbeke (1965) who measured the energy gap for the complete $Bi_2Te_3$–$Bi_2Se_3$ system and whose results are shown in figure 7.47, obtained a value of 0·145 eV for pure bismuth telluride. Figure 7.47 also shows the earlier results of Austin and Sheard (1957) whose results are in good agreement at the $Bi_2Te_3$-rich part of the system. Greenaway and Harbeke obtained a temperature dependence of the bandgap of $-0{\cdot}95 \times 10^{-4}$ eV $K^{-1}$ (between 300 and 77 K), this being in agreement with earlier results.

Mallinson *et al.* (1968) used approximate perturbation techniques to obtain a direct gap between conduction and valence bands with minima in the conduction band at the D points. This calculation gave only three conduction band minima but, with the limits of the calculation, it was possible for the minima to lie just off the D points, thus giving rise to the six minima necessary to explain the experimental data. Their de Haas–von Alphen experimental results showed ellipsoidal parameters with values rather different from those deduced from the magnetoresistance measurements. Figure 7.48 shows the ellipsoidal model deduced. The six ellipsoids are still centred on the reflection planes of the crystal. Two of the major axes of the ellipsoids lie in the reflection planes; one makes an angle of 25° with respect to the trigonal direction and the other makes the same angle with the bisectrix. Caywood and Miller (1970) reconsidered the earlier galvanomagnetic data and, by recalculating the effective mass parameters and no longer assuming complete degeneracy, obtained agreement with the data deduced from de Haas–van Alphen measurements.

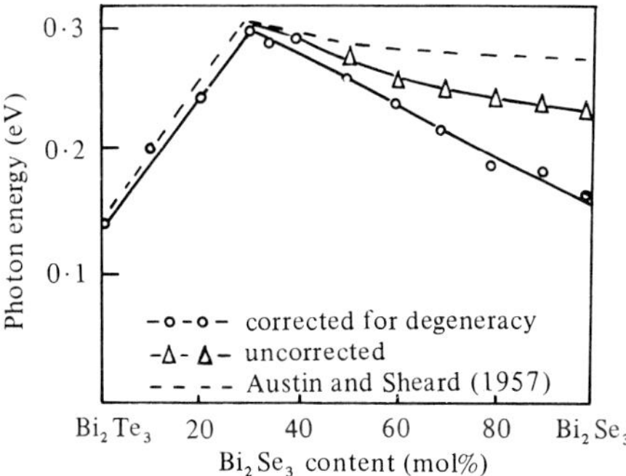


**Figure 7.47.** Absorption edge in $Bi_2Te_3$–$Bi_2Se_3$ alloys (Greenaway and Harbeke, 1965).

Measurements on samples with higher carrier concentrations ($>10^{25}$ m$^{-3}$) indicated the need for the presence of a second conduction band with carriers of high mass and low mobility and this has been confirmed (Ashworth *et al.*, 1971). This second band is less anisotropic than the first band. However, calculation of the energy bands by Togei and Miller (1971) who used a pseudopotential method showed the lower conduction band minima to lie between Γ and A and the valence band minima to lie between Γ and D, and that it is the upper conduction band minima that lie between Γ and D. Middendorff and Landwehr (1972) have deduced from magneto-Seebeck and Shubnikov–de Haas measurements on p-type material of hole concentration $1 \cdot 3 \times 10^{25}$ m$^{-3}$ the presence of a lower valence band, 15 meV below the upper one, with a cyclotron mass for the case of a magnetic field parallel to the hexagonal *c* axis of $0 \cdot 09 m_0$ and a density-of-states effective mass of $1 \cdot 25 m_0$.

Caywood and Miller, when interpreting their results on bismuth selenide assumed a six-valley model by analogy with bismuth telluride and from the data deduced that the constant energy surfaces are more spherical than in bismuth telluride. Köhler and Landwehr (1971) concluded from Shubnikov–de Haas measurements on n-type samples that the lowest conduction band consists of one valley not six, and Hyde *et al.* (1973, 1974) also deduced a single valley at the centre of the Brillouin zone, by comparing the number of carriers deduced from Hall and Shubnikov–de Haas measurements. The latter workers used samples of higher carrier concentration than previously reported, $9 \times 10^{25}$ m$^{-3}$, and found the Fermi surface to be elongated along the trigonal axis. Köhler (1973) from Shubnikov–de Haas measurements deduced a parabolic $E(k)$ dependence perpendicular to the $k_c$ axis, and parallel to the $k_c$ axis a dependence which becomes increasingly nonparabolic. An effective mass along the *c* axis of $0 \cdot 124 m_0$ was measured, whereas Hyde *et al.* obtained a value of $0 \cdot 25 m_0$ for their samples with higher carrier concentrations. Greenaway and Harbeke (1965) obtained a value of $0 \cdot 145$ eV for the bandgap when they had made full allowance for the Burstein effect in the degenerate material used. Although larger values had been obtained

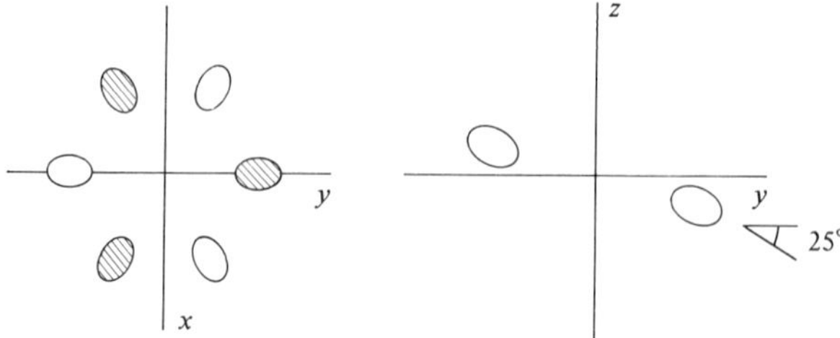


**Figure 7.48.** Schematic diagram of the ellipsoidal model for bismuth telluride fitted to de Haas–van Alphen data (Mallinson *et al.*, 1968). Location of the ellipsoids in the mirror planes has no significance.

previously, Greenaway and Harbeke considered this value to be more consistent with the fact that bismuth telluride and bismuth selenide are of similar structure.

### 7.6.3 Properties

#### 7.6.3.1 *Bismuth telluride*

Because of the form of the crystallographic structure and of the band structure, the galvanomagnetic and similar effects require characterisation by a considerable number of coefficients (see table 4.3). These have been measured extensively for bismuth telluride; for instance by Ashworth *et al.* (1971) and by Caywood and Miller (1970) who also studied bismuth selenide. Besides variation with orientation, the effects show considerable dependence on carrier concentration. Figure 7.49 shows the resistivity $\rho_{11}$ for both n- and p-type samples as a function of carrier concentration as measured along the binary direction. It is very difficult to measure $\rho_{33}$ (along the trigonal direction) because of the mechanical weakness inherent in samples suitably cut for this measurement. Caywood and Miller determined the carrier concentrations from the Hall effect by taking the

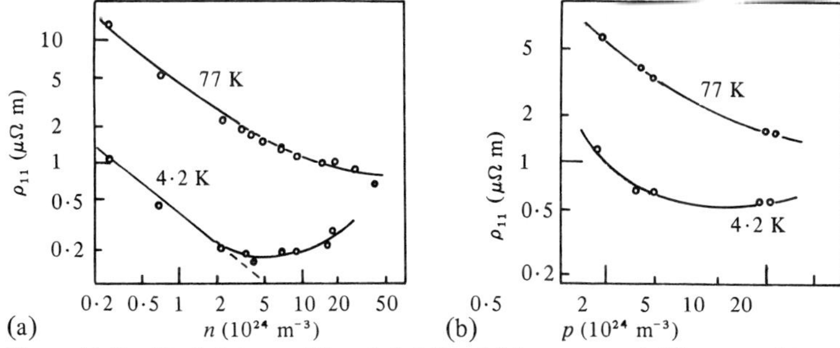


**Figure 7.49.** Resistivity at 77 and 4·2 K of (a) n-type and (b) p-type bismuth telluride as a function of carrier concentration. (From Ashworth *et al.*, 1971.)

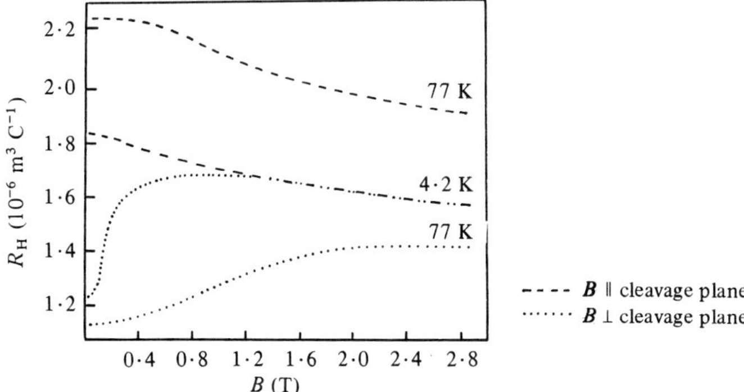


**Figure 7.50.** Typical Hall data for n-type $Bi_2Te_3$ showing the variation of $\rho_{123}$ and $\rho_{312}$ with magnetic field at 77 and 4·2 K (Mallinson *et al.*, 1968).

average of $\rho_{123}$ and $\rho_{312}$ at 4·2 K in a field of 3 T, where the coefficients have saturated to the same limiting value (see figure 7.50 for the Hall effect in a typical sample of carrier concentration $3{\cdot}7 \times 10^{24}$ m$^{-3}$). Of particular interest is the increase in resistivity with increasing carrier concentration for concentrations greater than $5 \times 10^{24}$ m$^{-3}$, thought to arise because of a reduction in total electronic mobility because of the presence of the second (heavy-mass) conduction band. In figure 7.49a, predicted behaviour from a single band model is shown by dashed lines. With the addition of impurities whose donor electrons are going into this second band, the ionised impurities so produced are not especially well screened, and this increased number of scattering centres reduces the mobility of the electrons in the lower band also. In consequence, the de Haas–van Alphen and the Shubnikov–de Haas effects are weak in samples of carrier concentration greater than $4 \times 10^{24}$ m$^{-3}$. Although other materials, such as grey tin and gallium antimonide, have this second heavy-mass band also, they do not show the same effect, their carrier mobilities being enhanced as electrons enter the higher band. The results of Drabble *et al.* (1958) and Goldsmid (1961) were obtained by doping with iodine; those of Caywood and Miller and Ashworth *et al.* were on tellurium-doped samples. The results for p-type material are similar, although fields of 5 T have been required to saturate the Hall effect to obtain the carrier concentrations. The two-band behaviour is less distinct here, probably because of the limited range of hole concentrations over which results have been obtained.

Another very clear indication of the presence of the second band is a plot of the ratio of the magnetoresistance coefficients $\rho_{1133}$ and $\rho_{1122}$ at 4·2 K. This ratio shows a distinct break at $4 \times 10^{24}$ m$^{-3}$ for both n- and p-type materials. Caywood and Miller have calculated from their own data and by recalculating the data of Drabble *et al.* the effective mass ratios $m_1/m_2$ and $m_3/m_2$ (where $m_1$, $m_2$ and $m_3$ are the coefficients of the effective mass tensor), and these also show a distinct discontinuity (figure 7.51). The scatter is too great to permit accurate determination of the mass parameters as a function of concentration, but as the horizontal lines are root-mean-square values for the high- and low-concentration ranges, and as the error bars are standard deviations for the two ranges, the discontinuity is very clear.

Bismuth telluride is very important as a thermoelectric material and figure 7.52 (Yim and Rosi, 1972) shows room-temperature data. The n-type data are for copper-doped samples and the p-type data are either for samples containing excess bismuth (Rosi *et al.*, 1959) or doped with lead (Rosi and Ramberg, 1960). In the high-conductivity region there is a decrease of the Seebeck coefficient due to the onset of degeneracy, whereas in the low conductivity region there is also a fall-off of the Seebeck coefficient, this time owing to the onset of intrinsic conduction.

The Seebeck coefficient for the n-type material is always higher than for the p-type at a specific conductivity. The product $m^{*3/2}\mu$, where $m^*$ is the density-of-states effective mass and $\mu$ the mobility, should be approximately the same in each sample, the higher mobility of the electrons being offset by their lower effective mass. However, addition of dopant, particularly excess bismuth which is required to several at%, considerably reduces the hole mobility in p-type material.

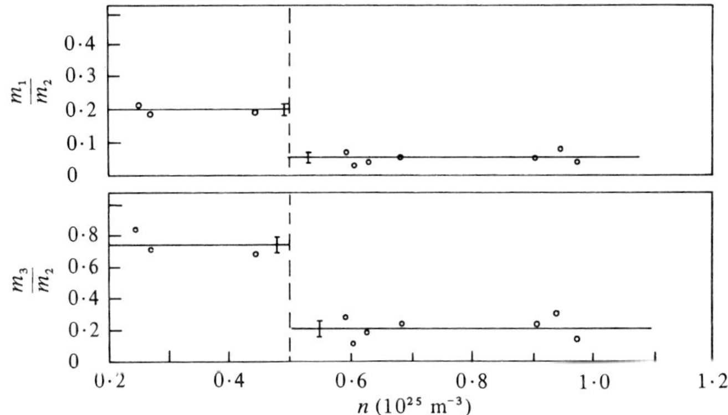


**Figure 7.51.** Variation in the ellipsoidal energy-surface shape for bismuth telluride as a function of electron carrier concentration at 76 K. (After Caywood and Miller, 1970.) The vertical line represents the carrier concentration above which second-band effects enter the de Haas–van Alphen data used.

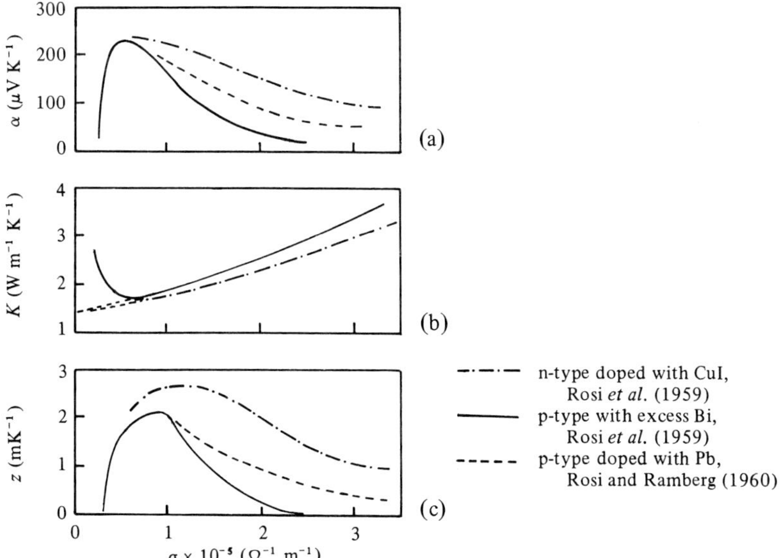


**Figure 7.52.** Thermoelectric properties of bismuth telluride as a function of electrical conductivity, $\sigma$: (a) the Seebeck coefficient, $\alpha$, (b) thermal conductivity, $K$, and (c) the figure of merit $z$.

The thermal conductivity, which includes electronic, lattice, and ambipolar contributions, shows a minimum. The rise at low carrier concentrations is due to the ambipolar effect, whereas over the remaining range of concentrations and electrical conductivities the thermal conductivity shows an almost linear relationship with electrical conductivity, as one would expect from $K_{el} = L_0 \sigma T$, where $K_{el}$ is the electronic contribution to the thermal conductivity and $L_0$ is the Lorentz number. In the intrinsic region the thermal conductivity of n-type material is lower than that of p-type because of additional scattering of phonons by the dopant.

From the two curves the dependence of the thermoelectric figure of merit on conductivity $\sigma$ is obtained; a maximum of $2 \cdot 2 \times 10^{-3}$ $K^{-1}$ occurs at a conductivity of $9 \times 10^4$ $\Omega^{-1}$ $m^{-1}$ for p-type material and $2 \cdot 6 \times 10^{-3}$ $K^{-1}$ at $1 \cdot 1 \times 10^5$ $\Omega^{-1}$ $m^{-1}$ for n-type. These results apply to measurements in the (111) plane. As has already been seen, there is a high anisotropy of the electrical conductivity [it is four times as great in the (111) plane as it is perpendicular to it], the anisotropy of the thermal conductivity is less (the corresponding factor is approximately two), whereas the Seebeck coefficient is almost isotropic. Hence these values of the figure of merit should be the highest attainable at this temperature.

7.6.3.2 *Bismuth selenide*

The galvanomagnetic coefficients of bismuth selenide were measured extensively by Hashimoto (1961), but more recent measurements by Caywood and Miller (1970) yielded much lower values. The earlier results may have been anomalous owing to diffusion of copper into the samples from the copper leads, the crystals may have contained cracks, or the sample geometry and/or field alignment may have been poor. Doping appears to cause only small changes of carrier concentration. Woollam *et al.* (1972) measured the galvanomagnetic effects in a p-type sample of carrier concentration $10^{25}$ $m^{-3}$ and found the samples to be highly degenerate and nearly compensated, with the electrons and holes having almost equal mobilities of $0 \cdot 055$ $m^2$ $V^{-1}$ at $4 \cdot 2$ K. Conductivity in the $x$ direction increases by a factor of four as samples are cooled from room temperature to $4 \cdot 2$ K. Thermoelectric measurements have been made by Middendorff *et al.* (1973) in strong magnetic fields up to 6 T. Besides confirmation of the nonparabolic one-valley model for the lowest conduction band from calculation of the Fermi levels in samples of different carrier concentrations, scattering exponents of $0 \cdot 5$ at 40 K and $0 \cdot 1$ at 65 K were obtained, suggesting a mixture of acoustic and ionised impurity scattering. However, the samples used had carrier concentrations of $10^{24}$ to $3 \times 10^{25}$ $m^{-3}$ and the scattering exponent remained independent of this concentration, so that Middendorff *et al.* were cautious as to what reliable conclusions can be drawn regarding the scattering exponent in bismuth selenide. In a typical sample with carrier concentration of

$2 \cdot 6 \times 10^{24}$ $m^{-3}$ the zero-field thermoelectric power was measured as approximately 52 $\mu V\ K^{-1}$ at 27·5 K rising to 82 $\mu V\ K^{-1}$ at 64 K. Higher values were obtained for samples with a reduced carrier concentration (up to nearly 90 $\mu V\ K^{-1}$ at 27·5 K for $n = 0 \cdot 81 \times 10^{24}$ $m^{-3}$) and lower values for samples with a higher carrier concentration. In a magnetic field oriented parallel to the trigonal axis there was a reduction in the thermoelectric power but this reduction levelled off.

## 7.7 Antimony telluride, $Sb_2Te_3$

Unlike antimony selenide which has a relatively large bandgap, antimony telluride has a small bandgap which may even be approaching a semimetallic value. The compound is of interest both in its own right and as a suitable binary to alloy with the previous two chalcogenide binaries.

### 7.7.1 Structure and preparation

Antimony telluride has a similar crystallographic structure to bismuth telluride and bismuth selenide. It is of point group $R\bar{3}m$ with the hexagonal unit cell having dimensions $a_0 = (4 \cdot 25 \pm 0 \cdot 02)$ Å and $c_0 = (29 \cdot 96 \pm 0 \cdot 10)$ Å (Semiletov, 1956), i.e. it is almost identical in size to bismuth telluride. The phase diagram for Sb–Te has been completed by Brown and Lewis (1962) and is shown in figure 7.53. From the diagram it can be seen that single-phase material exists in the range 11 to 60 at% Te, which is a wider range than was found for the Bi–Te system. The maximum melting point corresponds to a slightly antimony-rich composition of antimony telluride which on cooling tends to segregate tellurium in the grain boundaries. This may be why Abrikosov *et al.* (1959) in a previous determination of the phase diagram reported that at the stoichiometric $Sb_2Te_3$ composition, Sb–Te exists only as two phases, antimony telluride enriched with antimony, and tellurium.

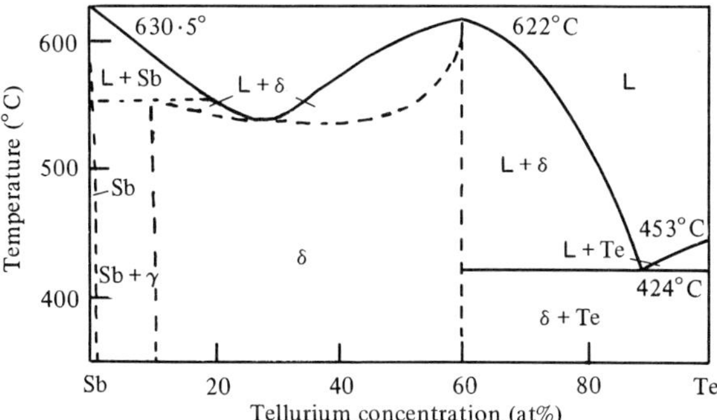


**Figure 7.53.** Phase diagram for the Sb–Te system (from Brown and Lewis, 1962).

Polycrystalline material can be obtained by heating the elements in a sealed silica tube to above the melting point. Single crystals with their cleavage planes parallel to the direction of growth (*c* axis perpendicular) can be grown by the Bridgman technique. Rönnlund *et al.* (1965) used a rate of 23 mm $h^{-1}$ and a temperature gradient of 60 K $cm^{-1}$. This is a very fast rate of descent. Horak *et al.* (1972) in obtaining iodine-doped samples used a somewhat slower rate of 4·5 mm $h^{-1}$ in a temperature gradient of 400 K in 5 cm. It was found that up to 2·5 at% iodine could be dissolved, the atoms probably occupying the vacant sites in the tellurium sublattice. To make the sample surfaces suitable for reflectivity measurements, they polished them with chromium oxide paste, thereafter etched with a 1 : 1 solution of nitric acid, and finished by chemical polishing with a mixture of bromine and glycerine.

#### 7.7.2 Band structure and properties

Black *et al.* (1957) from infrared transmission measurements deduced a bandgap of 0·3 eV whereas Airapetyants and Efimova (1958) determined that the bandgap is close to zero, probably with a slight overlap of the conduction and valence bands. More recently, Procarionne and Wood (1970) have determined the optical properties of a complete range of $Sb_2Se_3$–$Sb_2Te_3$ alloys. They did not determine the bandgap for 100% $Sb_2Te_3$ but deduced a decreasing bandgap from ~0·2 eV for 100% to 0·09 eV at 50% $Sb_2Te_3$ to 0 at 33% $Sb_2Te_3$. They took as the precise value for 100% $Sb_2Te_3$ the value of 0·23 eV obtained by Sehr and Testardi (1962) from their optical data. At less than 33% $Sb_2Te_3$, the structure changes to a mixture of hexagonal and orthorhombic phases. From measurements of electrical conductivity and the Seebeck coefficient on samples doped with lead, iodine, copper bromide, and tin, Rönnlund *et al.* (1965) deduced a two-valence-band model. By curve-fitting to a plot of Seebeck coefficient as a function of electrical conductivity, they deduced to an accuracy of a few per cent a value of separation of the valence bands of 0·23 eV; this could account for the value obtained by Sehr and Testardi for the bandgap.

Undoped material tends to be p-type with carrier concentrations of $\sim 1{\cdot}5 \times 10^{26}$ $m^{-3}$. Horak *et al.* (1972), for the interpretation of their results, used a six-valley model as for bismuth telluride and assumed that the constant-energy surfaces are rotational ellipsoids. For this analysis they used one type of hole and assumed the effective mass ratio for directions parallel and perpendicular to the *c* axis to be independent of carrier concentration. Scattering was taken to be acoustic lattice scattering. From the measurements they deduced an effective hole mass perpendicular to the *c* axis of $0{\cdot}20m_0$ and parallel to the *c* axis of $0{\cdot}59m_0$, and a mobility of 0·0168 $m^2$ $V^{-1}$ $s^{-1}$ perpendicular to the *c* axis. The density of states effective mass was obtained as $1{\cdot}08m_0$, $m_1 = 0{\cdot}20m_0$, $m_2 = 0{\cdot}89m_0$ and $m_3 = 0{\cdot}020m_0$; also $\sin^2\nu = 0{\cdot}853$, where $\nu$ is the angle of inclination of the constant energy ellipsoids to the $k_3$ axis, the $k_3$ axis coinciding with the crystallographic $c$ axis.

Figure 7.54 shows the dependence of the electrical conductivity and Seebeck coefficient on doping with lead as obtained by Rönnlund *et al.* The curves are somewhat surprising in that one might expect lead to substitute for antimony leading to a higher hole concentration and increased conductivity. In antimony telluride (and similarly in bismuth telluride), antimony atoms tend to occupy $Te^{[2]}$ sites (see page 182) and to a lesser extent $Te^{[1]}$ sites. Adding lead tends to restrict the occupation of the $Te^{[2]}$ sites by the antimony atoms, as $Pb^{++}$–$Te^{--}$ ionic bonds tend to be formed. At the optimum level of doping, the antimony and tellurium atoms should be occupying their respective sites to nearly 100%, although the precise mechanism is subject to some doubt.

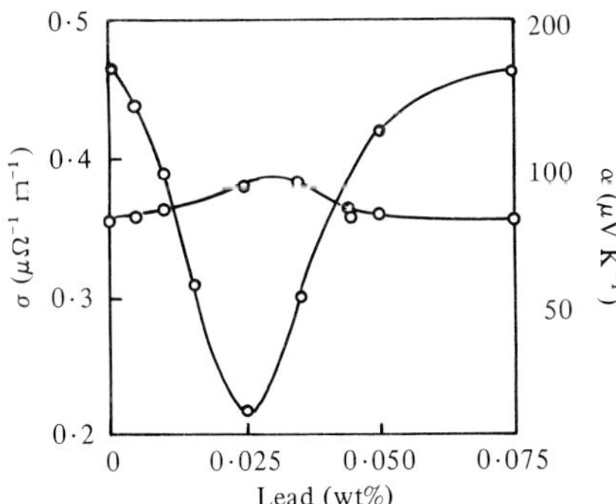


**Figure 7.54.** Seebeck coefficient $\alpha$ and electrical conductivity $\sigma$ for antimony telluride as functions of lead doping concentration (Rönnlund *et al.*, 1965).

## 7.8 Some alloy systems

### 7.8.1 $Bi_2Te_3$–$Bi_2Se_3$ system

The compounds are isomorphous and show complete solid solubility over the entire composition range (McHugh and Tiller, 1959). The undoped alloys are p-type in the $Bi_2Te_3$-rich region and n-type in the $Bi_2Se_3$-rich region, the changeover occurring at approximately 33% $Bi_2Se_3$. The compounds tend to be nonstoichiometric and from density measurements an antistructure type of disorder has been shown to exist (Miller and Li, 1965). In bismuth telluride excess bismuth occupies the tellurium sites to produce the p-type conductivity, whereas in bismuth selenide a vacancy or interstitial mechanism occurs to produce the n-type conductivity. As tellurium is replaced by selenium, the more electronegative selenium atoms replace the more weakly bound $Te^{[2]}$ atoms and by increasing the ionicity of the average Bi–$Te_x^{[2]}Se_{1-x}^{[2]}$ bond, increase the energy gap. When all the $Te^{[2]}$ sites are filled, $Te^{[1]}$ sites become occupied at random by selenium atoms. There is a tendency for charge to be attracted at the Bi–$Te_x^{[1]}Se_{1-x}^{[1]}$ bonds so that the ionicity of the Bi–$Se^{[2]}$ bonds is reduced and hence the energy gap reduced. The bandgap passes through a maximum at $x = 0{\cdot}33$ and, although it has been argued that this arises from the formation of the ordered compound $Bi_2Te_2Se$, Hyde *et al.* (1974) support the argument that it arises from a band-crossing effect involving the Γ–X and Γ–L extrema.

Measurements of the lattice component of the thermal conductivity show a decrease on alloying with either of the pure materials, but measurements by Rosi *et al.* (1959) and LaChance and Gardner (1961) show a peak in the middle of the system, this peak occurrring near 32% $Bi_2Se_3$, possibly because of ordering of the lattice. No superlattice has been identified for the system but there is some evidence of ordering from thermodynamic and x-ray data (Misra and Bever, 1964). The minimum in the lattice component of the thermal conductivity occurs at about 25% $Bi_2Se_3$, which is therefore a good composition at which to check the thermoelectric figure of merit as a function of resistivity. For p-type material, the figure of merit is small compared with that of p-type bismuth telluride, whereas for n-type material, $z$ shows a maximum of $2{\cdot}7 \times 10^{-3}$ $K^{-1}$ for $\rho \sim 1{\cdot}2 \times 10^{-5}$ Ω m, this value of $z$ being little better than for pure bismuth telluride at room temperature (Yim and Rosi, 1972). Imamuddin and Dupre (1972) have measured the figure of merit for pressed and sintered samples of n-type $Bi_2Te_3$–$Bi_2Se_3$ alloys in the temperature region 300 to 600 K. The maximum obtained was for a composition of $Bi_2Te_{2{\cdot}41}Se_{0{\cdot}59}$ doped with either antimony iodide or silver iodide and was $1{\cdot}6 \times 10^{-3}$ $K^{-1}$ in the temperature range 370 to 480 K.

### 7.8.2 $Bi_2Te_3$–$Sb_2Te_3$ system

These two compounds also form a continuous range of solid solutions (Smith *et al.*, 1962). When undoped, all the alloys are p-type with increasing hole concentration at the $Sb_2Te_3$-rich end of the system. The lattice thermal conductivity $K_L$ shows a minimum on alloying, this minimum occurring at 70% $Sb_2Te_3$. There have been widely differing estimates of the actual value depending on the extent of the degeneracy and the type of scattering assumed when calculating the Lorentz number from which the lattice thermal conductivity is derived. However, as at the minimum point, $K_L$ is reduced compared with the value for pure bismuth telluride, this material should exhibit an enhanced p-type figure of merit. Values of $z$ measured on 25% $Bi_2Te_3$–75% $Sb_2Te_3$ have been as follows: $3{\cdot}58 \times 10^{-3}$ $K^{-1}$ by Smirous and Stourac (1959) on a sample containing 4 wt% excess of tellurium and 0·05 wt% germanium, $3{\cdot}53 \times 10^{-3}$ $K^{-1}$ by Rosi *et al.* (1961) on a sample with 1·75 wt% excess selenium, $3{\cdot}2 \times 10^{-3}$ $K^{-1}$ on a selenium-doped sample of resistivity $1{\cdot}1 \times 10^{-5}$ Ω m and $3{\cdot}1 \times 10^{-3}$ $K^{-1}$ on a tellurium-doped sample of resistivity $8 \times 10^{-6}$ Ω m, these last two being by Yim and Rosi (1972). These values represent almost a 50% increase in the value of $z$

over that for pure bismuth telluride. Undoped alloys having a high percentage of antimony telluride are too degenerate to exhibit a high figure of merit. However, it is not thought that the addition of tellurium and selenium has an effect of producing donors to compensate for the high acceptor concentration; rather it is thought that $Bi_2Te_3$–$Sb_2Te_3$–$Sb_2Se_3$ alloys are being produced and this system will be considered next.

### 7.8.3 $Bi_2Te_3$–$Sb_2Te_3$–$Sb_2Se_3$ alloys

The three binaries form pseudo-ternary alloys over a large range of compositions, although not for large concentrations of antimony selenide; unlike the other two binaries, antimony selenide has an orthorhombic structure with $a_0 = 11 \cdot 62$ Å, $b_0 = 11 \cdot 77$ Å, and $c_0 = 3 \cdot 962$ Å (Wyckoff, 1964). Of the binaries, antimony selenide has not been discussed because its bandgap of 1·2 eV (Black *et al.*, 1957) lies outside the scope of this book. Addition of antimony selenide increases the bandgap of the alloy and this counteracts the decrease in bandgap which occurs when antimony telluride is added to bismuth telluride. Addition of about 5 mol% $Sb_2Se_3$ is found to be suitable. The figure of merit for an extrinsic semiconductor is related to the carrier mobility and the effective mass as well as the lattice component of the thermal conductivity, so if $(\mu/K)(m^*/m_0)^{3/2}$ is plotted as a function of mol% $Sb_2Te_3$ it is found to peak at 5%, 25%, and 70% $Sb_2Te_3$, these peaks being of increasing magnitude (Yim *et al.*, 1966). Consequently these three compositions are usually the ones investigated. Rosi *et al.* (1959) considered the undoped alloy for a composition corresponding to approximately the first peak and found that the undoped material showed some improvement over bismuth telluride and $(Bi_2Te_3)_{75}(Sb_2Te_3)_{25}$. However, the maximum value of $z$ obtained, $2 \cdot 8 \times 10^{-3}$ K$^{-1}$ at room temperature (see figure 7.55) is less than that quoted earlier for certain doped $Bi_2Te_3$–$Sb_2Te_3$ samples. Undoped the samples are p-type. When maximising $z$ for the composition corresponding to the third peak, it is necessary to raise the electrical resistivity by doping. Yim and Rosi (1972) have measured $z$ for samples

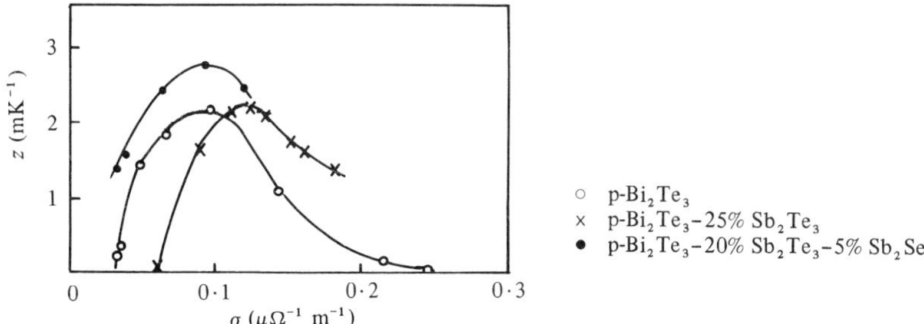


**Figure 7.55.** The figure of merit $z$ at 300 K for p-type $Bi_2Te_3$, $Bi_2Te_3$–25% $Sb_2Te_3$ and $Bi_2Te_3$–20% $Sb_2Te_3$–5% $Sb_2Se_3$ as a function of electrical conductivity, $\sigma$; all samples were doped with excess bismuth (Rosi *et al.*, 1959).

doped with both antimony iodide and tellurium and the largest value of $z$ that they obtained, $3\cdot4 \times 10^{-3}$ K$^{-1}$, was on a tellurium-doped sample with a resistivity of $1\cdot1 \times 10^{-5}$ Ω m.

n-Type samples can be obtained by doping. The best value of $z$ obtained by Yim and Rosi was $3\cdot2 \times 10^{-3}$ K$^{-1}$ on a $(Bi_2Te_3)_{90}(Sb_2Te_3)_5(Sb_2Se_3)_5$ sample doped with antimony iodide which had an electrical resistivity of $1\cdot1 \times 10^{-5}$ Ω m. Thus, the best values of $z$ obtained for p-type and n-type materials are very similar. The Hall mobility of the p-type alloy is $\sim 0\cdot03$ m$^2$ V$^{-1}$ s$^{-1}$ and that of the n-type $\sim 0\cdot02$ m$^2$ V$^{-1}$ s$^{-1}$. Thus the effective mass of the holes in the p-type material must be smaller than the effective mass of the electrons in the n-type, and this is the converse of the situation in pure bismuth telluride.

The figure of merit shows little change with temperature down to 200 K, but below this temperature falls rapidly because of the steep rise in the thermal conductivity. However, investigation indicates that the alloy doped to give the highest figure of merit at room temperature is also the most suitable at lower temperatures.

#### 7.8.4 Applications

##### 7.8.4.1 *Peltier cooling devices*

Bismuth telluride and bismuth selenide and their alloys have particular application for use in Peltier cooling devices. Table 7.7 summarises the figures of merit achieved for different alloys; as can be seen the $(Bi_2Te_3)(Sb_2Te_3)(Sb_2Se_3)$ pseudo-ternary alloy is particularly promising. Using a single-stage device made from n- and p-type alloys of the material, Yim *et al.* (1966) have achieved a cooling from room temperature by an amount of 77·6 K, which is thought to be the highest reproducible value for a single-stage device. A cooling of 141 K has been achieved in a six-stage device made of the alloys (mentioned by Yim and Rosi, 1972).

**Table 7.7.** Summary of thermoelectric figures of merit, $z$, achieved in bismuth and antimony telluride and selenide alloys at room temperature.

| Alloy | $z$ (K$^{-1}$) | Reference |
|---|---|---|
| **n-type** | | |
| $Bi_2Te_3$ + CuI dopant | $2\cdot6 \times 10^{-3}$ | Rosi *et al.* (1959) |
| $(Bi_2Te_3)_{75}(Bi_2Se_3)_{25}$ | $2\cdot7 \times 10^{-3}$ | Yim and Rosi (1972) |
| $(Bi_2Te_3)_{90}(Sb_2Te_3)_5(Sb_2Se_3)_5 + SbI_3$ dopant | $3\cdot2 \times 10^{-3}$ | Yim and Rosi (1972) |
| **p-type** | | |
| $Bi_2Te_3$ + excess Bi | $2\cdot2 \times 10^{-3}$ | Rosi *et al.* (1959) |
| $(Bi_2Te_3)_{25}(Sb_2Te_3)_{75}$ + 4% Te + 0·5% Se dopants | $3\cdot58 \times 10^{-3}$ | Smirous and Stourac (1959) |
| $(Bi_2Te_3)_{75}(Sb_2Te_3)_{20}(Sb_2Se_3)_5$ + excess Bi | $2\cdot8 \times 10^{-3}$ | Rosi *et al.* (1959) |
| $(Sb_2Te_3)_{72}(Bi_2Te_3)_{25}(Sb_2Se_3)_3$ + Te dopant | $3\cdot4 \times 10^{-3}$ | Yim and Rosi (1972) |

7.8.4.2 *Hall effect magnetometer*

Woollam *et al.* (1973) have investigated the use of n-type bismuth selenide crystals for Hall effect magnetometers. They are capable of repeated cycling between room temperature and liquid helium temperature, and for fields up to 11 T have been shown to deviate by less than ±0·8% for all temperatures from 1·1 K to 300 K. Also, when the Hall resistivity $\rho_{yx}$ is plotted against field $B$, it is found to vary by less than 2% from the linear dependence over the temperature range 1·1 to 78 K. For very low fields, thermoelectrically-generated voltages become significant and this means that the crystals must be used only in fields greater than 0·1 T. Disadvantages and advantages over indium arsenide probes are as follows:

*Disadvantages*

(i) A sensitivity which is down by a factor of $10^2$. This can be partly offset in that the lower resistivity of bismuth selenide means that three times the current can be used for the same power dissipation.

(ii) Although below 78 K the temperature coefficient of the Hall output is comparable with that of indium arsenide, above 78 K it becomes increasingly worse.

*Advantages*

(i) No thermal shock resistance has been noted even after 70 cycles.

(ii) Only small quantum oscillation effects occur. The maximum amplitude at 15 T was noted to be 1·8% peak to peak.

## 7.9 Other chalcogenide compounds and systems

### 7.9.1 $HgS_xSe_{1-x}$

The system has been investigated by Kharakhorin and Petrov (1967) who prepared their samples by the Bridgman method. The electron concentrations for the samples at 77 K were in the range (4 to 6) x $10^{23}$ $m^{-3}$ and the mobilities in the range 2 to 7 $m^2$ $V^{-1}$. The semimetal–semiconductor transition was found to occur at $x = 0{\cdot}18$. The forbidden energy gap was measured as changing from $-0{\cdot}07$ eV for pure mercury selenide to 0·10 eV as measured for a composition corresponding to $x = 0{\cdot}4$, but the measured overlap for pure mercury selenide is larger than the value discussed in section 7.1.2. As the sulphur content was increased, the electron effective mass increased from $0{\cdot}007m_0$ for $x = 0{\cdot}05$, $0{\cdot}0085m_0$ for $x = 0{\cdot}10$ to $0{\cdot}021m_0$ for $x = 0{\cdot}30$.

### 7.9.2 Calaverite, $AuTe_2$

The system $AuTe_2 \rightarrow AuAgTe_4 \rightarrow AgTe_2$ forms a group of three phases whose structure has been determined by Tunnel and Pauling (1952). In the calaverite structure the unit cell contains two molecules of $AuTe_2$, is of class $2/m$ and has dimensions $a_0 = 7\cdot19$ Å, $b_0 = 4{\cdot}40$ Å, and $c_0 = 5{\cdot}07$ Å, with an angle $\beta = 90°10'$ between the $c$- and $a$-axes.

(The *b*-axis is a two-fold rotation axis and the *ac* plane is a mirror plane.) Up to 5% of the gold can be replaced by silver without change of phase.

Sagar *et al.* (1971) have investigated the galvanomagnetic coefficients of the calaverite phase in synthetic crystals whose composition is hence controlled, unlike the composition of the naturally occuring minerals. Below 200 K the Hall coefficient was found to be independent of temperature with $R_H \approx -3 \times 10^{-8}$ m$^3$ K$^{-1}$. Hence calaverite is either a semimetal or a narrow-bandgap semiconductor with a net excess carrier concentration of $2 \times 10^{26}$ m$^{-3}$, equivalent to 0·018 per molecule. Above 200 K the Hall coefficient decreases with increasing temperature. So, if the phase is semiconducting, the bandgap is small and the extrinsic electron concentration is produced by vacancies on one of the lattice sites. However, the electron mobility is high for a material containing such a large defect concentration ($\mu_H \approx 0 \cdot 1$ m$^2$ V s$^{-1}$ at 77 K or 1 m$^2$ V s$^{-1}$ at 4·2 K). Sagar *et al.* have measured the tensor components of Hall coefficient, resistivity and magnetoresistance and suggest that the galvanomagnetic coefficients are best explained with the aid of a single-band model for the electrons. Zero-field resistivity was found to vary by at most a factor of 2 with direction and the transverse Hall coefficient less so.

### 7.9.3 Germanium telluride, GeTe

Germanium telluride has a face-centred rhombohedral structure with $a_0 = 5 \cdot 986$ Å and $\alpha = 88 \cdot 35°$ (Hansen and Anderko, 1958). However, this represents only a small distortion from the cubic NaCl structure so that the structure is not very dissimilar from those of lead telluride and tin telluride. Germanium telluride has a bandgap of 0·2 eV as measured from tunnelling experiments (Chang *et al.*, 1966) and is unusual in that it was the first semiconductor reported to undergo a transition to a superconducting state. It has a degenerate behaviour and this is thought to be caused by vacancies and antistructure disorder (Lewis, 1970). It is because of its more complicated crystal structure and complicated defect structure that germanium telluride has been studied less than the other two tellurides. In the phase diagram for Ge–Te, germanium telluride almost coincides with a peritectic melt at 725°C. However, McHugh and Tiller (1960) have shown that germanium telluride forms a eutectic with germanium at 49·85 at% Te at 723°C and that the germanium telluride compound forms from the melt at 724°C and lies at the off-stoichiometric composition of 50·61 at% Te. To prepare germanium telluride the elements are reacted at 850°C over 24 h so as to ensure thorough mixing and then the melt is slowly cooled in a vertical Bridgman furnace, although this tends to produce only polycrystalline samples (Lewis, 1969).

There remains some doubt as to whether the band structure involves two valence bands as in the case of tin telluride and lead telluride. Early results, for instance measurements of transport effects in bulk samples, suggested a two-band model with spherical bands, and a two-band valence model was supported by the optical and electrical measurements of Tsu *et al.* (1968). But some recent electrical measurements by Bahl and Chopra (1969, 1970) indicated a single nonparabolic band. The electronic band structure of germanium telluride as well as those of tin telluride and lead telluride have been measured by Tung and Cohen (1969) who used the empirical pseudopotential method and included spin–orbit interactions. Whereas all three have a valence maximum and a conduction minimum near to the L point, Tung and Cohen found a second valence maximum along the Σ axis for tin telluride and lead telluride but not for germanium telluride; the equivalent point was deduced to be a saddle point from a critical-point analysis. The value for the $L_6^+–L_6^-$ bandgap of 0·23 eV obtained by Tung and Cohen agrees well with the experimental value.

More recent experiments by Lewis (1973) of the optical properties of germanium telluride strongly indicate the existence both of a second valence band maximum and a second conduction band minimum, and the band model is shown in figure 7.56 together with the optical gap to be expected from such a model and the experimental values for the gap obtained for various carrier densities. The effective masses for the first set of bands are $m_{v1}$ (taken as $1·2m_0$) and $m_{c1}$ (which varies from $0·48m_0$ to $0·78m_0$). These bands are taken as being separated by an energy gap of 0·23 eV (from Tung and Cohen). The second pair of bands

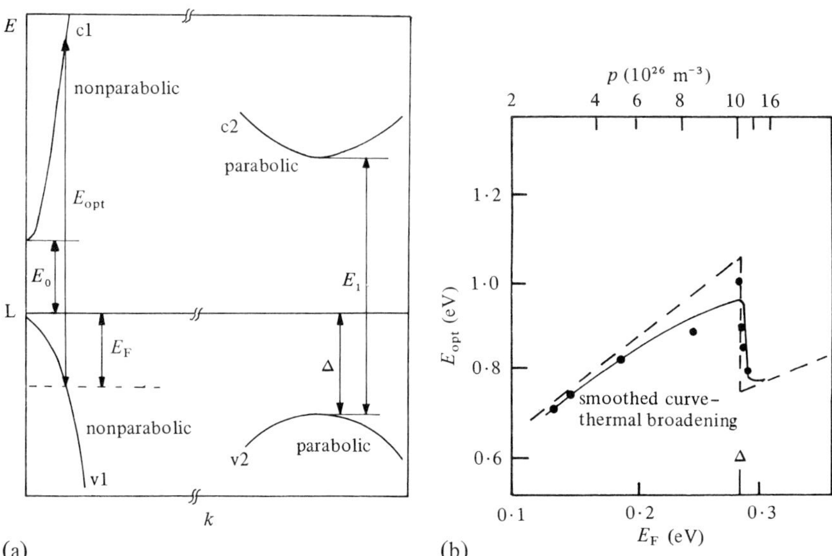


**Figure 7.56.** (a) Schematic band model for germanium telluride, and (b) the optical energy gap. (After Lewis, 1973.)

along the $\Sigma$ axis is found to be separated by energy $E_1 = 0{\cdot}75$ eV and the two bands have the single effective mass $m_2$ (not determined). The carrier density is normally such that the Fermi level is found within the valence band 1 as shown in the figure. In this case optical transitions would involve the bands at L only (the second set of bands are too widely spaced in $k$-space for there to be a large probability of phonon-assisted transitions between the pairs of bands) and the energy involved would be

$$E_{\text{opt}} = E_0 + \left(\frac{1+m_{\text{v1}}}{m_{\text{c1}}}\right) E_{\text{F}} , \qquad E_{\text{F}} < \Delta .$$

Once the carrier densities become large enough for the Fermi energy to enter the second valence band, the transitions would be confined to the second pair of bands and involve the energy

$$E_{\text{opt}} = E_0 + 2(E_{\text{F}} - \Delta) , \qquad E_{\text{F}} > \Delta .$$

The experimental results for the optical bandgap agree with the predicted values on this model, but carrier densities in germanium telluride have not been achieved with a sufficiently large value to produce the final increase of $E_{\text{opt}}$ with carrier density which the model predicts.

### 7.9.4 Tin telluride, SnTe

Tin telluride has the NaCl structure with $a_0 = 6{\cdot}313$ Å (Wyckoff, 1963). A feature it has in common with germanium telluride is that it also has a superconducting state. If grown perfectly pure and with exact stoichiometry, the compound would be an intrinsic semiconductor with a valence–conduction bandgap which is either direct or nearly so and which is approximately $0{\cdot}2$ eV at room temperature. From the phase diagram it can be shown that at equilibrium there is a tin deficiency which results in an extrinsic semiconductor with two holes per tin vacancy. The phase diagram is shown in figure 7.57 (Savage *et al.*, 1972), where the left boundary of the solidus field represents the limit of solubility of tin.

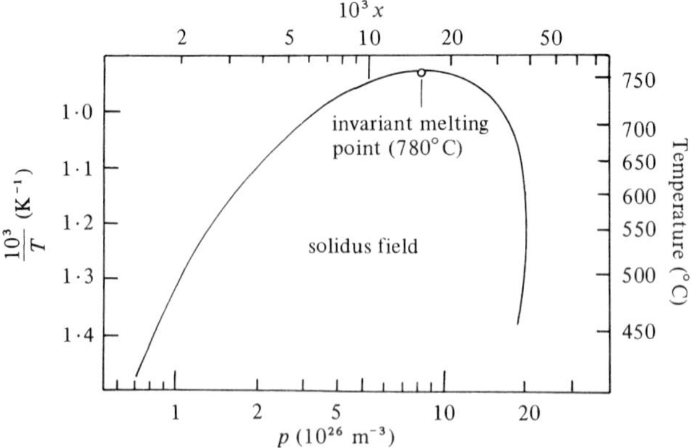


**Figure 7.57.** Phase diagram for $Sn_{1-x}Te$. (After Savage *et al.*, 1972.)

Hence, a range of hole concentrations from $7 \times 10^{26}$ to $2 \times 10^{27}$ $m^{-3}$ is possible. Large single crystals can be grown by the Czochralski technique, and this growth occurs at the invariant melting composition of $x = 0{\cdot}016$ corresponding to a carrier concentration of approximately $8 \times 10^{26}$ $m^{-3}$. Lower carrier concentrations can be achieved by subsequently diffusing tin into the samples; to achieve this, tin is electroplated onto the samples and is diffused in by a high-temperature anneal, the amount of tin being controlled by the extent of the electroplating or by controlling the temperature of anneal and thus the solubility limit of the tin. In the latter case, it is best to use a sequence of anneals, with higher-temperature anneals to obtain fast initial diffusion followed by lower-temperature anneals when the tin is more soluble. Savage *et al.* obtained samples ranging from a hole concentration of $8 \times 10^{26}$ $m^{-3}$ and a Hall mobility at 77 K of $0{\cdot}019$ $m^2$ $V^{-1}$ $s^{-1}$ to a hole concentration of $7{\cdot}2 \times 10^{25}$ $m^{-3}$ and a Hall mobility of $0{\cdot}32$ $m^2$ $V^{-1}$ $s^{-1}$.

Energy band models for tin telluride have been discussed by Ota and Rabii (1974) who carried out reflectance and transmittance measurements in the infrared region on samples grown epitaxially on sodium chloride substrates and having a range of carrier concentrations of $1{\cdot}8 \times 10^{25}$ to $4{\cdot}8 \times 10^{26}$ $m^{-3}$. Their results are in agreement with the band model proposed theoretically by Rabii (1969) and shown in figure 7.58. Two sets of valence-band maxima occur, one at the L and the other in the Σ direction. From $k \cdot p$ theory it is shown that the valence band rises steeply in the direction transverse to the Σ direction such that the valence band energy when expanded about the L point has a saddle shape such

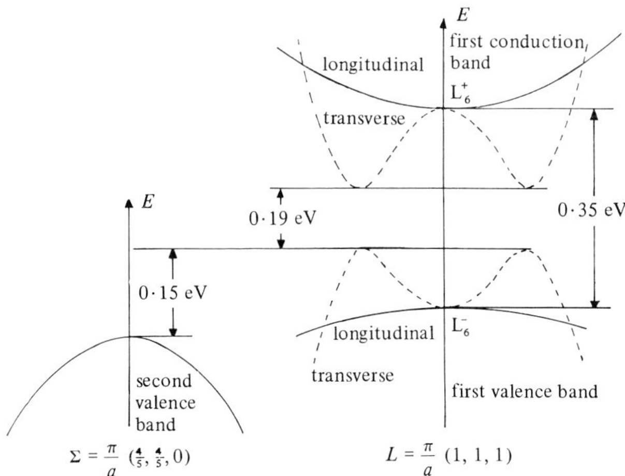


**Figure 7.58.** Band model for tin telluride as proposed by Rabii (1969). Room-temperature gaps are as determined experimentally (Ota and Rabii, 1974). The curvature of the second valence band is not indicative of its effective mass.

that there are two true maxima on the hexagonal face of the Brillouin zone on each side and close to the L point. The conduction band has an inverted but similar shape to that of the valence band. The constant energy surfaces are (ignoring nonparabolic effects) hyperloids of revolution about the [111] direction. The bandgap value and other energy values shown in figure 7.58 are the experimental values obtained by Ota and Rabii. The energy model agrees with that obtained by Tsang and Cohen (1971a) by an empirical pseudopotential method. Tsang and Cohen also showed that, rather unusually, the temperature coefficient for the bands in the Brillouin zone near the minimum energy gap is negative but is positive slightly away from the minimum gap. This is consistent with a negative temperature coefficient obtained from tunnelling experiments and a positive temperature coefficient obtained from optical measurements. Thus, although the direct gap is less than 0·2 eV at room temperature, it is rather larger at lower temperatures, assuming a value of 0·3 eV at 4·2 K. The significant difference from the lead chalcogenides (particularly from the point of view of alloying) is the reversal of the order of the $L_6^-$ and $L_6^+$ bands corresponding to the conduction and valence bands. Crossing of the bands away from the L point is forbidden because of symmetry requirements and it is thus, from the avoidance of crossing, that the valence-band maxima and conduction-band minima arise.

To observe the details of the L point valence bands, it is necessary to use material possessing a low carrier concentration (Ota and Rabii obtained films of carrier concentration $1{\cdot}8 \times 10^{25}$ $m^{-3}$). Higher concentrations indicate the presence of the second valence band along the [110] direction, the existence of this band being particularly consistent with piezoresistance measurements.

The band model largely accounts for the Hall data and thermoelectric power as measured by Allgaier and Houston (1962) and Brebrick and Strauss (1963) respectively. With increasing carrier density the Hall coefficient increases. There is also a rapid increase in carrier mobility with increasing temperature, and the thermoelectric power increases to a maximum with respect to carrier concentration. All these features arise from the presence of a second valence band with a lower carrier mobility than the first. However, Brebrick and Strauss could not account completely for their thermoelectric results. This is not surprising as Shubnikov–de Haas measurements (Savage *et al.*, 1972) have demonstrated the complicated form of the primary valence band by the observation of a large number of cross-sections. If the $\Sigma$ maxima are taken into account, then there are 20 valleys in all and it is probably the multi-valleyed structure that gives rise to the superconductivity observed in tin telluride. In addition, tin telluride has a very high dielectric constant. Savage *et al.* obtained cyclotron effective masses at a carrier concentration of $3{\cdot}6 \times 10^{26}$ $m^{-3}$ of $0{\cdot}125m_0$ for the $\langle 111 \rangle$ pockets and $0{\cdot}094m_0$ for the $\langle 100 \rangle$ pockets.

**References**


Abrikosov, N. Kh., Poretskaya, L. V., Ivanova, J. V., 1959, *Russ. J. Inorg. Chem.*, **4**, 1163.

Airapetyants, S. V., Efimova, B. A., 1958, *Sov. Phys. - Tech. Phys.*, **3**, 1632.

Allgaier, R. S., Houston, B. B., 1962, *Proceedings of the Sixth International Conference of Physics of Semiconductors, Exeter* (Institute of Physics, London), p.172.

Allgaier, R. S., Scanlon, W. W., 1958, *Phys. Rev.*, **111**, 1029.

Ashworth, H. A., Rayne, J. A., Ure, R. W., Jr., 1971, *Phys. Rev.*, **B3**, 2646.

Austin, I. G., Sheard, A. J., 1957, *Electron. Contr.*, **3**, 236.

Bahl, S. K., Chopra, K. L., 1969, *J. Appl. Phys.*, **40**, 4171, 4940.

Bahl, S. K., Chopra, K. L., 1970, *J. Appl. Phys.*, **41**, 2196.

Bernick, R. L., Kleinman, L., 1970, *Solid State Commun.*, **8**, 569.

Black, J., Conwell, E., Seigle, L., Spencer, C. W., 1957, *J. Phys. Chem. Solids*, **2**, 240.

Blair, J., Newman, R., 1961, *Metallurgy of Elemental and Compound Semiconductors*, **12**, Ed. R. O. Grobel (Interscience, New York), p.393.

Blue, M. D., Kruse, P. W., 1962, *J. Phys. Chem. Solids*, **23**, 577.

Borde, D., Albany, H. J., Vandevyver, M., 1966, *C. R. Acad. Sci., Ser. B.*, **262**, 123.

Brebrick, R. F., Strauss, A. J., 1963, *Phys. Rev.*, **131**, 104.

Brebrick, R. F., Strauss, A. J., 1965, *J. Phys. Chem. Solids*, **26**, 989.

Brown, A., Lewis, B., 1962, *J. Phys. Chem. Solids*, **23**, 1597.

Burke, J. R., Jr, Houston, B. B., Savage, H. T., 1970, *Phys. Rev.*, **B2**, 1977.

Butler, J. F., Harman, T. C., 1967, *Solid State Res. Rpts., Lincoln Lab., M.I.T.*, part 3, 3.

Butler, J. F., Harman, T. C., 1968, *Appl. Phys. Lett.*, **12**, 347.

Calawa, A. R., Harman, T. C., Finn, M., Youtz, P., 1968, *Trans. Metall. Soc. AIME*, **242**, 374.

Caywood, L. P., Jr., Miller, J. R., 1970, *Phys. Rev.*, **B2**, 3209.

Chang, L. L., Stiles, P. J., Esaki, L., 1966, *IBM J. Res. Dev.*, **10**, 484.

Chernik, I. A., Kaidanov, V. I., Vinogradova, M. I., Kolomoets, N. V., 1968, *Sov. Phys. - Semicond.*, **2**, 645.

Cohen-Solal, G., Marfaing, Y., Bailly, F., Rodot, M., 1965, *C. R. Acad. Sci.*, **261**, 931.

Crocker, A. J., Rogers, L. M., 1967, *Br. J. Appl. Phys.*, **18**, 563.

Dalven, R., 1969, *Infrared Phys.*, **9**, 141.

Dalven, R., 1971, *Phys. Rev.*, **B3**, 3359.

Dalven, R., 1973, *Solid State Phys.*, **28**, 179.

Delves, R. T., 1963, *J. Phys. Chem. Solids*, **24**, 885.

Delves, R. T., 1965, *Br. J. Appl. Phys.*, **16**, 343.

Delves, R. T., 1966, *Proc. Phys. Soc. London*, **87**, 809.

Delves, R. T., Lewis, B., 1963, *J. Phys. Chem. Solids*, **24**, 549.

Dimmock, J. O., 1971, *Proceedings of the International Conference on Semimetals and Narrow Gap Semiconductors, Dallas (1970), J. Phys. Chem. Solids*, **32**, supplement 1, 20.

Dimmock, J. O., Melngailis, I., Strauss, A. J., 1966, *Phys. Rev. Lett.*, **16**, 1193.

Dionne, G., Woolley, J. C., 1974, *Physics of IV-VI Compounds and Alloys*, Ed. S. Rabii (Gordon and Breach, London), p.20.

Dornhaus, R., Nimtz, G., Schlabitz, W., Zaplinski, P., 1974, *Solid State Commun.*, **15**, 495.

Drabble, J. R., Groves, R. D., Wolfe, R., 1958, *Proc. Phys. Soc. London*, **71**, 430.

Drabble, J. R., Wolfe, R., 1956, *Proc. Phys. Soc. London*, **B69**, 1101.

Dziuba, E. Z., Zakrzewski, T., 1964, *Phys. Status Solidi*, **7**, 1019.

Efimova, B. A., Kolomoets, L. A., 1965, *Sov. Phys. Solid State*, **7**, 339.

Elliott, R. P., 1965, *Constitution of the Binary Alloys,* first supplement (McGraw-Hill, New York).
Elliott, C. T., 1971, *J. Phys. D.,* **4**, 697.
Galazka, R. R., 1970, *Phys. Lett.,* **32A**, 101.
Giriat, W., 1964, *Br. J. Appl. Phys.,* **15**, 151.
Goldsmid, H. J., 1961, *J. Appl. Phys.,* **32**, 2198.
Greenaway, D. L., Harbeke, G., 1965, *J. Phys. Chem. Solids,* **26**, 1585.
Hamilton, D. R., 1958, *Br. J. Appl. Phys.,* **9**, 103.
Hansen, M., Anderko, K., 1958, *Constitution of the Binary Alloys* (McGraw-Hill, New York).
Harman, T. C., 1964, *J. Phys. Chem. Solids,* **25**, 931.
Harman, T. C., 1967, *Physics and Chemistry of II–VI Compounds,* Eds M. Aven, J. S. Prener (North-Holland, Amsterdam), p.767.
Harman, T. C., 1971, *Proceedings of the International Conference on Semimetals and Narrow Gap Semiconductors, Dallas (1970), J. Phys. Chem. Solids,* **32**, supplement, 363.
Harman, T. C., 1974, *Physics of IV–VI Compounds and Alloys,* Ed. S. Rabii (Gordon and Breach, London), p.141.
Harman, T. C., Honig, J. M., Trent, P., 1967, *J. Phys. Chem. Solids,* **28**, 1995.
Harman, T. C., Logan, M. J., Goering, H. L., 1958, *J. Phys. Chem. Solids,* **7**, 228.
Harris, J. J., Ridley, B. K., 1972a, *J. Phys. Chem. Solids,* **33**, 1455.
Harris, J. J., Ridley, B. K., 1972b, *J. Phys. C.,* **5**, 2746.
Harris, J. S., Longo, J. T., Gertner, E. R., Clarke, J. E., 1975, *J. Cryst. Growth,* **28**, 334.
Hashimoto, K., 1961, *J. Phys. Soc. Jpn.,* **16**, 1970.
Hoff, G. F., Dixon, J. R., 1972, *Solid State Commun.,* **10**, 433.
Horak, J., Tichy, L., Frumar, M., Vasko, A., 1972, *Phys. Status Solidi,* **A9**, 369; **A14**, 289.
Hyde, G. R., Beale, H. A., Spain, I. L., 1974, *J. Phys. Chem. Solids,* **35**, 1719.
Hyde, G. R., Dillon, R. O., Spain, I. L., Woollam, J. A., Sellmyer, D. J., 1973, *Solid State Commun.,* **13**, 257.
Imamuddin, M., Dupre, A., 1972, *Phys. Status Solidi,* **A10**, 415.
Kasai, I., Daniel, D. R., Maier, H., Wurzinger, H. -D., 1974, *J. Cryst. Growth,* **23**, 201.
Kharakhorin, F. F., Lutziv, R. V., Pashkovskii, M. V., Petrov, V. M., 1971, *Phys. Status Solidi,* **A5**, 69.
Kharakhorin, F. F., Petrov, V. M., 1967, *Sov. Phys. - Semicond.,* **1**, 112.
Köhler, H., 1973, *Phys. Status Solidi,* **B58**, 91.
Köhler, H., Landwehr, G., 1971, *Solid State Commun.,* **11**, 203.
Kohn, S. E., Yu, P. Y., Petroff, Y., Shen, Y. R., Tsang, Y. W., Cohen, M. L., 1973, *Phys. Rev.,* **B8**, 1477.
Kudman, I., 1972, *J. Mater. Sci.,* **7**, 1027.
LaChance, M. H., Gardner, E. E., 1961, *Adv. Energy Convers.,* **1**, 133.
Laugier, A., Cadoz, J., Faure, M., Moulin, M., 1974, *J. Cryst. Growth,* **21**. 235.
Lawson, W. D., 1951, *J. Appl. Phys.,* **22**. 1444.
Lawson, W. D., 1952, *J. Appl. Phys.,* **23**. 495.
Lewis, J. E., 1969, *Phys. Status Solidi,* **35**, 737.
Lewis, J. E., 1970, *Phys. Status Solidi,* **38**, 131.
Lewis, J. E., 1973, *Phys. Status Solidi,* **B59**, 367.
Li, C. Y., Ruoff, A. L., Spencer, C. W., 1961, *J. Appl. Phys.,* **32**, 1733.
Lin, P. J., Kleinman, L., 1966, *Phys. Rev.,* **142**, 478.
Long, D., 1968a, *Energy Bands in Semiconductors* (Interscience, New York), p.517.
Long, D., 1968b, *Phys. Rev.,* **176**, 923.

Long, D., Schmit, J. L., 1970, *Semiconductors and Semimetals,* **5**, Ed. R. K. Willardson, A. C. Beer (Academic Press, New York), p.517.
Machonis, A. A. Cardoff, I. B., 1964, *Trans. Metall. Soc. AIME,* **230**, 333.
Mallinson, R. B., Rayne, J. R., Ure, R. W., Jr., 1968, *Phys. Rev.,* **175**, 1049.
Martinez, G., 1973, *Phys. Rev.,* **B8**, 4678.
Martinez, G., Schlüter, M., Cohen, M. L., 1975, *Phys. Rev.,* **B11**, 651.
McHugh, J. P., Tiller, W. A., 1959, *Trans. Metall. Soc. AIME,* **215**, 651.
Melngailis, I., Harman, T. C., 1968, *Appl. Phys. Lett.,* **13**, 180.
Melngailis, J., Harman, T. C., Mavroides, J. G., Dimmock, J. O., 1971, *Phys. Rev.,* **B3**, 370.
Melngailis, J., Harman, T. C., Kernan, W. C., 1972, *Phys. Rev.,* **B5**, 2250.
Middendorff, A. von, Landwehr, G., 1972, *Solid State Commun.,* **11**, 203.
Middendorff, A. von, Köhler, H., Landwehr, R., 1973, *Phys. Status Solidi,* **B57**, 207.
Miller, G. R., Li, Che-Ya, 1965, *J. Phys. Chem. Solids,* **26**, 173.
Misra, S., Bever, M. B., 1964, *J. Phys. Chem. Solids,* **25**, 133.
Mitchell, D. L., Palik, E. D., Zemel, J. N., 1964, *Physics of Semiconductors, Proceedings of the Seventh International Conference, Paris,* Ed. M. Hulin (Academic Press, New York), p.325.
Nakajima, S., 1963, *J. Phys. Chem. Solids,* **24**, 479.
Ota, Y., Rabii, S., 1974, *Physics of IV-VI Compounds and Alloys,* Ed. S. Rabii (Gordon and Breach, London), p.113; also, 1973, *J. Nonmetals,* **1**, 117.
Overhof, H., 1971a, *Phys. Status Solidi,* **B43**, 221.
Overhof, H., 1971b, *Phys. Status Solidi,* **B45**, 315.
Parker, S. G., Pinnell, J. E., Johnson, R. E., 1974, *J. Electron. Mater.,* **3**, 731.
Pitt, G. D., McCartney, J. H., Lees, J., Wright, D. A., 1972, *J. Phys. D.,* **5**, 1330.
Procarionne, W., Wood, C., 1970, *Phys. Status Solidi,* **42**, 871.
Putley, E. H., 1955a, *Proc. Phys. Soc. London,* **68**, 22.
Putley, E. H., 1955b, *Proc. Phys. Soc. London,* **68**, 35.
Putley, E. H., 1965, *Materials Used for Semiconductor Devices,* Ed. C. A. Hogarth, (Interscience, New York), p.97.
Rabii, S., 1969, *Phys. Rev.,* **182**, 821.
Rabii, S., Lasseter, R. H., 1974, *Phys. Rev.,* **33**, 703.
Ravich, Ya. I., Efimova, B. A., Tamarchenko, V. I., 1971, *Phys. Status Solidi,* **B43**, 11, 453.
Ray, B., Spencer, P. M., 1967, *Phys. Status Solidi,* **22**, 371.
Riedl, H. R., 1962, *Phys. Rev.,* **127**, 162.
Rogers, L. M., 1967, *Br. J. Appl. Phys.,* **18**, 1227.
Rönnlund, B., Beckmann, O., Levy, H., 1965, *J. Phys. Chem. Solids,* **26**, 1281.
Rosi, F. D., Abeles, B., Jensen, R. V., 1959, *J. Phys. Chem. Solids,* **10**, 191.
Rosi, F. D., Hockings, E. F., Lindenblad, N. E., 1961, *RCA Rev.,* **22**, 82.
Rosi, F. D., Ramberg, E. G., 1960, *Thermoelectricity,* Ed. P. H. Egli (John Wiley, New York).
Sagar, A., Miller, R. C., Damon, D. H., 1971, *Proceedings of the International Conference on Semimetals and Narrow Gap Semiconductors, Dallas (1970), J. Phys. Chem. Solids,* **32**, supplement 1, 545.
Savage, H. T., Houston, B., Burke, J. R., Jr., 1972, *Phys. Rev.,* **B6**, 2292.
Scanlon, W. W., 1959, *Solid State Physics,* **9**, 83.
Scott, W., 1972, *J. Appl. Phys.,* **43**, 1055.
Sehr, R., Testardi, L. R., 1962, *J. Phys. Chem. Solids,* **23**, 1219.
Semiletov, S. A., 1956, *Kristallografiya,* **1**, 403.
Smirous, K., Stourac, L., 1959, *Z. Naturforsch.,* **14A**, 848.
Smith, M. J., Knight, R. J., Spencer, C. W., 1962, *J. Appl. Phys.,* **33**, 2186.
Spencer, P. M., 1964, *Br. J. Appl. Phys.,* **15**, 625.

Stankiewicz, J., Giriat, W., Bien, M. V., 1975, *Phys. Status Solidi,* **B68**, 485.
Stankiewicz, J., Giriat, W., Dobrowolski, W., 1974, *Phys. Status Solidi,* **B61**, 267.
Strauss, A. J., 1967, *Phys. Rev.,* **157**, 608.
Strauss, A. J., 1968, *Trans. Metall. Soc. AIME,* **242**, 354.
Strauss, A. J., Farrell, L. B., 1962, *J. Inorg. Nucl. Chem.,* **24**, 1211.
Tauber, R. N. Machonis, A. A. Cadoff, I. B., 1966, *J. Appl. Phys.,* **37**, 4855.
Togei, R., Miller, G. R., 1971, *Proceedings of the International Conference on Semimetals and Narrow Gap Semiconductors, Dallas (1970), J. Phys. Chem. Solids,* **32**, supplement 1, 349.
Tsang, Y. W., Cohen, M. L., 1971a, *Phys. Rev.,* **B3**, 1254.
Tsang, Y. W., Cohen, M. L., 1971b, *Solid State Commun.,* **9**, 261.
Tsu, R., Howard, W. E., Esaki, L., 1968, *Phys. Rev.,* **172**, 779.
Tung, Y. W., Cohen, M. L., 1969, *Phys. Rev.,* **180**, 823.
Tung, Y. W., Cohen, M. L., 1970, *Phys. Rev.,* **B2**, 1216.
Tunnel, G., Pauling, L., 1952, *Acta Cryst.,* **5**, 375.
Vérié, C., Decamps, E., 1965, *Phys. Status Solidi,* **9**, 797.
Wagner, J. W., Willardson, R. K., 1968, *Trans. Metall. Soc. AIME,* **242**, 366.
Wagner, J. W., Thompson, A. G., Willardson, R. K., 1971, *J. Appl. Phys.,* **42**, 2515.
Whitsett, C. R., 1961, *J. Appl. Phys.,* **32**, supplement, 2257.
Woollam, J. A., Beale, H. A., Spain, I. L., 1972, *Bull. Am. Phys. Soc.,* **17**, 304; *Phys. Lett.,* **41A**, 319.
Woollam, J. A., Beale, H. A., Spain, I. L., 1973, *Rev. Sci. Instrum.,* **44**, 434.
Wyckoff, R. W. G., 1963, *Crystal Structures,* second edition, **1** (Interscience, New York).
Wyckoff, R. W. G., 1964, *Crystal Structures,* second edition, **2** (Interscience, New York).
Yim, W. M., Rosi, F. D., 1972, *Solid-State Electron.,* **15**, 1121.
Yim, W. M., Fitzke, E. V., Rosi, F. D., 1966, *J. Mater. Sci.,* **1**, 52.
Zakrzewski, T., Dziuba, E. Z., 1972, *Phys. Status Solidi,* **52**, 665.
Zallen, R., Slade, M., 1970, *Solid State Commun.,* **8**, 1291.
Zhitinskaya, M. K., Kaidanov, V. I., Chernik, I. A., 1966, *Sov. Phys. – Solid State,* **8**, 246.

# 8

## Other compounds and mixed systems

### 8.1 Indium antimonide, InSb

#### 8.1.1 Introduction

This material has been very extensively investigated and all that will be attempted here is to outline its main electrical properties. It is of ZnS structure ($F\bar{4}3m$) having four molecules in the unit cell. It consists of a face-centred cubic lattice with the In at (000) and the Sb at ($\frac{1}{4}\frac{1}{4}\frac{1}{4}$). The lattice parameter is $a_0 = 6 \cdot 4782$ at 25°C (Wyckoff, 1963). Considerable attention has been given in chapter 3 to the symmetry of zinc blende and diamond structures. The lack of inversion symmetry in zinc blende structures is important for many of the properties. That the [111] direction is not equivalent to the [$\bar{1}\bar{1}\bar{1}$] direction is important in crystal growth and etching, and because of the lack of an inversion centre the materials show such effects as the piezoelectric effect which require this lack of inversion. Zinc blende structures cleave most easily along {110} planes in contrast to diamond-type materials which cleave along {111} planes. This is because in zinc blende materials the {111} planes are alternately composed of group III and group V atoms. In the {110} planes there are equal numbers of group III and group V atoms, and so charge neutrality is preserved on cleavage along these planes.

#### 8.1.2 Preparation

The phase diagram for the In–Sb system is shown in figure 8.1 (Hansen and Anderko, 1958). The compound can be produced by direct reaction of the elements under oxygen-free conditions. After synthesis it is usual to purify the compound by zone refining and high purity is attainable. However, it is difficult to remove zinc and tellurium (Harman, 1956) because the effective segregation coefficients for the two elements are 2·3

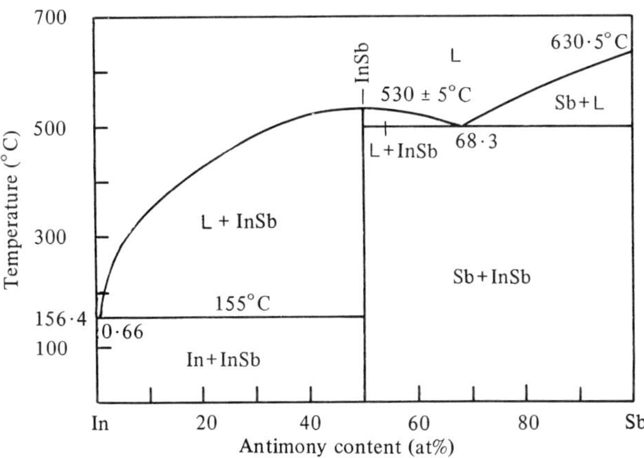


**Figure 8.1.** Phase diagram for the In–Sb system. (After Hansen and Anderko, 1958.)

and ~1 respectively. Cadmium removal can also be difficult because of redistribution via the vapour phase. Because indium antimonide expands 15% on freezing, the boat containing the ingot of material is tilted approximately 8° during zone refining. Quartz is the accepted material for the boats. Early work indicated that indium antimonide sticks to quartz but, as is often the case with other materials, this is due to the presence of oxide impurity. Quartz superseded carbon when it was found that carbon can be electrically active in III–V compounds. Zone refining is carried out in an atmosphere of pure hydrogen. Material which is intrinsic at room temperature can be prepared by zone refining; zone refining usually results in p-type material at the leading end and n-type material at the trailing end; n- or p-type material can be produced by adding dopants such as tellurium and zinc.

Single-crystals can be grown by Czochralski or horizontal Bridgman techniques. When the Czochralski method is used, crystals along all three major axes $\langle 111\rangle$, $\langle 110\rangle$, and $\langle 100\rangle$ are possible, although there is a strong tendency for crystals to grow in the $\langle 111\rangle$ direction; crystals in the $\langle 110\rangle$ and $\langle 100\rangle$ directions tend to twin. Also, when growing along the $\langle 111\rangle$ axis it is necessary to have the correct polarity ($\langle 111\rangle$ not $\langle\bar{1}\bar{1}\bar{1}\rangle$, Gatos *et al.*, 1960) otherwise twinning occurs here also. The Bridgman method tends to result in crystals in the $\langle 111\rangle$ direction, and for this orientation they can be grown at rates of 2 mm $min^{-1}$ without twinning, although slower rates, 0·1 mm $min^{-1}$ or less, produce lower dislocation densities. The growth of single-crystal indium antimonide is reviewed by Liang (1962).

### 8.1.3 Band structure

Indium antimonide is an approximately direct gap semiconductor with $E_g = 0{\cdot}235$ eV at 0 K; it has a conventional negative temperature coefficient such that the gap decreases to approximately 0·225 eV at 77 K and 0·18 eV at room temperature (Pidgeon and Brown, 1966; Mooradian and Fan, 1966). Thus it possesses one of the larger bandgaps among the materials considered in this text; other III–V compounds have larger bandgaps and are mostly excluded here. The conduction band minimum in indium antimonide is a single valley centred at $k = 0$ as shown in figure 8.2 (Long, 1968), which gives the $E(k)$ curves for the $\langle 111\rangle$ and $\langle 100\rangle$ directions close to the bandgap. The conduction band is nonparabolic and included in the figure as a dashed line is the $E(k)$ curve if a parabolic dependence is assumed. The conduction band and also the valence bands are labelled A, B, C and D, corresponding to the labelling in table 3.4. The lack of inversion symmetry in InSb and the effect of spin–orbit coupling lead to a complex structure at the valence band edge. As shown, the splitting as a result of spin–orbit interaction is 0·8 eV. Included is a lifting of the twofold degeneracy of the valence band A; there is also a much smaller lifting of degeneracy for C but this does not affect the experimental results. As can be seen, although the valence bands tend to maximise at $k = 0$, theory predicts that the true maxima are on the ⟨111⟩ axes, and hence there are eight of them. At most, they are 3% of the way to the Brillouin zone edge along the ⟨111⟩ directions and are $\sim 10^{-3}$ eV above the energy at $k = 0$. For band A very close to $k = 0$, the dependence on $k$ is linear rather than second power. Around the maxima the constant energy surfaces are spheroids with axes of revolution along the ⟨111⟩ directions (Kane, 1957). Valence band C as expected from $\boldsymbol{k} \cdot \boldsymbol{p}$ theory is nonparabolic and the valence bands, particularly the heavy-hole valence band A, are somewhat warped. Parameters for the different bands have been determined by many types of experiments, particularly from cyclotron resonance. The electron effective mass has been determined at zero temperature (and momentum) to be $0{\cdot}0138m_0$, and it has been shown by Koteles and Datars (1974) to go through a maximum 4·2% higher than this value and then to decrease with increasing temperature. The average effective masses of the holes in bands A and B are $0{\cdot}4m_0$ and $0{\cdot}016m_0$, respectively (Pidgeon and Brown, 1966). The first experimental evidence for inversion splitting of the conduction band was obtained by Seiler and Hathcox (1972) from measurement of the Shubnikov–de Haas effect under an applied stress.

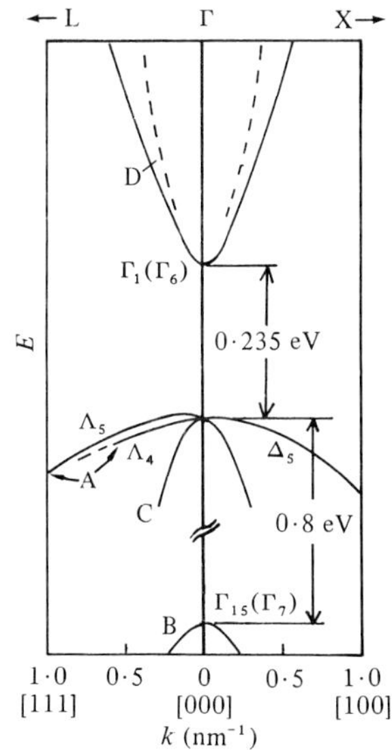


**Figure 8.2.** $E(k)$ curves for InSb in the ⟨111⟩ and ⟨100⟩ directions (Long, 1968).

### 8.1.4 Properties

Currently, indium antimonide is particularly important as a detector material and many of the properties have been discussed in the literature in connection with this application. In place of a detailed discussion of the properties, a selection of them are illustrated in figure 8.3; these are based mainly on curves given by Hilsum and Rose-Innes (1961) and also discussed by Kruse (1970). Indium antimonide crystals can be prepared with less than $10^{19}$ donors $m^{-3}$, but the information in figure 8.3 is for samples having concentrations greater than $10^{20}$ $m^{-3}$. Figure 8.3a shows the variation of intrinsic concentration with temperature and can be used up to 300 K (before degeneracy sets in) to deduce the approximate energy gap (by replotting $n_i T^{-3/2}$ versus $T^{-1}$). At lower temperatures the change of $n_i$ is too rapid for an exact value of the intrinsic concentration to be determined, but at 77 K it is approximately $10^{16}$ $m^{-3}$. Figures 8.3b

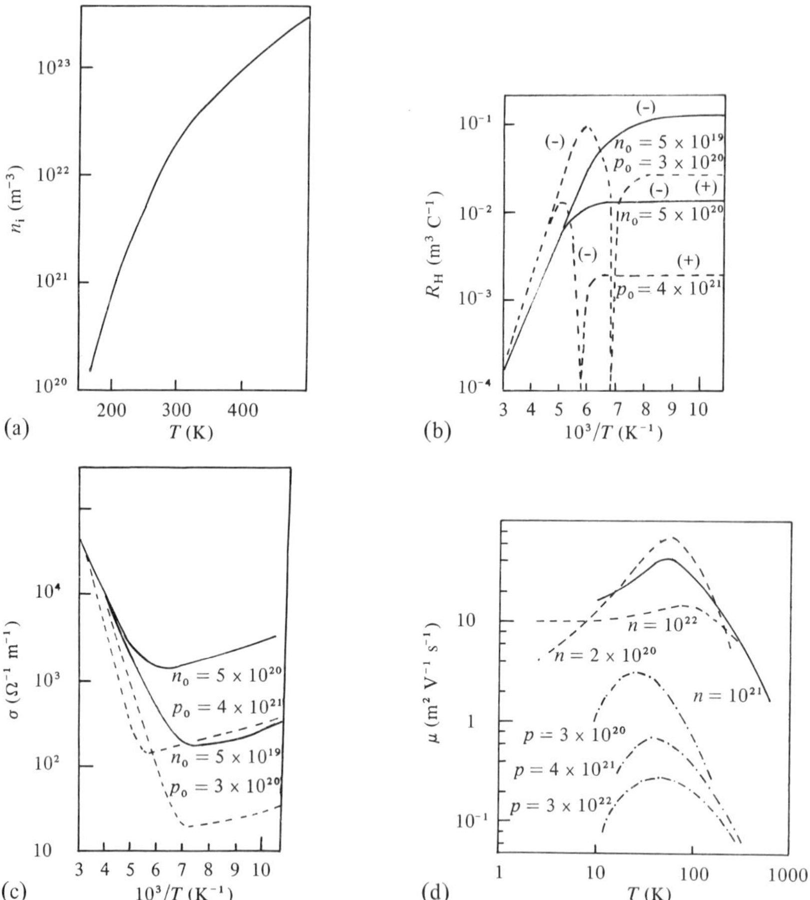


**Figure 8.3.** Some typical properties of InSb. Variation with temperature of (a) intrinsic concentration, $n_i$, (b) Hall coefficient, $R_H$, (c) electrical conductivity, $\sigma$, and (d) electron and hole mobilities. Facing page: The dependence on purity of (e) carrier mobility, $\mu$, and (f) electrical resistivity, $\rho$, at 77 K; (g) the dependence of magnetoresistance, $\Delta\rho/\rho_0$, on magnetic field, $B$, and hole concentration at room temperature, $p_0$. (After Hilsum and Rose-Innes, 1961, and Kruse, 1970).

and 8.3c show the variation of Hall coefficient and electrical mobility with temperature and it is from these curves that much of the data can be deduced and it is with these types of curves that data for other materials can be compared. Pure n-type samples become intrinsic above 150 K, and below 100 K there is little variation of Hall coefficient with temperature, any donor ionisation energies being very small. On the other hand, for the p-type samples there is indication of an ionisation energy of 0·0075 eV and impurity-band conduction. The transition temperature for these samples at which $R_H$ changes sign depends on the purity. The curves for p-type approach those for n-type at high temperatures. From the sets of curves the variation of electron and hole mobility with temperature can be deduced easily provided one type of carrier is involved (figure 8.3d). The purer samples exhibit an increase in mobility up to approximately 60 K after which the mobility decreases because of lattice scattering, the scattering being a combination of polar and electron–hole. As figure 8.3e shows, there is a systematic increase of carrier mobility with decrease of impurity concentration at 77 K, although it is found that at room temperature it is possible for a pure specimen to have a lower mobility than an impure one. Figure 8.3f shows how resistivity varies with impurity concentration at 77 K. The maximum resistivity does not occur when the material is intrinsic because the electron mobility is greater by a factor of ~$10^2$ than the hole mobility, giving rise to the

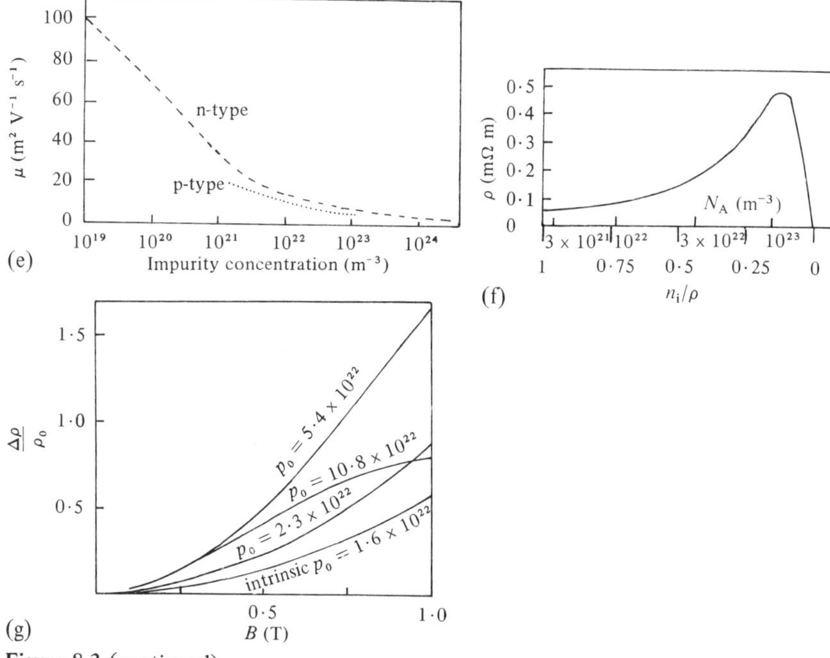

**Figure 8.3** (continued).

maximum occurring for a p-type material. The fact that the electron mobility is large and that there is a large electron-to-hole mobility ratio leads to a large transverse magnetoresistance in low magnetic fields and this is illustrated in figure 8.3g.

Optical properties which are of particular interest for use of the material in detectors have been reviewed by Kruse (1970). For electron concentrations greater than $5 \times 10^{23}$ $m^{-3}$ the Fermi level lies in the conduction band; because intrinsic absorption of photons requires sufficient energy to excite electrons from the top of the valence band to the Fermi level, the optical absorption edge shifts to shorter wavelengths as the free electron concentration increases above this value. The number of free electron–hole pairs produced per absorbed photon has been found by Tauc and Abraham (1959) to increase for energies above 0·5 eV owing to the production of additional hole–electron pairs by impact ionisation. Variation of recombination times with temperature and purity is important and this is also discussed by Kruse.

The Gunn effect in indium antimonide at atmospheric pressure was first observed by Smith *et al.* (1969) at 77 K; it requires a threshold value of the electric field of 600 V $cm^{-1}$. It had been thought previously that carrier multiplication effects prevented the occurrence of the Gunn effect [because the energy gap $E_g$ is less than the energy separation $\Delta$ between the lowest conduction band minimum and the higher lying (low mobility) minimum], but the presence of a region of negative differential mobility caused by intervalley electron transfer must be able to dominate this effect. Although a region of negative differential mobility could arise from a single valley, the intervalley transfer involving $\langle 111 \rangle$ minima is more likely.

## 8.2 Other antimonides

### 8.2.1 $InAs_xSb_{1-x}$

Another narrow-bandgap alloy system investigated recently is the alloy system between indium antimonide and indium arsenide. The bandgap corresponding to absolute zero temperature has been shown by Coderre and Woolley (1968, 1971) to decrease to a minimum for a mixed composition (figure 8.4a). There is considerable variation of bandgap with temperature as is shown in figure 8.4b for a typical alloy. For all compositions there is a temperature below the solidus temperature at which the bandgap passes through zero (figure 8.4c). This occurs because at all compositions the bandgap decreases with increasing temperature until the zero bandgap is reached and the band structure passes from a normal to an inverted form as shown. It is thought that the (000) $\Gamma_6$ conduction band interchanges with the $\Gamma_8$ light-hole valence band, the conduction band and the heavy-hole valence band being degenerate at the critical temperature $T_c$ giving a zero bandgap.

Samples were prepared by a number of methods. Horizontal directional freeze methods were chosen to prepare some of the samples. Growth was achieved by placing ingots in a temperature gradient of 5–10 K $cm^{-1}$ and the temperature lowered to give a growth of 1–2 mm $day^{-1}$; or by using a two-zone furnace with zones at 1000°C (above the melting point of indium arsenide) and 500°C (below the melting point of indium antimonide) with a steep temperature gradient between, and moving the ingot at 5 mm $day^{-1}$. The first of these methods gave satisfactory material for $0 < x < 0{\cdot}1$ and $0{\cdot}9 < x < 1{\cdot}0$, but the material in the middle range was multiphase; the second method gave homogeneous slices for most values of $x$ although there was a rapid change of composition in the

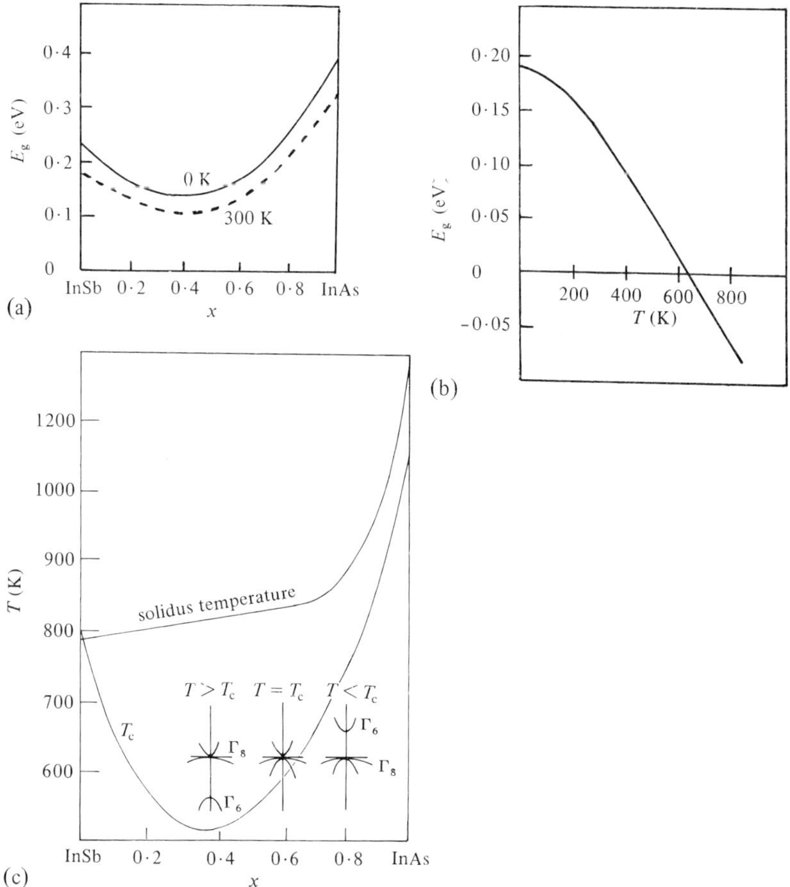


**Figure 8.4.** (a) Variation of energy gap of $InAs_xSb_{1-x}$ at 0 and 300 K with mole fraction $x$; (b) temperature variation of the energy gap for $x = 0{\cdot}68$; and (c) variation of the transition temperature and solidus temperature with $x$. (After Coderre and Woolley, 1968, 1971.)

middle where $0 \cdot 6 < x < 0 \cdot 8$. A third, but similar, method was zone recrystallisation in which a hot zone at 1000°C was passed down the ingot. Again it was difficult to obtain good samples with a middle range of $x$. However, even within this range it proved possible to obtain samples with a homogeneity of $\pm 1 \cdot 5\%$ in $x$ and samples outside the middle range had even higher homogeneity. All the samples possessed n-type extrinsic properties with carrier concentrations within the range $1 \times 10^{22}$ m$^{-3}$ to $2 \times 10^{23}$ m$^{-3}$. As indicated already, the absolute-zero bandgap was calculated over the range and showed a minimum at approximately 0·17 eV so that the minimum value was not as low as had been hoped for narrow-bandgap applications.

Amemiya *et al.* (1973) have made measurements on thin films, but because of problems of segregation and phase separation due to constitutional supercooling only samples in the composition range $0 < x < 0 \cdot 37$ were found to be free of composition variation. The samples were undoped and n-type, and below $x = 0 \cdot 14$ had carrier concentrations of $\sim 5 \times 10^{21}$ m$^{-3}$. In the range $x = 0 \cdot 14$ to $0 \cdot 37$ there was a decrease in grain size and mobility, and an increase in carrier concentration by a factor of 10. Similar data for the variation of bandgap with composition were determined as for the bulk material. The mobility showed a $T^{-1 \cdot 7}$ variation for all compositions, the same as for indium antimonide.

### 8.2.2 Platinum antimonide, $PtSb_2$

Platinum antimonide has a pyrites structure (*Pa*3; *m*3) which is equivalent to the NaCl structure with the $Cl^-$ ions replaced by anion pairs pointing in the ⟨111⟩ directions. The lattice parameter is $a_0 = 6 \cdot 428$ Å (Wyckoff, 1963). The material has a congruent melting point of 1226°C. Its bandgap was first reported from electrical transport properties to be 0·07 eV (Damon *et al.*, 1965) but the later value of Reynolds *et al.* (1968) of 0·11 eV (with little dependence on temperature up to room temperature) appears to be more consistent with recent measurements. This is a small bandgap considering the high melting point of platinum antimonide.

The compound can be grown by zone melting, Czochralski, and Bridgman techniques. By the second method, Damon *et al.* obtained crystals with extrinsic hole concentrations from 3 to $5 \times 10^{24}$ m$^{-3}$. On adding tellurium to the melt, hole concentrations less than $3 \times 10^{24}$ m$^{-3}$ and also n-type material could be obtained. Reynolds *et al.* used seed crystals of ⟨111⟩ and ⟨100⟩ orientation grown by the Czochralski technique. The part of the ingot freezing last was found to contain a second phase, a Pt–Sb eutectic of 24 wt% Sb, indicating that the melt becomes platinum-rich during growth. From their measurements, Reynolds *et al.* did not derive precise values for the effective masses, but they found the effective mass of the holes to be approximately half of that of the electrons ($0 \cdot 16 m_0 g^{1/2}$ compared with $0 \cdot 31 m_0 g^{1/2}$ where the factor $g$ takes a value between 1 and 1·8). Acoustic-mode phonon scattering limits the mobility in the temperature region 77–300 K with $\mu_n/\mu_p = 0 \cdot 5$, this ratio being independent of temperature.

Damon *et al.* proposed a multivalley valence band model, with ellipsoidal energy surfaces centred on the ⟨100⟩ axes in $\boldsymbol{k}$-space as being the simplest to satisfy the magnetoresistance and piezoresistance data. Emtage (1965) used the tight-binding approximation to calculate the band structure. He assumed that the conduction and valence bands are formed primarily from an incomplete set of 5d electrons on ionised platinum atoms, the bandgap arising from spin–orbit splitting. The interaction with 3s orbitals was included but interactions with antimony atoms were largely ignored. The calculated bands showed the measured symmetries and approximately the measured effective masses, but the bands were found to be very shallow. The band structure is shown in figure 8.5 and has eight conduction band minima on the ⟨111⟩ axes and six valence band minima on the ⟨101⟩ axes. For excitations greater than the barrier heights marked (0·007 eV in the case of the conduction bands and 0·011 eV in the case of the valence bands) the eight and six surfaces respectively effectively fuse into a single surface.

Abdullaev *et al.* (1973) have measured the galvanomagnetic and thermomagnetic properties and found the model assuming a bandgap of 0·11 eV to be self-consistent. The hole mobility at room temperature was obtained as ~0·05 $m^2\ V^{-1}\ s^{-1}$ and the electron mobility as 0·025 $m^2\ V^{-1}\ s^{-1}$.

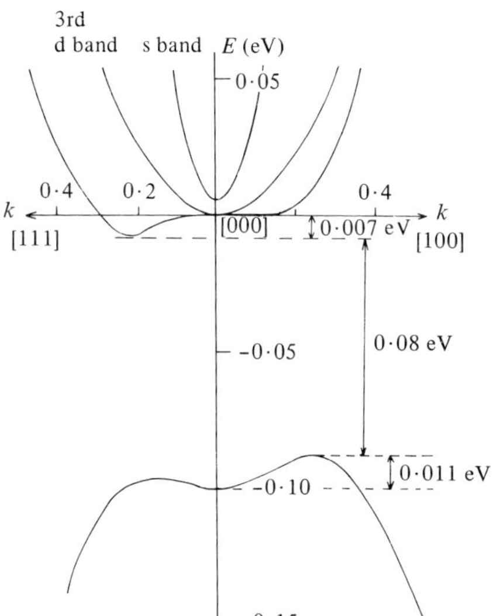


**Figure 8.5.** Band structure near $k = 0$ for platinum antimonide. The position of the s band is undetermined and need not come so close to the d bands as indicated (Emtage, 1965).

The hole mobility varied with temperature as $T^{-2 \cdot 2}$ and the electron mobility as $T^{-0 \cdot 33}$. Acoustic scattering was thought to be the main, but not the only, scattering mechanism and the different temperature dependences are due to different dependences of the effective masses. The transverse and longitudinal effective masses for the valence band were calculated on the basis of Emtage's band model as $0 \cdot 45 m_0$ and $0 \cdot 35 m_0$ at 77 K, and $0 \cdot 85 m_0$ and $0 \cdot 66 m_0$ at 310 K. The conduction-band effective masses were found to be approximately constant with temperature with $m_v^*/m_c^* \approx 0 \cdot 5 - 0 \cdot 6$ at around room temperature as measured from the thermoelectric effect. The thermal conductivity was measured also: at 77 K for a p-type sample of carrier concentration $8 \times 10^{23}$ m$^{-3}$ the thermal conductivity was $1 \cdot 13 \times 10^2$ W m$^{-1}$ K$^{-1}$ changing to 39 W m$^{-1}$ K$^{-1}$ at 300 K when the sample has become intrinsic; the lattice thermal conductivity gives the main contribution. Damon *et al.* obtained a maximum for the Seebeck coefficient of 320 $\mu$V K$^{-1}$ at 140 K.

### 8.3 Magnesium plumbide, $Mg_2Pb$

Magnesium plumbide is one of a group of intermetallic compounds of the form $Mg_2X$ where X is Si, Ge, Sn, or Pb. The first three of the series are semiconductors whose bandgaps decrease from $0 \cdot 77$ eV for $Mg_2Si$ to $0 \cdot 33$ eV for $Mg_2Sn$ (bandgaps at 0 K), whereas $Mg_2Pb$ is now known to be a semimetal. Each of the compounds crystallises in the antifluorite structure of $CaF_2$, *Fm*3*m* [i.e. a face-centred structure with the negative ion at (000) and the positive ions at $\pm(\frac{1}{4}\frac{1}{4}\frac{1}{4})$]. As the compounds show considerable solubility within each other, they provide a possible variable bandgap system with a range of bandgaps from $0 \cdot 77$ to 0 eV. The semiconducting properties for the overall system were predicted by Mott and Jones (1936) as the number of valence electrons is just sufficient to fill the fourth Brillouin zone of the fcc Bravais lattice. The lattice parameter for $Mg_2Pb$ is $a_0 = 6 \cdot 836$ Å (Wyckoff, 1963).

There was ambiguity in early work as to whether in fact magnesium plumbide should be classed as a semimetal or a semiconductor, but it has been shown by Stringer and Higgins (1970) that two condensed phases occur and that the phase corresponding to the semimetallic magnesium plumbide melts incongruently. Consequently, magnesium plumbide cannot be produced from a stoichiometric melt; rather it must be produced from the magnesium-rich side of the phase diagram (see figure 8.6). Although it is unusual to find the Bridgman–Stockbarger method employed for growing crystals which melt incongruently, it has proved possible to use it here because the composition at the peritectic point is only 2% different from that at the congruent melting point, and because the compound is of chemically invariant composition on the magnesium-rich side of the phase diagram (shown by Hall effect measurements). A nonstoichiometric melt was used by Stringer and Higgins in a horizontal setup incorporating a two-section double-walled graphite crucible and a stainless-steel furnace tube because of the reactivity of magnesium vapour towards other materials. The temperature gradient at the growth point in the furnace was enhanced by the use of an air-cooled heat exchanger. Pulling was carried out at 0·2 to 1 cm $h^{-1}$ and the furnace was tilted at 20° to the horizontal to allow the molten charge to flow into the tipped part of the crucible. A vertical arrangement was not used because of expansion on freezing. Early samples had inclusions taken in as the magnesium plumbide expanded, but this problem was overcome with the steep temperature gradient (31 K $cm^{-1}$ with the heat exchanger in operation).

Figure 8.7a sketches the probable band structure for magnesium plumbide, although that in figure 8.7b remains a possibility. These structures are based on experimental (Stringer and Higgins, 1971) and

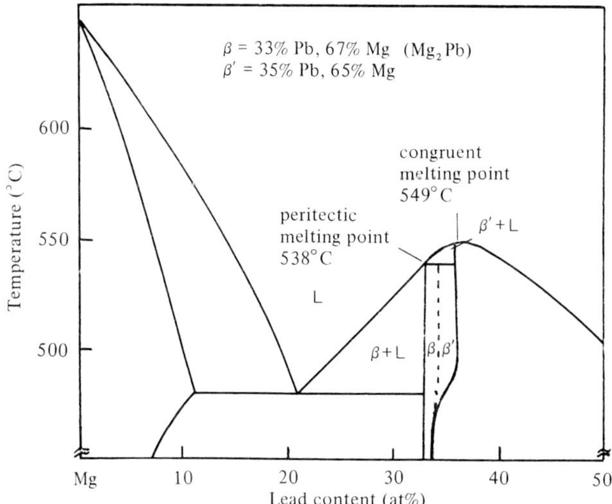


**Figure 8.6.** Part of the Mg–Pb phase diagram (from Stringer and Higgins, 1970).

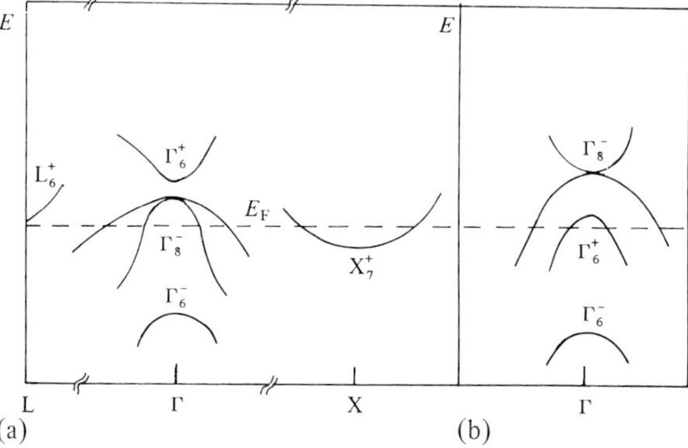


**Figure 8.7.** (a) Main features of the $Mg_2Pb$ energy bands and (b) an alternative band structure (Stringer and Higgins, 1971).

theoretical (Van Dyke and Herman, 1970) deductions. The latter workers have carried out a relativistic orthogonalised-plane-wave calculation of the structure. The band structure is similar to the widely-investigated p-type germanium structure, but there is a difference in symmetry. In magnesium plumbide the centre of the symmetry is located at an atomic site and the valency-band edge is $\Gamma_8^-$ ($\Gamma_{15}$ without spin–orbit splitting) whereas in germanium it is $\Gamma_8^+$ ($\Gamma_{25}$).

The material is almost compensated, the slight departure being caused by the fact that the compound is prepared from a nonstoichiometric melt. Stringer and Higgins (1970, 1971) have used a three-carrier model (heavy and light holes, and electrons) for explaining the electrical measurements. They deduced for their samples the following concentrations and effective masses: (a) light holes $5 \cdot 4 \times 10^{25}$ m$^{-3}$ (classical analysis) to $7 \cdot 9 \times 10^{25}$ m$^{-3}$ (from Shubnikov–de Haas effect), $m^* = 0 \cdot 04 m_0$; (b) heavy holes $3 \cdot 6 \times 10^{25}$ m$^{-3}$ (classical analysis) to $4 \cdot 2 \times 10^{25}$ m$^{-3}$ (Shubnikov–de Haas effect), $m^* = 0 \cdot 35 m_0$ (perpendicular to 100), $m^* = 0 \cdot 45 m_0$ (perpendicular to 111); (c) electrons $4 \cdot 1 \times 10^{25}$ m$^{-3}$ (from high-field experiments aimed at measuring the lack of compensation). The effective masses are averaged values. The curvature of the heavy-hole band is independent of energy but the light-hole band is highly nonparabolic, although the effective mass corresponding to any particular energy is almost isotropic.

## 8.4 Cadmium arsenide, $Cd_3As_2$

Cadmium arsenide is a narrow-bandgap semiconductor, or possibly a zero-bandgap material, which exhibits high mobility of its carriers at relatively high carrier concentrations. Despite extensive investigation, many of its important properties remain subject to doubt and, in particular, its band structure is not known with certainty.

### 8.4.1 Structure and preparation

Cadmium arsenide has a melting temperature of 721°C and undergoes at least one solid–solid phase change, reported to be at 578°C (Jayaraman *et al.*, 1966), 615°C (Hiscocks, 1969), and 595°C (Pietraszko and Łukaszewicz, 1973), the variation possibly being due to hysteresis. There is expansion on cooling through the phase change, and this leads to cracking of crystals grown above this phase transition temperature. To avoid this cracking, the crystals are usually grown from the vapour phase below this temperature. The crystal structure at room temperature has been determined by Steigmann and Goodyear (1968). The unit cell (space group $I4_1cd$) is tetragonal with $a_0 = 12 \cdot 67 \pm 0 \cdot 01$ Å and $c_0 = 25 \cdot 48 \pm 0 \cdot 02$ Å with thirty-two formula units per unit cell. The arsenic ions are approximately cubic close-packed in their arrangement (variation of the As–As bond-lengths shows that there is considerable distortion) and the cadmium ions are tetrahedrally coordinated. The structure is related to the fluorite structure with $As^{3-}$ replacing $Ca^{2+}$, and $Cd^{2+}$ replacing $F^-$, but with two cube-diagonally opposite sites vacant and with these vacant sites situated systematically through the entire unit cell made up from sixteen of these fluorite cells.

Pietraszko and Łukaszewicz report a further two solid–solid phase changes and the variation of lattice parameters with temperature and the unit cells are shown in figure 8.8. The $\beta$ phase with lattice constants $a_0' = b_0' = c_0'$ is a fluorite structure and has the simplest unit cell. Each of the other phases is composed of slightly deformed cells of this type. Although $a_0'$, $b_0'$, and $c_0'$ are taken as existing in the same directions throughout the different phases, the fourfold axis changes its direction in the transition from $\alpha'$ to $\alpha''$ and the $c$-directions in the two phases are perpendicular to each other. These lower transitions do show hysteresis, particularly the change which occurs in the temperature range 410 to 460°C. The relationships between $a$, $b$, and $c$ for the three phases and the units $a_0'$, $b_0'$, and $c_0'$ of the fluorite-type structure are shown in table 8.1.

The phase diagram for the As–Cd system is shown in figure 8.9 (Hansen and Anderko, 1958). The method of preparation consists of heating stoichiometric quantities of cadmium and arsenic gradually to above the melting temperature of 721°C (usually on up to 800°C) and

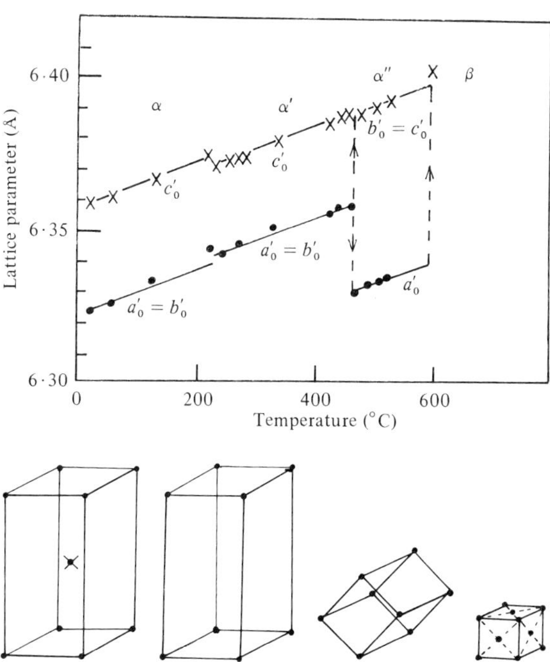


**Figure 8.8.** Variation of lattice parameters of cadmium arsenide with temperature, and the unit cells for different phases. (After Pietraszko and Łukaszewicz, 1973.)

then rapidly cooling the compound. Although crystals have been grown by the Stockbarger–Bridgman method, single-crystal growth is usually carried out from the vapour either in an enclosed system or in a flow of an inert gas. Using the former system, Koltirine and Chaumereuil (1966) have grown crystals of up to 2 cm length and 0·4 cm diameter, the largest single crystals reported. They used a two-zone furnace with the zones at 565 and 555°C with a very steep temperature gradient at the interface between the two zones. As each crystal grew, it was withdrawn from the interface at a rate of 1 mm day$^{-1}$ (figure 8.10). Using the system involving the flow of an inert gas (argon), Lovett (1972) has grown crystals of various habits and sizes, but in particular single crystal platelets which possessed particularly low carrier concentrations (see later).

**Table 8.1.** Relationship of $a, b, c$ to $a_0', b_0', c_0'$ for the different phases of cadmium arsenide (Pietraszko and Łukaszewicz, 1973).

| Phase | $a, b, c$ in units of $a_0', b_0', c_0'$ | $a, b, c$ (Å) |
|---|---|---|
| $\alpha$ | $a = 2a_0'$  $b = 2b_0'$  $c = 4c_0'$ | $a = b = 12{\cdot}65$  $c = 25{\cdot}44$ (23°C) |
| $\alpha'$ | $a = 2a_0'$  $b = 2b_0'$  $c = 4c_0'$ | $a = b = 12{\cdot}69$  $c = 25{\cdot}49$ (250°C) |
| $\alpha''$ | $a = 2^{1/2}c_0'$  $b = 2^{1/2}b_0'$  $c = 2a_0'$ | $a = b = 9{\cdot}039$  $c = 12{\cdot}67$ (500°C) |
| $\beta$ | $a = a_0' = b_0' = c_0'$ | $a = 6{\cdot}403$ (600°C) |

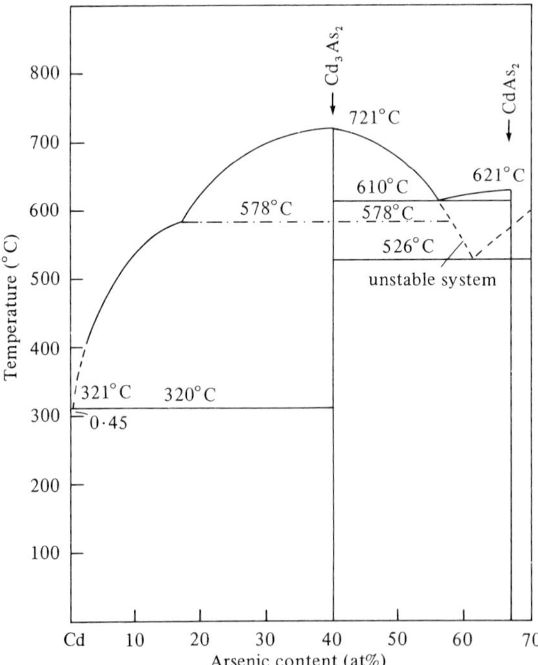


**Figure 8.9.** Phase diagram for the Cd–As system. (After Hansen and Anderko, 1958.)

The platelets grew at temperatures around 400°C whereas larger crystals grew at higher temperatures, ~500 to 540°C. The argon was passed over cadmium heated to just below the melting temperature to prevent the cadmium arsenide containing excess arsenic.

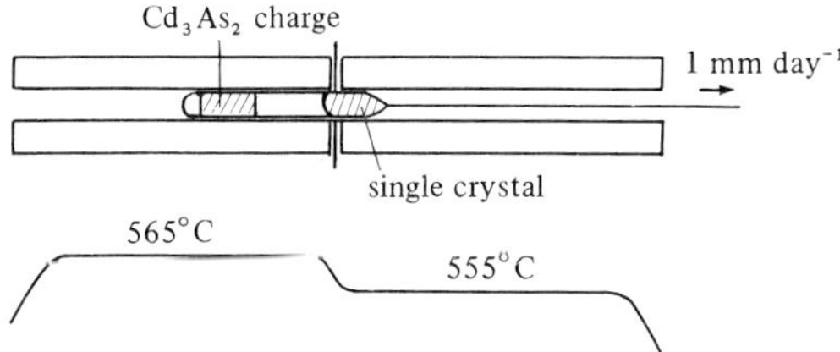


**Figure 8.10.** Growth of cadmium arsenide from the vapour (Koltirine and Chaumereuil, 1966).

### 8.4.2 Properties

All samples of cadmium arsenide have been n-type and most possess carrier concentrations in excess of $10^{24}$ m$^{-3}$, although crystals with carrier concentrations of 3 to 6 x $10^{23}$ m$^{-3}$ have been grown by Lovett (1972) and Ugai and Zyubina (1965). Figures 8.11a and 8.11b show the variation of Hall coefficient and resistivity with temperature for samples of these different carrier concentrations. Even for samples of the lower carrier concentrations, decomposition or change of phase sets in before sample temperature can be raised to that corresponding to the intrinsic region.

To explain the electrical properties a number of band structure models have been suggested. Sexer (1966, 1967) suggested a two-type electron model with the two minima in the conduction band separated by 0·15 eV and suggested that for low carrier concentrations electrons exist only in the narrow band at $k = 0$ whereas for high concentrations they exist in both bands (figure 8.12a). A similar model has been suggested by Radautsan *et al.* (1973) but with $m_2$ much smaller ($0{\cdot}018m_0$). Wagner *et al.* (1969) suggested an α-Sn-like zero-bandgap model and this model or one with a very small gap as shown in figure 8.11b is supported by the theoretical paper of Lin-Chung (1969). The exceptionally complicated crystal structure of cadmium arsenide makes a theoretical study difficult: Lin-Chung used a pseudo-potential method and considered the crystal structure as a fluorite crystal structure with cadmium-ion vacancies distributed periodically throughout the crystal. With experimental results from interband magnetooptical transitions, Wagner *et al.* assumed a Kane model for the light conduction and valence bands and assumed an energy gap $E_g$ of 0·38 eV. They also suggested that a weak optical edge of 0·2 eV and a strong optical edge of 0·6–0·7 eV, measured by optical transmission on thin films by Zdanowicz (1967), arise respectively from transitions from heavy valence to conduction band, and from light valence to conduction band, both shifted by the Burstein effect.

[In this latter case, the Burstein shift is twice the Fermi energy (≈0·15 eV) if the light-hole mass is comparable with the heavy-hole mass, thus giving a transition of 0·68 eV.]

Rogers *et al.* (1971) discussed two possible band models, shown in figures 8.12c and 8.12d, to explain the measured transport and optical properties. Both models involve an indirect gap, this having been deduced from the form of the wavelength dependence of the absorption coefficient, although it is possible that there is actual overlap of the principal valence and conduction bands. They used a value of 0·13 eV

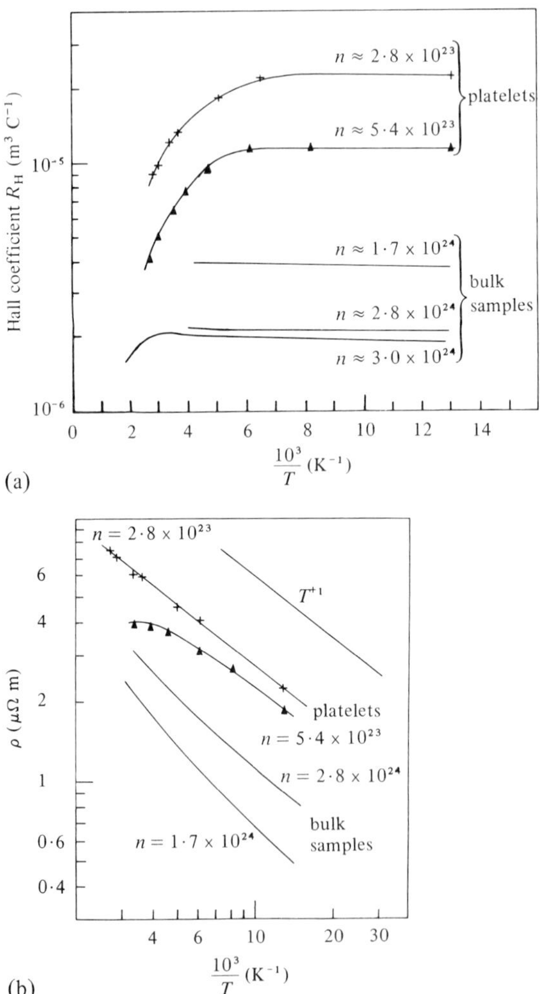


**Figure 8.11.** (a) Hall coefficient, $R_H$, and (b) resistivity, $\rho$, in cadmium arsenide platelets and bulk samples (Lovett, 1972).

for the energy gap $E_g$, whereas Lovett (1972) tested values between 0·20 and 0·50 eV with varying effective masses to fit the Hall data of figure 8.11a. A choice of $E_g$ = 0·38 eV and a heavy-hole effective mass of $0{\cdot}12m_0$ would fit the data and give an electron effective mass at the bottom of the band of approximately $0{\cdot}02m_0$. The extra conduction band shown in figure 8.12d may be necessary to explain certain discrepancies in the results and a second conduction minimum would seem necessary to explain a variation of the Hall coefficient with applied magnetic field which was measured in highly degenerate material (Lovett, 1969). However, Cisowski and Zdanowicz (1973), in making electrical measurements under hydrostatic pressure up to 12 kbar, explained their results on the basis of a single Kane-type conduction band and suggested that any second conduction-band minimum must lie at least 0·4 eV above the first. Aubin *et al.* (1970) deduced from magneto-Seebeck and magnetoresistance measurements that a second minimum lies 0·4 eV above the first conduction-band minimum and obtained an electron effective mass of $0{\cdot}012m_0$ at the bottom of the first conduction band.

As can be seen from figures 8.11a and 8.11b, the resistivity of cadmium arsenide over most of the temperature range is approximately proportional to absolute temperature because the electron mobility is inversely proportional to temperature and the carrier concentration remains constant. Samples with lower carrier concentrations exhibit

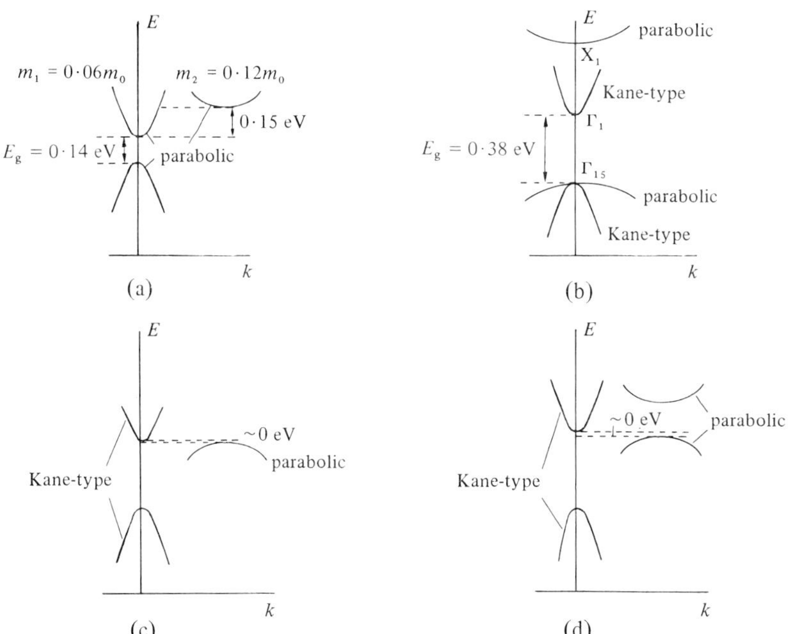


**Figure 8.12.** Various band-structure models proposed for cadmium arsenide.

higher mobilities. For samples possessing the usual carrier concentration of $\sim 3 \times 10^{24}$ m$^{-3}$, mobilities are around 0·7 to 1·0 m$^2$ V$^{-1}$ s$^{-1}$ at room temperature and 2 to 3 m$^2$ V$^{-1}$ s$^{-1}$ at 80 K, or slightly higher for good single crystals, whereas for samples with the lower carrier concentrations, mobilities are higher reaching a value of almost 10 m$^2$ V$^{-1}$ s$^{-1}$ at 80 K for a sample of carrier concentration less than $3 \times 10^{23}$ m$^{-3}$.

Thin films do not necessarily exhibit the same properties as the bulk material. Extensive measurements have been made by Zdanowicz (1968) on thin-film elements which were 0·1 to 2·0 $\mu$m thick and obtained by evacuation on to glass or mica substrates. Here the electrical properties depend primarily on the temperature of the substrate. If the substrate was held at less than 100°C, an amorphous film was obtained with resistivity in the range $10^{-4}$ to $10^{-2}$ $\Omega$ m and a carrier concentration of $10^{23}$ m$^{-3}$, these values corresponding to the rather low mobility value of 0·02 m$^2$ V$^{-1}$ s$^{-1}$. To obtain crystalline films, it was necessary for the substrate to be between 160° and 180°C giving lower resistivities of $10^{-5}$ to $7 \times 10^{-5}$ $\Omega$ m and electron mobilities of 0·1 to 0·3 m$^2$ V$^{-1}$ s$^{-1}$. Both resistance and Hall voltage show decreases in value with increase of temperature from room temperature upwards, this effect setting in at lower temperatures than for the bulk samples.

Magnetoresistance values show considerable variation from sample to sample and remain unexplained in detail. Iwami *et al.* (1971) obtained almost zero longitudinal magnetoresistance coefficients for samples of $3 \times 10^{24}$ m$^{-3}$ carrier concentration but the coefficients were not zero for carrier concentrations of $1 \times 10^{24}$ and $1 \times 10^{25}$ m$^{-3}$. Low-field transverse magnetoresistance measurements on low-carrier-concentration material (Lovett, 1972) and high-carrier-concentration material (Lovett, 1967) both show larger low-field coefficients than for material of concentration $3 \times 10^{24}$ m$^{-3}$.

Rosenman (1969) measured the Shubnikov–de Haas effect in single crystals of $Cd_3As_2$ over a temperature range of 4·2 to 40 K. From the oscillatory magnetoresistance results, he obtained a transverse cyclotron mass varying from 0·042 to 0·065$m_0$ when the carrier concentration varied from 2·6 to $8{\cdot}3 \times 10^{24}$ m$^{-3}$, hence consistent with a nonparabolic conduction band. He proposed a Fermi surface of one ellipsoid prolate in the fourfold axis direction with an anisotropy axis ratio of 1·18. Quantum oscillations in thermoelectric power and thermal conductivity have also been measured (German *et al.*, 1972).

Thermoelectric and thermomagnetic measurements have been made by many workers. At room temperature the thermoelectric power is −60 $\mu$V K$^{-1}$ and increases to a maximum of approximately −85 $\mu$V K$^{-1}$ at 500 K (Zdanowicz, 1961) for a carrier concentration of $3 \times 10^{24}$ m$^{-3}$. The Nernst effect is small because of the extrinsic nature of the samples and there is considerable difference between adiabatic and isothermal values, making accurate determination of the thermomagnetic coefficients difficult.

This difference arises because cadmium arsenide possesses perhaps the largest Righi–Leduc effect measured in any material. Results of Lovett and Ballentyne (1967) for a typical sample are shown in figure 8.13 and may be compared with those for mercury selenide in section 7.1.3.

The thermal conductivity of cadmium arsenide shows certain anomalies. Extensive thermal conductivity measurements on cadmium arsenide and its alloys with zinc arsenide were made by Spitzer *et al.* (1966) who showed that for samples having high electrical conductivity (greater than $10^5\ \Omega^{-1}\ m^{-1}$) it can be as much as 30% below that calculated from the Wiedemann–Franz law plus the extrapolated lattice thermal conductivity. Because of the lack of correlation of the thermal conductivity with carrier concentration, Spitzer *et al.* postulated that the anomaly lies with the lattice component, which appears to be as low as 0·03 to 0·04 W $m^{-1}$ $K^{-1}$ for some samples. This involves a very high phonon scattering cross-section and Spitzer *et al.* suggested that phonon scattering may involve the transition of localised electrons from one type of arsenic vacancy site to another, there being three types of arsenic sites in all. The localised electrons associated with these types of sites might have energy levels differing by only a few hundredths of an electron volt. The largest carrier concentrations do not necessarily correspond to the smallest lattice conductivities. [Another possibility suggested by Lovett (1972) is an antistructure disorder with arsenic substituting for cadmium.] Figure 8.14 shows total thermal conductivity as a function of electrical conductivity for a number of doped and undoped samples. For most of the samples, Spitzer *et al.* found little variation of total conductivity with temperature over the temperature range −40 to 140°C, whereas Ballentyne and Lovett (1968) found an increase in total thermal conductivity with temperature

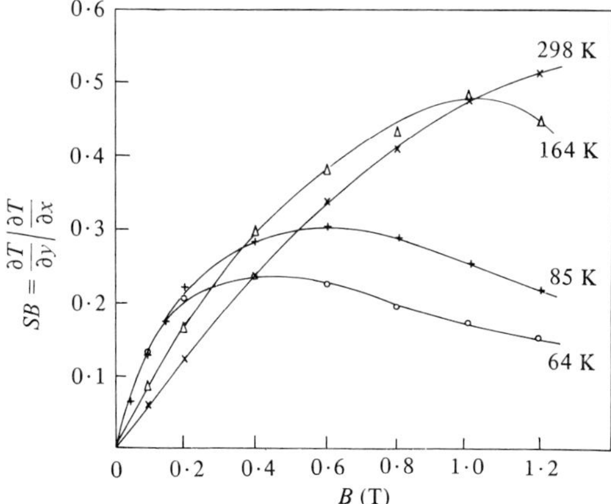


**Figure 8.13.** Variation of the dimensionless Righi–Leduc coefficient for cadmium arsenide with magnetic field (Lovett and Ballentyne, 1967).

from 150 K upwards, which they were unable to explain theoretically, and obtained room temperature values of total thermal conductivity of just under 5 W $m^{-1}$ $K^{-1}$. Also, Armitage and Goldsmid (1969) measured an anomalously high value of thermal conductivity of 8 W $m^{-1}$ $K^{-1}$ for a sample of exceptionally high carrier concentration ($1 \cdot 7 \times 10^{25}$ $m^{-3}$) and Lovett (1967) measured a value of over 16 W $m^{-1}$ $K^{-1}$ on a sample with a high carrier concentration which exhibited variation of Hall coefficient with magnetic field. Armitage and Goldsmid obtained values of 3 W $m^{-1}$ $K^{-1}$ for their other samples. These samples had a lattice conductivity falling continuously with temperature to approximately 1 W $m^{-1}$ $K^{-1}$ at 300 K and they deduced a low Lorenz number arising from inelastic scattering of the electrons. They suggested that there may be a number of competing scattering processes dominating at low carrier concentrations. In any case, further investigation is desirable to relate the variation of thermal conductivity and other properties to the band structure and carrier concentrations.

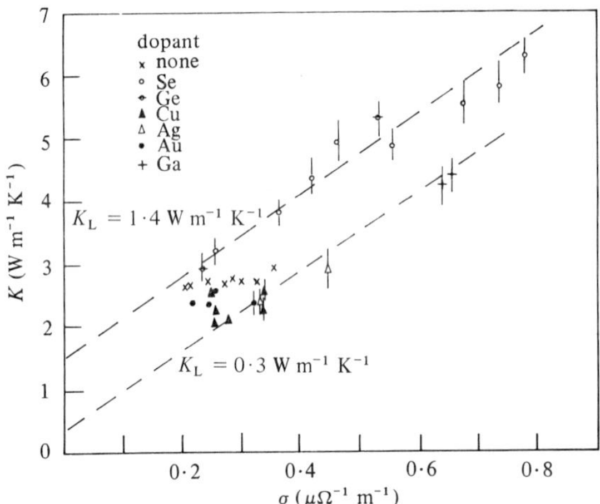


**Figure 8.14.** Thermal conductivity plotted against electrical conductivity for cadmium arsenide samples at room temperature, each point representing a different sample. (After Spitzer *et al.*, 1966.) The dashed lines are calculated with the use of the values of thermoelectric power and $K_L = 1 \cdot 4$ and $0 \cdot 3$ W $m^{-1}$ $K^{-1}$.

### 8.4.3 Solid solutions and eutectics involving cadmium arsenide

The $Cd_{3-x}Zn_xAs_2$ system has been investigated extensively. If one takes band structure models for cadmium arsenide shown in figures 8.12c and 8.12d, then the energy separation between the principal valence and conduction bands which is approximately zero for cadmium arsenide rapidly becomes positive with the addition of zinc (Rogers *et al.*, 1971). The direct energy gap also becomes larger [although an alternative model with it going to zero at $x = 2 \cdot 6$ has been suggested by Wagner *et al.* (1969) on the basis of model 8.12b]. Although by alloying with zinc it is possible to produce intrinsic and p-type material (both n-type $Cd_{1\cdot2}Zn_{1\cdot8}As_2$ and p-type $Cd_{1\cdot8}Zn_{1\cdot2}As_2$ were found to be intrinsic at room temperature), the samples have large band gaps, viz 0·5 eV and 0·7 eV in the two examples cited. Figure 8.15 shows the direct and indirect bandgaps as deduced by Rogers *et al.* on the basis of models 8.12c or 8.12d for the complete range $x = 0$ to $x = 3$. The lattice parameters $a_0$ and $c_0$ decrease linearly in value over the complete range. The highest Seebeck coefficient measured was $-850\ \mu V\ K^{-1}$ on a sample of carrier concentration less than $10^{19}\ m^{-3}$ and containing also selenium, with the composition $Cd_{1\cdot2}Zn_{1\cdot8}As_2Se_{0\cdot02}$, whilst a $Cd_{1\cdot2}Zn_{1\cdot8}As_2$ sample of carrier concentration $1\cdot1 \times 10^{22}\ m^{-3}$ had a Seebeck coefficient of $-820\ \mu V\ K^{-1}$. Undoped alloys with $x < 1\cdot5$ are always n-type; alloys with $x > 1\cdot5$ are p-type.

Similarly, cadmium arsenide has been alloyed with cadmium phosphide (Masumoto and Isomura, 1967). The alloying produced increasing resistivity as a result of decreasing carrier mobility and little change of carrier concentration. The absolute value of the Seebeck coefficient (which is negative) gradually increased.

In contrast, cadmium arsenide has been grown as a eutectic with nickel arsenide (Hiscocks, 1969) in such a way that rods of the metallic nickel arsenide are contained within a matrix of cadmium arsenide. The eutectic line occurs at 700 K and this intersects the two liquidus curves at 9 mol% NiAs. The eutectic crystals were grown by the Bridgman–Stockbarger technique but involving liquid encapsulation as used for crystal pulling. The encapsulant was $B_2O_3 : Na_3AlF_3$, the 1% of $Na_3AlF_3$ being added to lower the viscosity of the encapsulant. Charge and encapsulant were contained within a graphite crucible which was lowered at 1 to 2 cm $h^{-1}$.

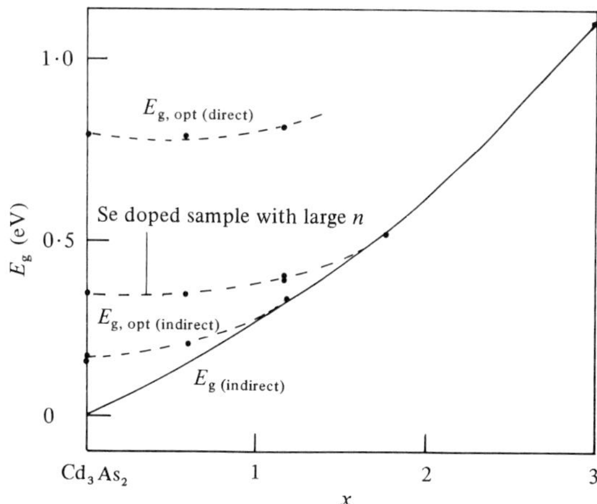


**Figure 8.15.** The direct and indirect optical energy gaps of $Cd_{3-x}Zn_xAs_2$ at 300 K as functions of alloy composition $x$. The solid curve shows the true energy gap obtained by correcting the optical gap for Burstein shift. (After Rogers *et al.*, 1971.)

To maintain thermal symmetry the crucible was rotated at 10 rev $min^{-1}$. Samples were obtained with a continuous eutectic along most of the grown length of 3 to 4 cm without the banding which would be associated with fluctuations of temperature or lowering rate. The lengths of the nickel arsenide rods were in the direction of growth of the ingot but tended to splay out towards the outside of the ingot. Because of the solid–solid phase change near 600°C, the ingots were polycrystalline but less cracked than in the case of pure cadmium arsenide.

### 8.4.4 Applications

Much of the earlier work on cadmium arsenide was concerned with the possibility of its use in a thermomagnetic device particularly for cooling; it has the advantage of high carrier mobility at high carrier concentrations and a low lattice thermal conductivity. However, as mentioned, it did not prove possible to reduce the electron carrier concentration and produce intrinsic material especially below room temperature.

Zdanowicz (1968) investigated the properties of thin film samples as possible Hall effect elements. She found that the sensitivity was greater than that of elements produced from bulk indium arsenide, indium antimonide, and alloys of indium arsenide with indium phosphide; and that the input resistance, nominal values of input current, and nominal output power are comparable with those of germanium Hall elements. Covered with a protective layer, the elements have very long term stability. Their disadvantages are rather large temperature coefficients of electrical resistivity and Hall voltage and relatively low mobility, these parameters being rather lower in thin films than in bulk material.

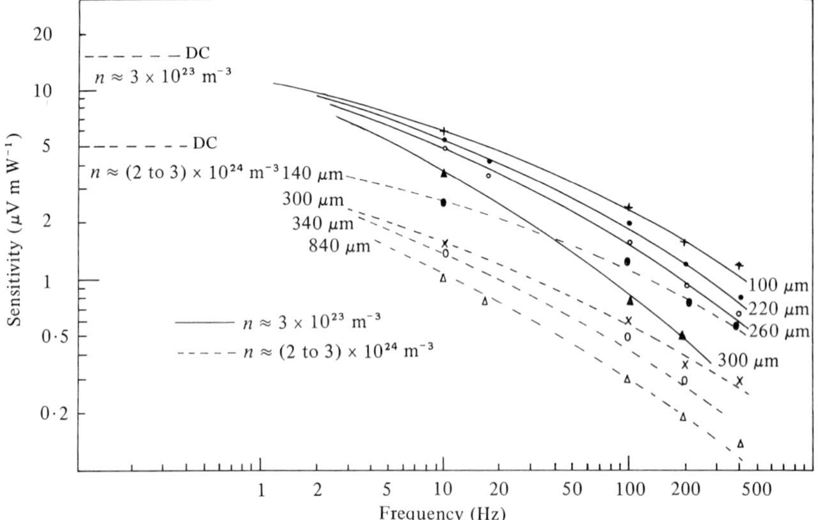


**Figure 8.16.** Nernst detector sensitivity as a function of frequency and thickness for cadmium arsenide with two ranges of carrier concentration; $B = 0{\cdot}3$ T (Lovett, 1973).

Another possible application is as a thermal radiation detector based on the Nernst effect (see section 5.6). Goldsmid *et al.* (1972) investigated the $Cd_3As_2$–NiAs eutectic for such a device and found that the inclusions of nickel arsenide adversely affect performance when the heat flow is parallel to the rods. Even so, this eutectic is better for such a detector than the InSb–NiSb eutectic. Lovett (1973) carried out similar measurements on cadmium arsenide single crystal platelets including measurements on the material with low carrier concentration, and the sensitivity curves are shown in figure 8.16. The sensitivities peaked at magnetic fields of approximately 0·8 T, the precise field value depending on the carrier mobility, and hence the carrier concentration, of the particular sample.

## 8.5 Metal–nonmetal transitions

A section should be devoted perhaps to compounds which show a metal–nonmetal transition where the nonmetal phase is either semimetallic or a narrow-bandgap semiconductor. A variety of compounds fall within the category, mainly transition-metal oxides and sulphides.

The metal–nonmetal transition has been reviewed by Mott (1968), Mott and Davis (1971), and others. The description of the difference between metals and insulators given by Wilson (1931) in terms of a model of noninteracting electrons does not work, for instance, for cupric oxide which should be a metal by the model. So the electron–electron interaction ($e^2/r_{12}$) was introduced and this led to the idea that at low densities a free-electron gas would 'crystallise' into a nonconducting state. Mott (1949) introduced the idea of a lattice parameter $d$ between the array of electrons; for large $d$ the array would constitute an insulator, for small $d$ a metal. At a particular value of $d$ there might be a phase transition, although not necessarily, as the transition need be sharp only at $T = 0$. However, not all transitions are due to electron–electron interactions. Some can be explained in terms of a noninteracting electron gas model. A change of crystal structure may give rise to a transition (this probably accounts for the transitions in the vanadium oxides) or a nonmetal–metal transition can occur at the Néel point (i.e. the temperature at which the spins of the neighbouring ions in an antiferromagnetic material cease to be ordered antiparallel as the temperature is increased), as occurs for nickel sulphide.

### 8.5.1 Nickel sulphide, NiS

Nickel sulphide occurs in two crystallographic forms, an orthorhombic form and a hexagonal form which is semimetallic (and antiferromagnetic) at low temperatures. At higher temperatures the hexagonal form is metallic (and paramagnetic) and there is an abrupt change of conductivity at the transition temperature [figure 8.17 (Koehler and White, 1973)]. The hexagonal structure normally exists only above 620 K but it can be maintained at lower temperatures by quenching.

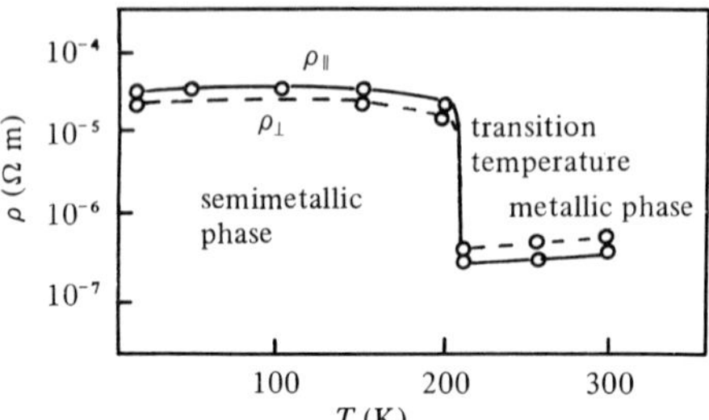


**Figure 8.17.** Resistivity as a function of temperature for single crystals of nickel sulphide. (After Koehler and White, 1973.)

8.5.1.1 *Preparation*

Preparation of polycrystalline single crystals is discussed in detail by Koehler *et al.* (1972). They prepared the hexagonal form of nickel sulphide by direct synthesis from 99·999% pure nickel wire and 99·9999% pure sulphur powder in a quartz ampoule. The sulphur was additionally purified by repeated remelting in a low vacuum to remove adsorbed gases. During synthesis the ampoule was held at a uniform temperature of 400°C for 7 days for completion of the chemical reaction. Next the temperature was raised to 100°C (i.e. above the melting point) and held for 4 days to enhance homogenisation, and then cooled to 560°C, whereupon the sample was quenched in water at 0°C to retain the hexagonal form. The quenched material was polycrystalline but contained several large grains. Some material was used directly in this form but other material was used for the preparation of single crystals.

For this a modified Bridgman technique was used. The polycrystalline material was placed in a quartz tube with conical bottom, and the tube evacuated to $10^{-5}$ Torr. It was placed in the upper zone of the furnace, heated to above the melting temperature (975–1025°C) with the tip at about 975°C, and lowered into the lower zone at 3 mm $h^{-1}$ with a temperature gradient of 25 K $cm^{-1}$. After lowering for 15 cm, the crystals were quenched in water at 0°C. Single crystals of up to 5 cm length were obtained. The first-grown regions of the crystals were found to be sulphur-rich with a composition of approximately $Ni_{0 \cdot 94}S$ and while sufficient sulphur remained present this composition continued. When the deficiency of sulphur became large enough, the composition approached NiS and then became nickel-rich.

The transition temperature was found to be highly dependent on the stoichiometry but very little dependent on the presence of substitutional impurities (figures 8.18a and 8.18b). Samples were annealed at 600°C to reduce the sulphur content and obtain $Ni_xS$ with $x = 1$. On annealing at 700°C $x$ became greater than 1; below 600°C, $x$ remained less than 1. One end of the ampoule was held at room temperature during the anneal so that excess sulphur was condensed out.

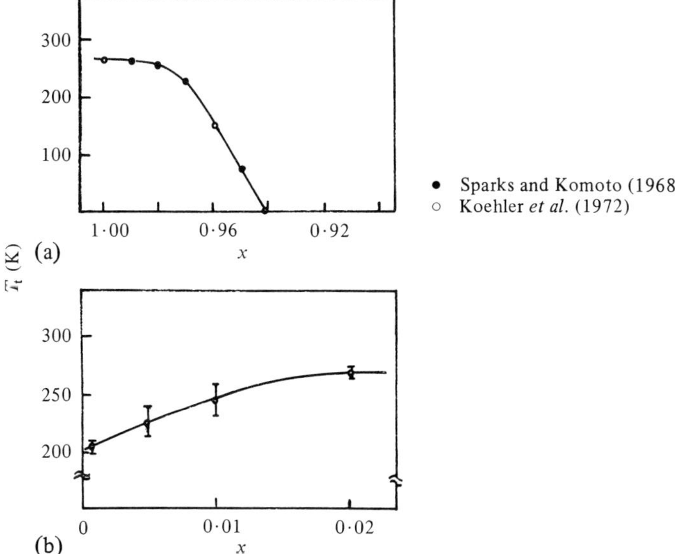


**Figure 8.18.** Transition temperature $T_t$, plotted as a function of $x$ for (a) the composition $Ni_xS$ and (b) the composition $Ni_{0\cdot 99-x}Cr_xS$ (brackets indicate the transition width).

8.5.1.2 *Properties*

Electrical and other properties have been investigated in particular by Townsend *et al.* (1971) who reported the transition as a metal–semiconductor transition rather than a metal–semimetal transition. Townsend *et al.* found a considerable difference in resistivity between resistivity measured parallel and perpendicular to the *c* axis. [Nickel sulphide has the noncentrosymmetric space group $P6_3/mc$ below the transition and the centrosymmetric space group $P6_3/mmc$ above the transition temperature (Trahan *et al.*, 1970).] As has been indicated, this was not found by Koehler and White and the high resistivity measured parallel to the *c* direction by Townsend *et al.* probably arises from cracking along the cleavage planes perpendicular to *c* as the sample passes through the transition temperature.

Measurements of the Hall effect have enabled calculation of the carrier concentration and mobility, but there have been even larger discrepancies between measurements here than for resistivity. The results of Townsend *et al.* above 165 K (and below the transition temperature) indicate Hall mobilities of $0\cdot 1$ $m^2$ $V^{-1}$ $s^{-1}$ and carrier concentrations of $10^{23}$ to $10^{24}$ $m^{-3}$ (and an activation energy of approximately $0\cdot 17$ eV suggesting an energy gap of $\approx 0\cdot 4$ eV). These compare with values obtained by Ohtani *et al.* (1970) of mobilities of $3 \times 10^{-4}$ $m^2$ $V^{-1}$ $s^{-1}$ and carrier concentrations of $10^{27}$ $m^{-3}$. Both groups obtained a thermoelectric power of a few tens of $\mu$V K$^{-1}$ indicating p-type material. The results of magnetic susceptibility measurements reported by Koehler and White are shown in figure 8.19; they are in broad agreement with those obtained by Townsend *et al.*

From the experimental results, nickel sulphide is found to have a density of states below the transition temperature approximately half that for the metallic state. Diagrams for the density-of-states models as used by Koehler and White above and below the transition temperature are shown in figure 8.20. The Ni(4s) bands are considered as lying well above the others and are unoccupied. The Ni(3d) and S(3p) overlap, and the Ni(3d) band is split into two bands by the cubic crystalline field and it is the higher of these which lies near the Fermi surface and is shown in the diagram; it splits below the transition temperature owing to crystal distortion and Hubbard interaction. If the Fermi level lies between the

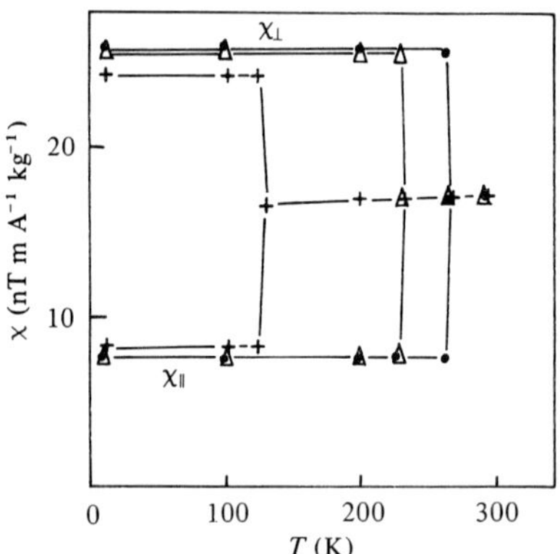


**Figure 8.19.** Magnetic susceptibility, $\chi$, as a function of temperature for single crystals of nickel sulphide with transition temperatures of 130, 228, and 267 K. (After Koehler and White, 1973.)

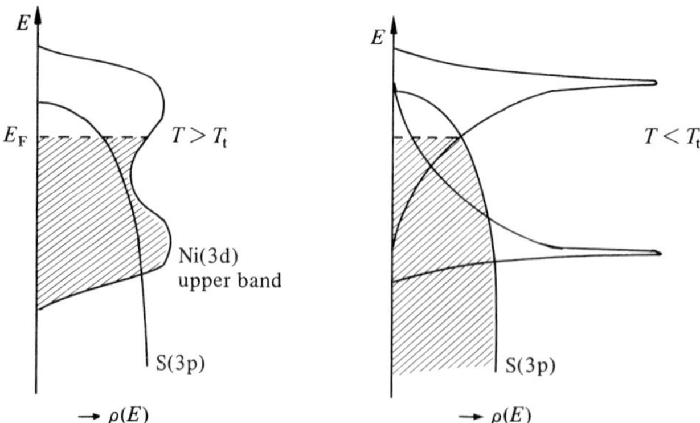


**Figure 8.20.** Density-of-states models for nickel sulphide. (After Koehler and White, 1973.)

two split portions of the band, then the band is half full. The model is consistent with metallic properties above the transition temperature, and below the transition temperature conductivity is due mainly to holes in the 3p band. The electrons in the copper Ni(3d) band have a large effective mass, thus affecting participation in conduction processes. The magnetic properties are also consistent. The model ignores hybridisation of the various degenerate electron bands but nevertheless supports the suggestion of a semimetallic rather than a semiconducting phase below the transition temperature.

### 8.5.2 Vanadium oxides

There are a number of vanadium oxides and they show a metal–nonmetal transition where the nonmetal is a semiconductor of relatively narrow bandgap. Adler (1968) has reviewed the available data on these oxides.

Vanadium trioxide, $V_2O_3$, has rhombohedral symmetry but the lattice can be considered as distorted hexagonal with the metal ions lying in the basal planes. Below 150 K there is a transition to monoclinic symmetry with a 2% shifting of the $V^{3+}$ ions. The conductivity varies from $10^5\ \Omega^{-1}\ m^{-1}$ above the phase transition to $10^{-3}\ \Omega^{-1}\ m^{-1}$ below the phase transition (Feinleib and Paul, 1967). An activation energy of approximately 0·15 eV was measured for the low-temperature phase and optical experiments indicated a drop in transmission at around 0·3 eV which is probably therefore the bandgap value. Different models have been used to explain the data. To account for the large difference in conductivity it is necessary to explain a mobility difference of a factor of $10^5$ between the metallic and semiconducting states. A crystalline distortion model will account for the energy gap but the large change of carrier mobility requires either a large correlation effect or the effect of self-trapping of the free carriers. However, another possibility is the occurrence of antiferromagnetism combined with the monoclinic distortion; this would widen the energy gap and lead to a change in carrier concentration which would account for the entire discontinuity in the conductivity. Ashkenazi and Weger (1973) have given a detailed model for the transition based on the electronic band structure. The model is treated in the Hartree–Fock approximation at zero and finite temperatures and the model investigates change of the $c_0/a_0$ ratio, creation of magnetic moments, changes in covalency, the effect of pressure and the order of the transition.

Vanadium dioxide, $VO_2$, has a structure which is a slight modification of rutile above the phase transition of 340 K (the $V^{4+}$ ions form a body-centred tetragonal lattice), whereas below 340 K the structure is monoclinic and resembles $MnO_2$. This monoclinic phase is an 8% distortion from tetragonal. The electrical conductivity changes by a factor of approximately $10^5$ through the transition from $10^6\ \Omega^{-1}\ m^{-1}$ above the transition to $10\ \Omega^{-1}\ m^{-1}$ below the transition. The activation energy increases from $\sim 0 \cdot 15$ eV at low temperatures to just below $0 \cdot 5$ eV near the transition temperature. The mobility of $1 \cdot 3 \times 10^{-5}$ m$^2$ V$^{-1}$ s$^{-1}$ indicates a very narrow-band model, but indications are that the energy gap is $\approx 0 \cdot 8$ eV, i.e. considerably greater than that for the trioxide.

Vanadium monoxide, VO, has a transition temperature of 126 K, activation energy of $0 \cdot 14$ eV, and a jump in the conductivity by a factor of $10^6$ at the transition. The conductivity of stoichiometric vanadium monoxide at 100 K has been measured as $10^5$ $\Omega^{-1}$ m$^{-1}$ (Kawano *et al.*, 1966). One possible explanation of the change of properties is that vanadium monoxide is metallic at all temperatures but below the transition temperature it decomposes into two phases, $V_3O$ and $V_3O_4$, with high resistivity due to the phase boundaries (or VO with $V_2O_3$ regions). However, similarity in properties between VO and $V_2O_3$ would suggest that explanations for their properties should also be similar. $V_6O_{13}$ has a transition of 149 K and an activation energy below the transition temperature of $0 \cdot 12$ eV. A crystalline distortion model predicts a change of conductivity of a factor of $10^4$ and an energy gap of $0 \cdot 14$ eV.

Other phases of vanadium oxide arising from packing faults in the $VO_2$ structure also exist and some of these probably exhibit transition temperatures.

### 8.5.3 Other transition-metal oxides and sulphides

The transition in titanium trioxide, $Ti_2O_3$, is also reviewed by Adler (1968). The compound has the so-called corundum structure at all temperatures; that is, it is rhombohedral with space group $R\overline{3}c$ with $a_0 = 5 \cdot 431$ Å and $\alpha = 56°36'$ (Wyckoff, 1964), or, if a hexagonal cell is used, $a_0' = 5 \cdot 148$ Å and $c_0' = 13 \cdot 636$ Å. The resistivity of titanium trioxide is found to drop by a factor of 50 as the temperature is raised from 300 K to 750 K compared with a change of approximately 8 for vanadium trioxide (Honig, 1968). Yahia and Frederikse (1961) determined a transition temperature of 450 K for their samples, whereas Abrahams (1963) found a transition temperature of approximately 650 K. There being no change in crystal symmetry, it had been suggested that the change in resistivity arose from splitting of the conduction bands into subbands as antiferromagnetic order set in. However, evidence for antiferromagnetic order is inconclusive and the transition has been explained by Van Zandt *et al.* (1968) in terms of a gradual change of band structure, the material changing from a narrow-bandgap semiconductor with a bandgap $\sim 0 \cdot 06$ eV to a semimetal; there is probably a direct correlation between the change of bandgap and distortion of the unit cell with change of temperature. The resistivity changes from approximately $8 \times 10^{-4}$ $\Omega$ m below the transition temperature to $1 \cdot 5 \times 10^{-3}$ $\Omega$ m above the transition. At the same time the resistivity is changing rapidly with temperature.

As with the V–O system, so also the Ti–O system forms Magnéli phases of the type $Ti_nO_{2n-1}$ where $n = 3, 4, 5$, etc. These also tend to exhibit temperature transitions. $Ti_3O_6$ exhibits a sharp change of magnetic susceptibility at 460 K, this probably being a semiconductor–metal transition, and similarly $Ti_5O_9$ exhibits a change at 130 K (see Adler, 1968).

Numerous other materials show transitions of the metal–nonmetal type and some involve narrow-bandgap or semimetal phases. Niobium dioxide, $NbO_2$, has a semiconducting phase at room temperature, but a somewhat wide bandgap (0·5 eV). Below its temperature transition of 119 K, black iron oxide, $Fe_3O_4$, has electrical behaviour consistent with a semiconductor of bandgap 0·25 eV. Both chromium sulphide and iron sulphide are semiconductors in the low-temperature phases and, having small activation energies, are probably relatively narrow-bandgap semiconductors.

This last section concerning transition-metal oxides and sulphides is distinct from the rest of the book in that the materials have been less widely investigated and differ greatly in many of their properties from materials considered earlier. Nonmetal–metal transitions are expected to be common also among a large range of transition-metal selenides and halides, and it is an area where much interest and research is likely to be centred in the future, even though these transitions do not have much practical importance at present.

**References**

Abdullaev, A. A., Angelova, L. A., Kuznetsov, V. K., Ormont, A. B., Pashintsev, Ya. I., 1973, *Phys. Status Solidi,* **A18**, 459.
Abrahams, S. C., 1963, *Phys. Rev.,* **130**, 2230.
Adler, D., 1968, *Rev. Mod. Phys.,* **40**, 714.
Amemiya, Y., Terao, H., Sakai, Y., 1973, *J. Appl. Phys.,* **44**, 1625.
Armitage D., Goldsmid, H. J., 1969, *J. Phys. C.,* **2**, 2138.
Ashkenazi, J., Weger, M., 1973, *Adv. Phys.,* **22**, 207.
Aubin, M., Brizard, R., Messa, J. P., 1970, *Can. J. Phys.,* **48**, 2215.
Ballentyne, D. W. G., Lovett, D. R., 1968, *Br. J. Appl. Phys. (J. Phys. D.), Ser.2,* **1**, 585.
Cisowski, J., Zdanowicz, W., 1973, *Acta Phys. Pol.,* **A43**, 293.
Coderre, W. M., Woolley, J. C., 1968, *Can. J. Phys.,* **46**, 1207.
Coderre, W. M., Woolley, J. C., 1971, *Proceedings of the International Conference on Semimetals and Narrow Gap Semiconductors, Dallas (1970), J. Phys. Chem. Solids,* **32**, supplement 1, 535.
Damon, D. H., Miller, R. C., Sagar, A., 1965, *Phys. Rev.,* **138**, A636.
Emtage, P. K., 1965, *Phys. Rev.,* **138**, A246.
Feinleib, J., Paul, W., 1967, *Phys. Rev.,* **155**, 841.
Gatos, H. C., Moody, P. L., Lavine, M. C., 1960, *J. Appl. Phys.,* **31**, 312.
German, Ph., Calecki, D., Coste, G., 1972, *J. Phys. Chem. Solids,* **33**, 69.
Goldsmid, H. J., Savvides, N., Uher, C., 1972, *J. Phys. D.,* **5**, 1352.
Hansen, M., Anderko, K., 1958, *Constitution of the Binary Alloys* (McGraw-Hill, New York).
Harman, T. C., 1956, *J. Electrochem. Soc.,* **103**, 128.

Hilsum, C., Rose-Innes, A. C., 1961, *Semiconducting III-V Compounds* (Pergamon Press, Oxford).
Hiscocks, S. E. R., 1969, *J. Mater. Sci.*, **4**, 773.
Honig, J. M., 1968, *Rev. Mod. Phys.*, **40**, 748.
Iwami, M., Matsunami, H., Tanaka, T., 1971, *J. Phys. Soc. Jpn.*, **31**, 768.
Jayaraman, A., Anantharaman, T. R., Klement, W., 1966, *J. Phys. Chem. Solids*, **27**, 1605.
Kane, E. O., 1957, *J. Phys. Chem. Solids*, **1**, 249.
Kawano, S., Kosuge, K., Kachi, S., 1966, *J. Phys. Soc. Jpn.*, **21**, 2744.
Koehler, R. F., Jr., White, R. L., 1973, *J. Appl. Phys.*, **44**, 1682.
Koehler, R. F., Jr., Feigelson, R. S., Swartz, H. W., White, R. L., 1972, *J. Appl. Phys.*, **43**, 3127.
Koltirine, B., Chaumereuil, M., 1966, *Phys. Status Solidi*, **13**, K1.
Koteles, E. S., Datars, W. R., 1974, *Phys. Rev.*, **B9**, 568.
Kruse, P. W., 1970, *Semiconductors and Semimetals*, **5**, Eds R. K. Willardson, A. C. Beer (Academic Press, New York), p.15.
Liang, S. C., 1962, *Compound Semiconductors*, Eds R. K. Willardson, H. L. Goering (Reinhold, New York), p.227.
Lin-Chung, P. J., 1969, *Phys. Rev.*, **188**, 1272.
Long, D., 1968, *Energy Bands in Semiconductors* (Interscience, New York), p.101.
Lovett, D. R., 1967, Thesis, London University.
Lovett, D. R., 1969, *Phys. Lett.*, **30A**, 90.
Lovett, D. R., 1972, *J. Mater. Sci.*, **4**, 733.
Lovett, D. R., 1973, *Phys. Status Solidi*, **A17**, K123.
Lovett, D. R., Ballentyne, D. W. G., 1967, *Br. J. Appl. Phys.*, **18**, 1399.
Masumoto, K., Isomura, S., 1967, *Trans. Jpn. Inst. Metals*, **8**, 139.
Mooradian, A., Fan, H. Y., 1966, *Phys. Rev.*, **148**, 873.
Mott, N. F., 1949, *Proc. Phys. Soc. London, Sect. A*, **62**, 416.
Mott, N. F., 1968, *Rev. Mod. Phys.*, **40**, 677.
Mott, N. F., Davis, E. A., 1971, *Electronic Processes in Non-Crystalline Materials* (Oxford University Press, Oxford).
Mott, N. F., Jones, H., 1936, *The Theory of the Properties of Metals and Alloys* (Oxford University Press, Oxford).
Ohtani, T., Kosuge, K., Kachi, S., 1970, *J. Phys. Soc. Jpn.*, **28**, 1588.
Pidgeon, C. R., Brown, R. N., 1966, *Phys. Rev.*, **146**, 575.
Pietraszko, A., Łukaszewicz, K., 1973, *Phys. Status Solidi*, **A18**, 723.
Radautsan, S. I., Arushanov, E. K., Chuiko, G. P., 1973, *Phys. Status Solidi*, **A20**, 221.
Reynolds, R. A., Brau, M. J., Chapman, R. A., 1968, *J. Phys. Chem. Solids*, **29**, 755.
Rogers, L. M., Jenkins, R. M., Crocker, A. J., 1971, *J. Phys. D.*, **4**, 793.
Rosenman, I., 1969, *J. Phys. Chem. Solids*, **30**, 1385.
Seiler, D. G., Hathcox, K. L., 1972, *Phys. Rev. Lett.*, **29**, 647.
Sexer, N., 1966, *Phys. Status Solidi*, **14**, K43.
Sexer, N., 1967, *Phys. Status Solidi*, **21**, 225.
Smith, J. E., Jr., Nathan, M. I., McGroddy, J. C., Porowski, S. A., Paul, W., 1969, *Appl. Phys. Lett.*, **15**, 242.
Sparks, J. T., Komoto, T., 1968, *Rev. Mod. Phys.*, **40**, 752.
Spitzer, D. P., Castellion, G. A., Haacke, G., 1966, *J. Appl. Phys.*, **37**, 3795.
Steigmann, G. A., Goodyear, J., 1968, *Acta Cryst.*, **B24**, 1062.
Stringer, G. A., Higgins, R. J., 1970, *J. Appl. Phys.*, **41**, 489.
Stringer, G. A., Higgins, R. J., 1971, *Phys. Rev.*, **B3**, 506.

Tauc, J., Abraham, A., 1959, *Czech. J. Phys.,* **9**, 95.
Townsend, M. G., Tremblay, R., Horwood, J. L., Ripley, L. J., 1971, *J. Phys. C.,* **4**, 598.
Trahan, J., Goodrich, R., Watkins, S., 1970, *Phys. Rev.,* **2**, 2859.
Ugai, Ya. A., Zyubina, T. A., 1965, *Izv. Akad. Nauk. SSSR, Neorg. Mater.,* **1**, 860.
Van Dyke, J. P., Herman, F., 1970, *Phys. Rev.,* **B2**, 1644.
Van Zandt, L. L., Honig, J. M., Goodenough, J. B., 1968, *J. Appl. Phys.,* **39**, 594.
Wagner, R. J., Palik, E. D., Swiggard, E. M., 1969, *Phys. Lett.,* **30A**, 175.
Wilson, A. H., 1931, *Proc. R. Soc. London, Sect. A,* **133**, 458.
Wyckoff, R. W. G., 1963, *Crystal Structures,* second edition, **1** (Interscience, New York).
Wyckoff, R. W. G., 1964, *Crystal Structures,* second edition, **2** (Interscience, New York).
Yahia, J., Frederikse, H. P. R., 1961, *Phys. Rev.,* **123**, 1257.
Zdanowicz, W., 1961, *Acta Phys. Pol.,* **20**, 647.
Zdanowicz, L., 1967, *Phys. Status Solidi,* **20**, 473.
Zdanowicz, L., 1968, *Solid-State Electron.,* **11**, 429.

# Appendix

Selected properties of elements and compounds constituting (or used in) semimetals or narrow-bandgap semiconducting systems considered in this book.

| Material | Crystal structure | Lattice parameters (lengths in Å) | Melting point (°C) | Specific gravity | Appearance | Bandgap (eV) | Carrier mobilities at 300 K ($m^2 V^{-1} s^{-1}$)[e] | |
|---|---|---|---|---|---|---|---|---|
| | | | | | | | electrons | holes |
| As | rhombohedral $R\bar{3}m$ | $a_0 = 4{\cdot}131$<br>$\alpha = 54°10'$ | 613[a]<br>817 (28 atm) | 5·727 | grey metallic | −0·5 | | |
| $AuTe_2$ | monoclinic $C2/m$ | $a_0 = 7{\cdot}19$<br>$b_0 = 4{\cdot}40$<br>$c_0 = 5{\cdot}07$<br>$\beta = 90°10'$ | 472 | 8·2–9·3 | | small or zero | 0·1[f] | |
| Bi | rhombohedral $R\bar{3}m$ | $a_0 = 4{\cdot}746$<br>$\alpha = 57°14{\cdot}2'$ | 271·3 | 9·80 | silver white or reddish metallic | −0·037 | 3 | 0·8 |
| $Bi_2Se_3$ | rhombohedral $R\bar{3}m$ | $a_0 = 9{\cdot}841$<br>$\alpha = 24°16'$ | 706 | 6·82 | black | −0·0145 | 0·018 | 0·018 |
| $Bi_2Te_3$ | rhombohedral $R\bar{3}m$ | $a_0 = 10{\cdot}418$<br>$\alpha = 24°12{\cdot}7'$ | 585 | 7·7 | grey | 0·145 | 0·12 | 0·05 |
| C(graphite) | hexagonal $P6mc$ | $a_0 = 2{\cdot}456$<br>$c_0 = 6{\cdot}696$ | 3652–3697[a]<br>4300 | 2·25 | black | −0·03 | 1 (basal plane) | 1 (basal plane) |
| $Cd_3As_2$ | tetragonal $I4_1cd$ | $a_0 = 12{\cdot}67$<br>$c_0 = 25{\cdot}48$ | 721 | 6·21 | grey metallic | ⩽0·14 | 1·5 | |
| CdSe | ZnO $C6mc$ | $a_0 = 4{\cdot}309$<br>$c_0 = 7{\cdot}021$ | 1258 | 5·81 | grey brown or red | 1·84 | 0·065 | |
| CdTe | ZnS $F\bar{4}3m$ | $a_0 = 6{\cdot}480$ | 1041 | 6·20 | white | 1·605 | 0·12 | 0·005 |
| GeTe | face-centred rhombohedral (distorted NaCl) | $a_0 = 5{\cdot}986$<br>$\alpha = 88°21'$ | 724 | 5·3 | | 0·23 | | 0·013 |
| HgS | hexagonal (distorted NaCl) $C3_12$ | $a_0 = 4{\cdot}149$<br>$c_0 = 9{\cdot}495$ | 583·5[a]<br>2020 | 8·10 | black opaque or reddish | large | | |

| | | | | | | | | |
|---|---|---|---|---|---|---|---|---|
| HgSe | ZnS $F\bar{4}3m$ | $a_0 = 6{\cdot}08$ | 500[a] 799 | 8·266 | grey | small | 2 | |
| HgTe | ZnS $F\bar{4}3m$ | $a_0 = 6{\cdot}44$ | 670 | 8·17 | grey | −0·001 | 2·5 | 0·035 |
| InAs | ZnS $F\bar{4}3m$ | $a_0 = 6{\cdot}036$ | 943 | 5·68 | dark grey with metallic lustre | 0·32 | 3 | 0·04 |
| InSb | ZnS $F\bar{4}3m$ | $a_0 = 6{\cdot}478$ | 530 | 5·78 | grey with metallic lustre | 0·235 | 7·8 | 0·075 |
| $Mg_2Pb$ | antifluorite $Fm3m$ | $a_0 = 6{\cdot}836$ | 549[b] 538[c] | 6·2 | | small overlap | 10[g] | 0·4[g] |
| MnTe | ZnO $C6mc$ | $a_0 = 4{\cdot}087$ $c_0 = 6{\cdot}701$ | | 6·2 | black | ~1·0 | | |
| PbS | NaCl $Fm3m$ | $a_0 = 5{\cdot}936$ | 1127 | 7·5 | black | 0·286 | 0·08 | 0·10 |
| PbSe | NaCl $Fm3m$ | $a_0 = 6{\cdot}124$ | 1081 | 7·5 | grey | 0·165 | 0·15 | 0·15 |
| PbTe | NaCl $Fm3m$ | $a_0 = 6{\cdot}462$ | 924 | 8·16 | grey | 0·190 | 0·16 | 0·075 |
| $PtSb_2$ | pyrite $Pa3$ | $a_0 = 6{\cdot}428$ | 1226 | 10·5 | | 0·11 | 0·025 | 0·04 |
| Sb | rhombohedral $R\bar{3}m$ | $a_0 = 4{\cdot}507$ $\alpha = 57°6{\cdot}5'$ | 630·7 | 6·68 | silver white metallic | −0·16 | 0·028 | 0·35 |
| $Sb_2Se_3$ | orthorhombic $Pbnm$ | $a_0 = 11{\cdot}62$ $b_0 = 11{\cdot}77$ $c_0 = 3{\cdot}962$ | 611 | | grey | 1·2 | 0·01 | |
| $Sb_2Te_3$ | hexagonal $R\bar{3}m$ | $a_0 = 4{\cdot}25$ $c_0 = 29{\cdot}96$ | 622 | 6·50 | grey | small overlap | | 0·03 |
| αSn | diamond $Fd3m$ | $a_0 = 6{\cdot}49$ | 13[d] | 5·76 | dark grey | 0 | 0·25 | 0·24 |
| SnSe | NaCl $Fm3m$ | $a_0 = 6{\cdot}313$ | 861 | 6·13 | steel grey | 0·88 | | |
| SnTe | NaCl $Fm3m$ | $a_0 = 6{\cdot}313$ | 780 | 6·43 | grey | 0·2 | | 0·32[f] |

[a] Sublimation temperature; [b] congruent point; [c] peritectic point; [d] transformation temperature; [e] approximate values; [f] at 77 K; [g] at 4 K.

## Author index

# Subject index

# Materials index

Main entries are in bold print